Fundamentals of Statistics for Aviation Research

This is the first textbook designed to teach statistics to students in aviation courses. All examples and exercises are grounded in an aviation context, including flight instruction, air traffic control, airport management, and human factors.

Structured in six parts, this book covers the key foundational topics relative to descriptive and inferential statistics, including hypothesis testing, confidence intervals, z and t tests, correlation, regression, ANOVA, and chi-square. In addition, this book promotes both procedural knowledge and conceptual understanding.

Detailed, guided examples are presented from the perspective of conducting a research study. Each analysis technique is clearly explained, enabling readers to understand, carry out, and report results correctly. Students are further supported by a range of pedagogical features in each chapter, including objectives, a summary, and a vocabulary check.

Digital supplements comprise downloadable data sets and short video lectures explaining key concepts. Instructors also have access to PPT slides and an instructor's manual that consists of a test bank with multiple choice exams, exercises with data sets, and solutions.

This is the ideal statistics textbook for aviation courses globally, especially in aviation statistics, research methods in aviation, human factors, and related areas.

Michael A. Gallo, Ph.D., is a Professor Emeritus in the College of Aeronautics at Florida Institute of Technology, USA.

Brooke E. Wheeler, Ph.D., is an Associate Professor in the College of Aeronautics at Florida Institute of Technology, USA.

Isaac M. Silver, Ph.D., is the President of Energy Management Aerospace, USA.

Aviation Fundamentals

Series Editor: Suzanne K. Kearns

Aviation Fundamentals is a series of air transport textbooks that incorporate instructional design principles to present content in a manner that is engaging to the learner, at an accessible level for young adults, allowing for practical application of the content to real-world problems via cases, reflection questions, and examples. Each textbook will be supported by a companion website of supplementary materials and a test bank. The series is designed to help facilitate the recruitment and education of the next generation of aviation professionals (NGAP), a task that has been named a "Global Priority" by the ICAO Assembly. It will also support education for new air transport sectors that are expected to rapidly evolve in future years, such as commercial space and the civil use of remotely piloted aircraft. The objective of *Aviation Fundamentals* is to become the leading source of textbooks for the variety of subject areas that make up aviation college/university degree programs, evolving in parallel with these curricula.

Fundamentals of International Aviation Law and Policy
Benjamyn I. Scott and Andrea Trimarchi

Fundamentals of Airline Operations
Gert Meijer

Fundamentals of International Aviation
Suzanne K. Kearns

Fundamentals of Airline Marketing
Scott Ambrose and Blaise Waguespack

Fundamentals of Statistics for Aviation Research
Michael A. Gallo, Brooke E. Wheeler, and Isaac M. Silver

For more information about this series, please visit: www.routledge.com/Aviation-Fundamentals/book-series/AVFUND

Fundamentals of Statistics for Aviation Research

Michael A. Gallo, Brooke E. Wheeler, and Isaac M. Silver

Routledge
Taylor & Francis Group

LONDON AND NEW YORK

Designed cover image: © Getty Images

First published 2023
by Routledge
4 Park Square, Milton Park, Abingdon, Oxon OX14 4RN

and by Routledge
605 Third Avenue, New York, NY 10158

Routledge is an imprint of the Taylor & Francis Group, an informa business

British Library Cataloguing-in-Publication Data
A catalogue record for this book is available from the British Library

Library of Congress Cataloging-in-Publication Data
Names: Gallo, Michael A., author. | Wheeler, Brooke E., author. |
Silver, Isaac M., author.
Title: Fundamentals of statistics for aviation research / Michael A. Gallo,
Brooke E. Wheeler and Isaac M. Silver.
Description: Abingdon, Oxon ; New York, NY: Routledge, 2023. |
Series: Aviation fundamentals | Includes bibliographical references and index. |
Identifiers: LCCN 2022054896 (print) | LCCN 2022054897 (ebook) |
ISBN 9781032311494 (hbk) | ISBN 9781032311463 (pbk) | ISBN 9781003308300 (ebk)
Subjects: LCSH: Aeronautics—Study and teaching. | Aeronautics—Research.
Classification: LCC TL560 .G35 2023 (print) | LCC TL560 (ebook) |
DDC 629.130071—dc23/eng/20230106
LC record available at https://lccn.loc.gov/2022054896
LC ebook record available at https://lccn.loc.gov/2022054897

ISBN: 9781032311494 (hbk)
ISBN: 9781032311463 (pbk)
ISBN: 9781003308300 (ebk)

DOI: 10.4324/9781003308300

Typeset in Myriad Pro
by codeMantra

Access the Support Material: www.routledge.com/9781032311494

To the memory of my first statistics teacher,
Robert B. Nenno
whose nearly 50-year relationship with me evolved
from teacher to colleague to coauthor to friend.
M.A.G.

To my parents, for always supporting my
academic endeavors and encouraging lifelong learning.
B.E.W.

To my uncle, Dr. Michael Detty,
who encouraged me to pursue academic
studies in the sciences.
I.M.S.

Research is a lot like hot air ballooning: The glamour of hot air ballooning is you never know where you are going to land; the glamour of research is you never know what you are going to find.

Contents

List of Figures   xi
List of Tables   xvii
Preface   xxi

PART A
Research and Statistics Basics   **1**

1 The General Nature of Research and Statistics   3
Student Learning Outcomes 3
1.1 Chapter Overview 3
1.2 The Research Process 3
1.3 The Population-Sample Relationship and Sampling Error 10
1.4 Quantitative Research Methodologies 11
1.5 Variables and Measurement Scales 17
Chapter Summary 21
Vocabulary Check 22
Review Exercises 23

2 Organizing and Summarizing Data   26
Student Learning Outcomes 26
2.1 Chapter Overview 26
2.2 Examining Distributions 27
2.3 Measures of Central Tendency 34
2.4 Measures of Dispersion 37
2.5 Measures of Position: Percentiles and Quartiles 43
Chapter Summary 44
Vocabulary Check 45
Review Exercises 45

PART B
Making Reasonable Decisions about a Population   **49**

3 Z Scores, the Standard Normal Distribution, and Sampling Distributions   51
Student Learning Outcomes 51
3.1 Chapter Overview 51
3.2 z Scores 52
3.3 The Standard Normal Distribution 53
3.4 The Sampling Distribution of Sample Means 63
Chapter Summary 71
Vocabulary Check 72
Review Exercises 72

4 The Concept of Statistical Inference   77
Student Learning Outcomes 77
4.1 Chapter Overview 77
4.2 Parameter Estimation 77
4.3 The Nature of Hypothesis Testing 86
4.4 Hypothesis Tests Involving the Mean (σ Known) 98
4.5 Confidence Intervals vs. Hypothesis Tests 104
Chapter Summary 105
Vocabulary Check 106
Review Exercises 106

PART C
Analyzing Research Data Involving a Single Sample   **111**

5 Single-Sample *t* Test   113
Student Learning Outcomes 113
5.1 Chapter Overview 113
5.2 The t Test Statistic and t Distribution 113
5.3 Confidence Intervals and the t Distribution 117
5.4 Hypothesis Testing and the t Distribution 119
5.5 Using the Single-Sample t Test in Research: A Guided Example 121
Chapter Summary 126
Vocabulary Check 127
Review Exercises 127

6 Examining Bivariate Relationships: Correlation   131
Student Learning Outcomes 131
6.1 Chapter Overview 131
6.2 Correlation Fundamentals 131
6.3 Quantifying Relationships: The Pearson r 137
6.4 Statistical Aspects of Correlation 142

6.5 *Statistical Inferences Involving Pearson r* 145
6.6 *Using Bivariate Correlation in Research: A Guided Example* 150
Chapter Summary 155
Vocabulary Check 155
Review Exercises 156

7 Examining Bivariate Relationships: Regression   160
Student Learning Outcomes 160
7.1 *Chapter Overview* 160
7.2 *The Regression Equation* 160
7.3 *Statistical Aspects of Regression* 171
7.4 *Statistical Inferences Involving the Regression Coefficient* 173
7.5 *Using Bivariate Linear Regression in Research:*
       A Guided Example 178
Chapter Summary 184
Vocabulary Check 185
Review Exercises 185

PART D
Analyzing Research Data Involving Two Independent Samples   **189**

8 Independent-Samples *t* Test   191
Student Learning Outcomes 191
8.1 *Chapter Overview* 191
8.2 *The Concept of Independent Samples* 191
8.3 *Statistical Inferences Involving Independent Samples* 195
8.4 *Using the Independent-Samples t Test in Research:*
       A Guided Example 201
Chapter Summary 206
Vocabulary Check 207
Review Exercises 207

9 Single-Factor ANOVA   212
Student Learning Outcomes 212
9.1 *Chapter Overview* 212
9.2 *The Concept of ANOVA* 212
9.3 *The ANOVA Summary Table and the F Distribution* 218
9.4 *Statistical Inferences Involving the Single-Factor ANOVA:*
       Hypothesis Testing 225
9.5 *Using the Single-Factor ANOVA in Research:*
       A Guided Example 229
Chapter Summary 234
Vocabulary Check 235
Review Exercises 235

10 Factorial ANOVA

240

Student Learning Outcomes 240
10.1 Chapter Overview 240
10.2 The Concept of Factorial Designs 240
10.3 The Logic and Structure of Factorial ANOVA 247
10.4 Statistical Inferences Involving Factorial ANOVA:
        Hypothesis Testing 252
10.5 Using Factorial ANOVA in Research:
        A Guided Example 255
Chapter Summary 264
Vocabulary Check 265
Review Exercises 265

PART E
Analyzing Research Data Using a Within-Groups Design

273

11 Repeated-Measures *t* Test

275

Student Learning Outcomes 275
11.1 Chapter Overview 275
11.2 The Concept of Repeated-Measures 275
11.3 Statistical Inferences Involving the Repeated-Measures t Test 279
11.4 Using the Repeated-Measures t Test in Research:
        A Guided Example 283
Chapter Summary 289
Vocabulary Check 289
Review Exercises 290

PART F
Nonparametric Statistics: Working with Frequency Data

295

12 The Chi-Square Statistic

297

Student Learning Outcomes 297
12.1 Chapter Overview 297
12.2 One-Way Chi-Square: The Test for Goodness of Fit 297
12.3 Using the Chi-Square Test for Goodness of Fit in Research:
        A Guided Example 302
12.4 Two-Way Chi-Square: The Test for Independence 307
12.5 Using the Chi-Square Test for Independence in Research:
        A Guided Example 312
Chapter Summary 318
Vocabulary Check 318
Review Exercises 319

Appendix: Statistics Tables 325
Answers to Part A Review Exercises 335
Index 337

Figures

1.1   The relationship between research and statistics   4
1.2   The research process   5
1.3   A sample is a subset of the population   7
1.4   A comparison of various sampling strategies   8
1.5   Comparing a sample statistic to the corresponding population
      parameter. The circled items represent randomly selected flight times   12
1.6   Common types of research methodologies used for answering
      quantitatively based research questions   13
1.7   Comparison between sample selection vs. sample assignment   14
1.8   Flowchart for classifying variables coupled with their corresponding
      measurement scales   18
2.1   The concept of organizing and summarizing data   27
2.2   Bar graph that corresponds to Table 2.1, which shows the distribution
      of AMTs' highest level of education   29
2.3   Histogram of the number of runway incursions reported at ATL for a
      30-day period. The graph corresponds to the frequency distribution
      given in Table 2.4   30
2.4   Histogram of ages of 100 people who attended the Oshkosh Air Show.
      The graph corresponds to the grouped frequency distribution given in
      Table 2.5   32
2.5   Common shapes of distributions partitioned into two main categories:
      symmetrical and skewed. (Adapted from Kachigan, 1991, p. 31.)   33
2.6   Comparing the mode, median, and mean with respect to the shape of
      a distribution   37
2.7   Demonstration of why it is not sufficient to describe a distribution
      solely by its shape and mean. Information about the amount of spread
      there is in the data also is needed   38
2.8   The interquartile range is the distance between the first quartile (Q_1)
      and the third quartile (Q_3) and represents the middle 50%
      of a distribution   39
2.9   General illustration of a boxplot   41
2.10 Visualizing the concept of standard deviation using different targets
      where each target is represented by concentric circles with the mean
      (M) in the center, and scores are represented by "dots." As scores
      become more clustered about the mean (and each other), the average

distance scores are from the mean, which is standard deviation, becomes smaller | 41
2.11 Chapter 2/Exercise A-10 | 47
3.1 Demonstration of how the standard deviation of a distribution can affect how far a score is from the mean | 52
3.2 Frequency distribution tables and corresponding histograms and frequency polygons of the distributions for the number of "heads" in (a) 3 flips, (b) 5 flips, and (c) 10 flips of a fair coin | 54
3.3 As the number of flips of a coin increases, the distribution approaches a bell-shaped curve as shown by the given frequency polygon, which represents the number of "heads" in 20 flips of a coin | 55
3.4 A *frequency distribution graph* for the number of runway incursions reported at ATL for a 30-day period. The data correspond to the frequency distribution in Table 2.4 | 56
3.5 Demonstration of how standardizing a distribution with z scores does not affect the shape of the distribution relative to the raw data. The figure is with respect to the complacency scores from the data set Ch_3 Standard Normal Distribution | 58
3.6 In (a), are the approximate percentages of scores that lie within 1, 2, and 3 standard deviations of the mean in a normal distribution. In (b), the normal distribution is applied to the complacency scores. | 59
3.7 Examples of how to find area under the standard normal curve using the z table (Table 1/Appendix A) | 61
3.8 For Example 3.2(a) | 62
3.9 For Example 3.2(b) | 63
3.10 Concept of sampling distribution of sample means | 64
3.11 Illustration of the concept of a sampling distribution of sample means | 66
4.1 Comparison of a distribution for a single sample (a) vs. a sampling distribution of sample means (b), and the application of the sampling distribution of sample means to the AOPA example where 25 random samples each of size $n = 5$ were selected (c) | 79
4.2 Shown are the 95% confidence intervals for the 25 randomly selected samples from the hypothetical AOPA population data of flight hours given in Figure 3.10(a) in Chapter 3. The CIs are represented by horizontal lines, the lower and upper bounds are presented as "end" bars, and the corresponding sample means are pictured as closed circles. (Adapted from Kachigan, 1991, p. 97.) | 84
4.3 Conceptual view of hypothesis testing involving the mean | 87
4.4 Summary and comparison of research vs. statistical hypotheses | 90
4.5 A summary of the progression from purpose statement to statistical hypotheses of a research study relative to the BIM IRA exam scores example | 91
4.6 Critical z boundary values for a one-tailed test to the left and right, and for a two-tailed test with respect to a significance level of $\alpha = .05$ | 93
4.7 The z boundary values for the critical regions relative to four commonly used significance levels: $\alpha = .10, .05, .01,$ and $.001$ | 94
4.8 Comparison of the alpha-level approach to significance (a) vs. the p-value approach to significance (b) | 95

4.9 Contrasting views of the alpha-level approach (a) vs. the *p*-value approach (b) to significance based on the BIM example 97

4.10 Contrasting views of the alpha-level approach (a) vs. the *p*-value approach (b) to significance based on CFIs' complacency scores from the file Ch_4 Complacency Scores for Hypothesis Testing Example 99

4.11 The treatment effect of the 15-hour workshop designed to reduce CFIs' level of complacency "shifted" the untreated population's distribution four-tenths of a standard deviation to the left. This is considered a medium effect 103

5.1 A comparison of different *t* distributions relative to the normal distribution 116

5.2 Illustration of a one-tailed *t* test to the left for Example 5.1 117

5.3 Illustration of a two-tailed *t* test for Example 5.2 117

5.4 Illustration of *t* test for Example 5.4 120

5.5 Illustration of a two-tailed test for Example 5.4 121

5.6 Critical regions for Chapter 5's Guided Example 123

5.7 Illustration of results of *t* test for Chapter 5's Guided Example 124

6.1 General forms of various bivariate relationships 134

6.2 Bivariate scatter plot of the raw data given in Table 6.1. (*Note*: The scatter plot does not show duplicate entries.) 135

6.3 Bivariate scatter plot of the raw data given in Table 6.1 with the "line of best fit" inserted 135

6.4 Example of a nonlinear relationship (parabolic). As IQ scores increase, career satisfaction scores also increase but up to a certain point where they level off and then begin to decrease 136

6.5 Scatter plot of altitude versus fuel consumption relative to jet aircraft. The line of best fit "falls," which signifies a negative, or inverse, relationship 136

6.6 Scatter plots of six different data sets showing different degrees of linear relationship 138

6.7 Bivariate scatter plot for the data given in Table 6.2 140

6.8 Illustration of critical *r* values for a two-tailed test of the correlation coefficient, *r*, based on a sample size of $n = 25$, $df = 23$, and $\alpha = .05$ (a), and illustration of testing the significance of the correlation coefficient for the hypothetical data given in Table 6.2 (b) 143

6.9 The effect of outliers on correlation. Outliers can either inflate significance (a) or mask significance (b) 144

6.10 An illustration of using Table 4/Appendix A for sample size planning for Chapter 6's Guided Example 152

6.11 Scatterplot and line of best fit confirm linearity assumption for Chapter 6's Guided Example 153

6.12 Illustration of using Table 4/Appendix A to determine the power of Chapter 6's Guided Example 154

7.1 Illustration of the general characteristics of a linear equation given in slope-intercept form (a), and a specific application of the slope-intercept form of a linear equation (b) 161

7.2 Illustration that shows the deviation between observed points and their corresponding points on the regression line 163

7.3 Graph of the regression line $\hat{y} = -1.3x + 12.9$. The black points represent the observed scores from the given chart, and the three red points were used to construct the regression line 165

7.4 Illustration of a regression line that highlights there are always two points that correspond to any given x value: the observed score (x, y), and the predicted score $(x, \hat{y})$ 166

7.5 Scatter plot of the salary–career satisfaction data from the file Ch_6 Salary–Career Satisfaction Data (see also Table 7.1). The regression equation is $\hat{y} = 0.0041233x - 209.6475$ 166

7.6 Scatter plot of the fuel cost–air fare data for Example 7.1. 170

7.7 Critical region for one-tailed t test to the right for $\alpha = .05$ with $df = 28$ $(n = 30)$ 176

7.8 Illustration of the critical t values for Chapter 7's Guided Example (a), and the location of the calculated t statistic (b) with respect to the critical values 180

7.9 Residual plot that examines (and confirms) the linearity assumption for Chapter 7's Guided Example 181

7.10 Residual plot that examines the independence of the residuals assumption for Chapter 7's Guided Example. The residual scores of Y are placed on the vertical axis and the case numbers (1–62) on the horizontal axis. The result is a scatter plot that depicts no discernible pattern 181

7.11 Scatterplot from the regression analysis for Chapter 7's Guided Example. The endpoints of the regression line represent the mean annual salary for each group, respectively 182

8.1 Illustration of the critical t values for the time to PPL data given in Table 8.2 (a), and the location of the calculated t statistic with respect to the critical t values (b) 200

8.2 Illustration of critical t values for Chapter 8's Guided Example (a), and the location of the calculated t value with respect to the critical t values (b) 203

9.1 Differences in critical region boundary values between a preset alpha level of $\alpha = .05$ vs. the Bonferroni alpha level of $\alpha = .05/6 = .0083$, which is based on six pairwise comparisons 215

9.2 The structure of ANOVA and its nine components as denoted by the circled numbers in bold. (Adapted from Gravetter & Wallnau, 2017, p. 377.) 220

9.3 Various F distributions for different degrees of freedom 224

9.4 The critical regions for $F(2, 12)$ with respect to $\alpha = .05$ and $\alpha = .01$ from Table 5/Appendix A 224

9.5 Illustration of the critical F value for the data set from Ch_9 IRA Exam Data for ANOVA Hypothesis Test Example (a), and the location of the calculated F value with respect to the critical F value (b) 228

9.6 Illustration of the critical F value for Chapter 9's Guided Example (a), and the location of the calculated F value with respect to the critical F value 231

10.1 Matrix structure of a 2×2 factorial design 241

10.2 Matrix structure of a 4×2 factorial design for Example 10.1 242

10.3   The cells of the 2 × 2 factorial design for the weather knowledge study
       are completed with group means using the data from the file Ch_10
       Weather Knowledge Data   242
10.4   Descriptive illustration of the three effects of the 2 × 2 factorial design
       related to the weather knowledge study (see also Figure 10.3)   243
10.5   Graphical illustrations of the three different situations involving interactions 245
10.6   Numerical and graphical illustrations of a zero interaction   245
10.7   Constructing an interaction plot from the group-means table given in
       Figure 10.3 and replicated here for the convenience of the reader   246
10.8   Comparing the partitioning of total variance between single-factor and
       factorial ANOVAs. (Adapted from Gravetter & Wallnau, 2017, p. 377.)   248
10.9   Group-means table for the regional airline pilots' salary study   250
10.10 Interaction plot of the regional airline pilots' salary study   251
10.11 Group-means table for Chapter 10's Guided Example   259
10.12 Interaction plot for Chapter 10's Guided Example   260
10.13 Interaction plot for Chapter 10 Review Exercises Part A9   269
10.14 Interaction plot for Chapter 10 Review Exercises Part A10   270
11.1   (a) Illustration of critical t value for the data in Table 11.1, and (b)
       location of calculated t value with respect to the critical t value   282
11.2   (a) Illustration of critical t value for Chapter 11's Guided Example, and
       (b) the location of the calculated t value with respect to the critical t value 286
12.1   Illustration of the chi-square (χ^2) distribution   299
12.2   Illustration of how to use Table 6/Appendix A   300
12.3   Illustration of a critical χ^2 value vs. the corresponding calculated χ^2 value 301
12.4   (a) Illustration of the critical chi-square value for Chapter 12's Guided
       Example of goodness of fit test, and (b) the location of the calculated
       chi-square value with respect to the critical value   304
12.5   (a) Illustration of the critical chi-square value for Chapter 12's Guided
       Example of test for independence, and (b) the location of the
       calculated chi-square value with respect to the critical value   315
A1   Figure for Table 1   325
A2   Figure for Table 2   327
A3   Figure for Table 6   332

Tables

2.1	Frequency Distribution of Participants' Highest Level of Education	28
2.2	Relative Frequency Distribution of Participants' Highest Level of Education	28
2.3	Cumulative Relative Frequency Distribution of Participants' Highest Level of Education	29
2.4	Frequency Distribution of Number of Runway Incursions Reported at ATL During a 30-Day Period	30
2.5	Grouped Frequency Distribution of 100 Ages	31
2.6	Five-Number Summary for One-Way Commute Distances to the Airport	40
2.7	Calculating Standard Deviation	42
3.1	Standard Deviation Calculations for the Population {1, 3, 5, 7} and the Distribution of Sample Means Based on All Possible Samples of Size $n = 2$ Randomly Selected from the Same Population	68
4.1	Possible Outcomes and Decisions Relative to Testing the Null Hypothesis	100
5.1	Comparison of Confidence Intervals Relative to the Status of σ	118
6.2	Hypothetical Data Used to Demonstrate Raw Scores Formula for Calculating Pearson r	140
6.3	Summary Data Used for the Raw Scores Formula for Calculating Pearson r	141
7.1	Observed Scores, Predicted Scores, and Residuals for the Airport Managers' Annual Salary–Career Satisfaction Scores Data Set	167
7.2	Inputs for Calculating the Standard Error of Estimate ($SE_{Residuals}$) for the Hypothetical Data Involving Number of Type Ratings and Number of Pilot Deviations	173
7.3	Inputs for Calculating the Standard Error of the Regression Coefficient (SE_B) for the Hypothetical Data Involving Number of Type Ratings and Number of Pilot Deviations	174
7.4	Inputs for Calculating the Standard Error of Estimate ($SE_{Residuals}$) and the Standard Error of Regression (SE_B) for the Annual Salary–Career Satisfaction Data	177
8.1	Pooled and Separate Variances Formulas for the Independent-Samples t Test	194
8.2	Time to PPL Data Used for Constructing the 95% CI for the Difference in Sample Means	197
8.3	Analysis of Variance (ANOVA) Table for Independent-Samples t Test Guided Example	205
8.4	Summary Chart for Chapter 8 Exercises, Part B–Section C	210
9.1	Small Within-Groups Variability but Large Between-Groups Variability	216

9.2 Large Within-Groups Variability but Small-Between Groups Variability 216
9.3 Replica of Table 2.7 from Chapter 2 for Calculating Standard Deviation 219
9.4 Basic Structure of the ANOVA Summary Table 220
9.5 ANOVA Summary Table for IRA Exam Scores Example 222
9.6 ANOVA Summary Table for BIM-GWS IRA Exam Study from Ch_8 Guided Example Data 225
9.7 ANOVA Summary Table for Ch_9 IRA Exam Data Used in ANOVA Hypothesis Test Example 228
9.8 Summary of Tukey's HSD Post Hoc Comparisons for Ch_9 IRA Exam Data Used in ANOVA Hypothesis Test Example 229
9.9 ANOVA Summary Table for Chapter 9's Guided Example 232
10.1 Generic Structure of the Factorial ANOVA Summary Table Using Letters of the Alphabet to Represent Cells 249
10.2 General Structure of the Factorial ANOVA Summary Table with Proper Notations 249
10.3 Factorial ANOVA Summary Table for the Salary Example 251
10.4 Factorial ANOVA Summary Table for the Weather Knowledge Study Data 254
10.5 Factorial ANOVA Summary Table for the Guided Example 258
10.6 Factorial ANOVA Summary Table for the Guided Example Extended to Include Effects Sizes 262
11.1 Fictitious Data to Demonstrate the Computation of the Repeated-Measures t Test 280
12.1 Number of Runway Incursions per Human Factor Errors 298
12.2 Observed and Expected Number of Runway Incursions per Human Factor Errors Based on a Hypothesized Even Distribution 299
12.3 Observed and Expected Number of Runway Incursions per Human Factor Errors Based on a Hypothesized Even Distribution and Configured to Follow the Chi-Square Test Statistic Formula 301
12.4 Frequency Distribution Table for the Guided Example Involving Chi-Square Test for Goodness of Fit 305
12.5 Observed and Expected Frequencies for the Guided Example Involving Chi-Square Test for Goodness of Fit 306
12.6 Observed and Expected Frequencies for Airline Seat Preferences for the Guided Example Involving Chi-Square Test for Goodness of Fit 306
12.7 Contingency Table for FAA 1500-Hour Requirement Study 308
12.8 Contingency Table for FAA 1500-Hour Requirement Study with Proportions (p) 310
12.9 Cohen's Effect Sizes for Chi-Square Test for Independence 312
12.10 Contingency Table for the Guided Example for Chi-Square Test for Independence Involving Weather Conditions and Runway Incursions with Observed Frequencies 313
12.11 Complete Contingency Table for the Weather Conditions–Runway Excursions Study for the Guided Example for the Chi-Square Test for Independence 316
1 Area Under the Standard Normal Curve Between the *Mean* and Z 325
2 Critical Values of Student's t Distribution, which Equate to a Particular Proportion of the Area Under the Curve for a One-Tailed Test 327
3 Critical Values for Pearson r 328

4   Power Table for Pearson *r*   329

5a   Critical *F* Values (α = .05)   330

5b   Critical *F* Values (α = .01)   331

6   Critical Values of the Chi-Square Distribution   332

7a   Sample Size (*N*) for χ^2 Test for Independence to Detect Effect Size (*V*)
    for α = .05 with Corresponding *df* and Contingency Table of Size (*R* × *C*)   333

7b   Sample Size (*N*) for χ^2 Test for Independence to Detect Effect Size (*V*)
    for α = .01 with Corresponding *df* and Contingency Table of Size (*R* × *C*)   334

Preface

This book is a traditional first-level statistics textbook grounded exclusively in an aviation context. We wrote the book to make the study of statistics more meaningful and accessible for aviation students by employing prose, examples, and exercises that are aligned to aviation research studies. To facilitate this approach, we augmented the standard statistical strategies of a first-year statistics course with guided examples structured into three parts: pre-data analysis, data analysis, and post-data analysis. Pre-data analysis includes writing purpose statements, operational definitions, research questions and research hypotheses, and engaging in sample size planning via an a priori power analysis. Data analysis includes conducting a hypothesis test relative to the targeted statistical strategy. Post-data analysis includes calculating and interpreting effect sizes, confidence intervals, post hoc tests (if relevant), and post hoc power. We also discuss plausible explanations for the results.

This book was written for both undergraduate and graduate aviation students and is appropriate for introductory statistics courses; capstone courses; research courses where students learn to plan, conduct, and report research; and as a reference or supplement for students pursuing independent study and/or conducting thesis or dissertation research. The concepts discussed in the textbook are explained fully and simply in an uncomplicated manner. Furthermore, by grounding the discussion in an aviation research context, aviation students will be able to better understand how statistics and research are applied to their discipline and real-world situations. The reader is cautioned that this book is *not* a research design textbook, but instead is a statistics textbook that integrates fundamental principles of research methodology applied to aviation.

We organized the material in this book into six major parts.

- Part A, *Research and Statistics Basics*, presents basic methodological concepts of research and descriptive statistics and consists of two chapters. Chapter 1 introduces the research process, population-sample relationship, sampling strategies, common research methodologies, variables, and measurement scales. Chapter 2 presents frequency distribution tables and graphs for organizing data, and measures of central tendency, dispersion, and position for summarizing data.
- Part B, *Making Reasonable Decisions about a Population*, begins the transition from descriptive statistics to inferential statistics and consists of two chapters. Chapter 3 introduces z scores, the standard normal distribution, and the concept of a sampling distribution, and Chapter 4 introduces the concept of statistical inference from two perspectives: directly via parameter estimation and indirectly via hypothesis testing.

- Part C, *Analyzing Research Data Involving a Single Sample*, presents statistical strategies for analyzing data from research studies involving a single sample and consists of three chapters. Chapter 5 introduces the single-sample *t* test, Chapter 6 introduces bivariate correlation, and Chapter 7 introduces bivariate linear regression.
- Part D, *Analyzing Research Data Involving Two Independent Samples*, presents statistical strategies for analyzing data from research studies involving two independent samples and contains three chapters. Chapter 8 introduces the independent-samples *t* test, Chapter 9 introduces the single-factor ANOVA, and Chapter 10 introduces factorial ANOVA.
- Part E, *Analyzing Research Data Using a Within-Groups Design*, consists of a single chapter, Chapter 11, which focuses on within-group designs, and introduces the repeated-measures *t* test.
- Part F, *Nonparametric Statistics: Working with Frequency Data*, consists of a single chapter, Chapter 12, which introduces two nonparametric strategies: the chi-square test for goodness of fit, and the chi-square test for independence.

Each chapter closes with a *chapter summary*, which reviews the salient aspects of the chapter, a *vocabulary check*, which lists the keywords or phrases introduced in the chapter, and *review exercises*, which contain two parts. Part A, Check Your Understanding, consists of 10 multiple-choice items. Part B, Apply Your Knowledge, consists of a research study and corresponding data for students to apply the three components of the guided examples—pre-data analysis, data analysis, and post-data analysis. The book's end matter consists of various statistics tables and an index.

This textbook is designed for use in a one-semester course. For undergraduate students, we suggest focusing on Chapters 1–8, and (optionally) Chapter 12 or another additional chapter as appropriate for a selected class project. For graduate students, all 12 chapters are appropriate. Additional resources are also available to facilitate instruction. These include PowerPoint slides by chapter; a separate PDF document with copies of the figures and tables; copies of all the data sets in Excel format; and an instructor's manual that contains solutions to the Part B chapter exercises, chapter tests, which consist of a multiple-choice items test bank and Apply Your Knowledge exercises with data sets, and solution to the test items.

Many people contributed considerably to the preparation of this material. We begin first with the editorial staff at Routledge. A warm "thank you" is extended to Dr. Suzanne Kearns, editor of the Aviation Fundamentals series, for considering and then recommending our proposal be a part of the series. We also are grateful to Natalie Tomlinson, former commissioning editor, who guided the revision of our initial proposal, and then successfully pitched it to the Routledge editorial board. A note of gratitude also is extended to Andrew Harrison, who was Natalie's successor; Helena Parkinson, our editorial assistant; and Routledge's production staff, particularly Kelly Cracknell. This book also benefited from the reviewers of the manuscript, whose constructive criticisms strengthened the quality of the book. We also acknowledge the contributions of our former students in the College of Aeronautics at Florida Institute of Technology, whose comments and research projects inspired us to write this textbook and served as the basis for the textbook's structure. Last but not least, personal gratitude is extended to Janie and our respective families for their support, patience, and understanding throughout this entire project.

Part A

Research and Statistics Basics

1 The General Nature of Research and Statistics

Student Learning Outcomes

After studying this chapter, you will be able to do the following:

1. Determine an appropriate research objective for a research topic by writing a corresponding purpose statement and research question, and operationally defining all key terms/phrases.
2. Describe the population, sample, and sampling strategy of a given research study.
3. Determine if a sample is representative of its parent population.
4. Explain the concept of sampling error.
5. Determine the appropriate research methodology for conducting a study.
6. Determine the variable(s) of a study, the variable(s) type, and the corresponding measurement scale.

1.1 Chapter Overview

Given the focus of this book, which is using statistics to answer research questions, we begin our study of statistics by introducing the concepts of research and statistics. As illustrated in Figure 1.1, research involves collecting data (called *raw scores*) to further our understanding of something, and statistics provides a bridge between the collected data and meaningful information by converting raw scores to a *derived score* such as a percent. In this chapter, we introduce the major concepts associated with any research endeavor that warrants a statistical analysis of the collected data. The major topics include the research process, the relationship between a population and sample, commonly used research methodologies for collecting numerical data, the different types of variables associated with research, and the measurement scales on which variables are measured.

1.2 The Research Process

There are two general types of research studies: quantitative and qualitative. *Quantitative studies* involve collecting numerical, or *empirical data*, and use statistics for data analysis. *Qualitative studies* involve collecting non-numerical, or *contextual data*, and do not use statistics for data analysis. In this textbook, we restrict our discussion to quantitative research studies and present various statistical strategies used to analyze empirical data.

Regardless of the approach, though, research may be thought of as a formal process of inquiry that involves collecting and analyzing data in a systematic manner to help provide dependable and useful answers to meaningful questions. Although there will

DOI: 10.4324/9781003308300-2

RESEARCH

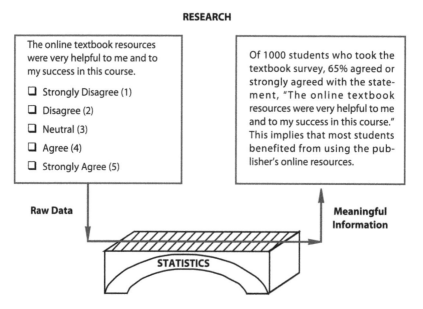

Figure 1.1 The relationship between research and statistics.

be some variation with respect to how this is done, this process generally involves the four steps shown in Figure 1.2.

Step 1: The research objective. Researchers first identify a research problem they want to investigate. This problem generally begins as a topic of interest, which is then refined and expressed as a specific objective called a purpose statement. Thus, a purpose statement describes the primary focus of a study. For example, if our topic of interest is runway incursions and we are interested in understanding what factors are associated with runway incursions, then the corresponding purpose statement would be:

> *The purpose of this study is to determine what factors are associated with runway incursions.*

This research objective also may be expressed as a research question (RQ):

> *What factors are associated with runway incursions?*

Note that a purpose statement is expressed in declarative form—it ends with a period—and the corresponding RQ is a rewording of the purpose statement and is expressed in interrogative form—it ends with a question mark.

When specifying the research objective, it often is necessary to also include operational definitions of key terms or phrases. With respect to the runway incursions study, there are several terms or phrases that need to be defined:

- What type of *pilots* will be targeted? For example, will we focus on general aviation (GA) pilots or airline transport pilots (ATPs)?

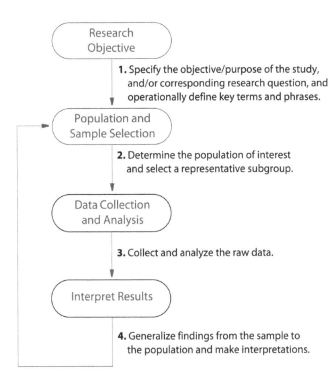

Figure 1.2 The research process.

- What *type of runway incursions* will be targeted? For example, will we focus on operational errors (OEs), which are air traffic control (ATC)-related incursions, pilot deviations (PDs), or pedestrian-vehicle deviations (P/VDs)?
- What *factors* will be targeted? For example, will we focus on the physical features of runways such as airfield signage, markings, and lighting; airport policies; or pilot demographics such as age, years as pilot-in-command (PIC), biological sex, highest level of education, total flight hours, and number of ratings and certificates held? (*Note:* Following the guidelines of the American Psychological Association (APA, 2020, p. 139), we use the term *biological sex* to mean "sex assigned at birth," which functionally reflects a binary assignment of female or male.

By defining key terms and phrases using operational definitions, we are informing the reader exactly what these terms and phrases mean relative to the context of the given study. Operational definitions also mitigate any ambiguity, and they facilitate replication studies. For our running example, we might define pilots as GA pilots, restrict the study to PD runway incursions, and focus on the pilot demographics and experiences listed above. Observe that this is how *we* choose to define these terms. Someone else might choose to define these terms differently. Thus, there are no correct or incorrect definitions, but instead definitions are relative to the researcher's objective.

Operational definitions also enable us to empirically measure psychological or social *constructs*, which are not directly observable. Unlike tangible characteristics such as age, total flight hours, and education level, constructs are abstractions that cannot be observed directly. Examples of constructs in aviation science research include attitudes, risk perception, motivation, anxiety, complacency, and self-efficacy. To examine a construct, we operationally define it by identifying an instrument that was designed specifically to measure the targeted behavior or construct.

For example, let's assume we want to examine certified flight instructors' (CFIs) attitudes toward flight instruction. To do this, we identify an instrument specifically designed to measure CFIs' "attitudes." One such instrument is Dunbar's (2015) Attitudes toward Flight Instruction Inventory (ATFII), which measures attitudes using a traditional 5-point *Likert response scale*. This type of scale consists of a set of statements reflecting the construct, and respondents assess their level of agreement with each statement—Strongly Disagree, Disagree, Neutral/Undecided, Agree, and Strongly Agree. Each response is given a weighted score of 1, 2, 3, 4, and 5, respectively, with higher aggregate scores reflecting a more positive tendency toward the targeted construct. (*Note:* Similar response scales that do not use a 5-point scheme are referred to as *Likert-type* scales.) Similarly, if we endeavor to examine GA pilots' level of risk perception, we could operationally define "risk perception" as scores on Hunter's (2002) Risk Perception-Other instrument, which measures risk perception on a scale from 1 to 100, where higher scores reflect a greater ability to perceive risky situations. Thus, operationally defining a construct from the perspective of a measuring instrument and its corresponding scores enables us to transform something that is not directly measurable into empirical data.

Example 1.1: *Purpose Statement, RQ, and Operational Definitions*

Let's assume you are interested in examining the relationship between pilots' diet and their flight performance. Write a corresponding purpose statement, RQ, and operationally define all key terms and phrases.

Solution

- **Purpose statement**. The purpose of the study is to determine the relationship between pilots' diet and flight performance.
- **Research question**. What is the relationship between diet and flight performance?
- **Operational definitions**. The terms that need to be defined are "pilots," "diet," and "performance." We will define pilots as Part 121 pilots. We will define diet as the average number of calories consumed daily. We will define performance, which is a construct, as an aggregate score based on the number of errors a pilot makes relative to maintaining a correct altitude, heading, and airspeed. Performance will be measured in a simulator, and lower aggregate scores reflect fewer errors and thus better performance.

Step 2: Population and sample selection. Note from the runway incursions example given in Step 1 that it would be cost-prohibitive, if not impossible, to solicit information from every GA pilot in the world. To make research studies more manageable and efficient, what we do instead is select a representative *subgroup* of the group we are interested in studying. Using the language of research, we call the targeted group of interest the population, we refer to the subgroup as the sample, and we call the process of selecting a sample from a population sampling. Thus, as illustrated in Figure 1.3, a population is the *universal set* because it refers to all the individuals or items we are interested in studying, and a sample is a *subset* because it is a subgroup of the population.

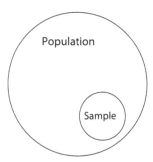

Figure 1.3 A sample is a subset of the population.

By selecting a sample from the population, we can concentrate our efforts on the sample instead of the entire population. Thus, we learn about the population's characteristics by studying the sample's characteristics, which makes conducting the study more feasible with respect to data collection efforts, time, money, and other resources. With respect to our running example involving runway incursions, the population might be all U.S. GA pilots, and the sample might be 200 records acquired from the runway incursions database maintained by the U.S. Federal Aviation Agency (FAA). Notice that we also could have defined the population to be all GA pilots in the world or all GA pilots in the Canadian province of Ontario. How broad or narrow the population is defined is up to the researcher, and the size of a sample is determined by a formal process called ***power analysis***, which we discuss in Chapter 2.

Because the primary purpose for selecting a sample is to learn about the population, it is extremely important that a sample accurately reflects the population from which it is selected, which we call the parent population. This makes samples very important in any research study because if the sample does not accurately represent its parent population, then the inferences we make will be spurious. Therefore, it is incumbent for us to select a representative sample, which means that the sample is reasonably typical of its parent population and is reasonably large enough to represent the parent population. If a sample is representative of its parent population, then we can generalize the results from the sample back to the parent population with reasonable confidence. If, on the other hand, a sample is not representative of its parent population, then it is considered a biased sample, and the findings from the sample cannot be generalized to the parent population with any degree of confidence.

There are many ways to select a sample from a population. As illustrated in Figure 1.4, which contains a summary of commonly used approaches, sampling strategies fall into one of two categories: probability vs. non-probability. Probability sampling approaches involve *random selection* and are more likely to produce a representative sample. The concept of random sampling involves selecting a sample from the population in such a manner that each person or object in the population has an *equal and independent chance* of being selected. An example of a random selection process is the air-mix lotto machine, which uses air to blow around numbered ping-pong balls in an enclosed chamber to determine winning lottery numbers. In simple cases that involve small sample sizes, we can select a random sample by flipping a coin, rolling dice, or picking numbers out of a hat. Other approaches to random sampling include using a random number table, using the random number feature from a statistics software package, and consulting an online random number generator such as random.org (2022).

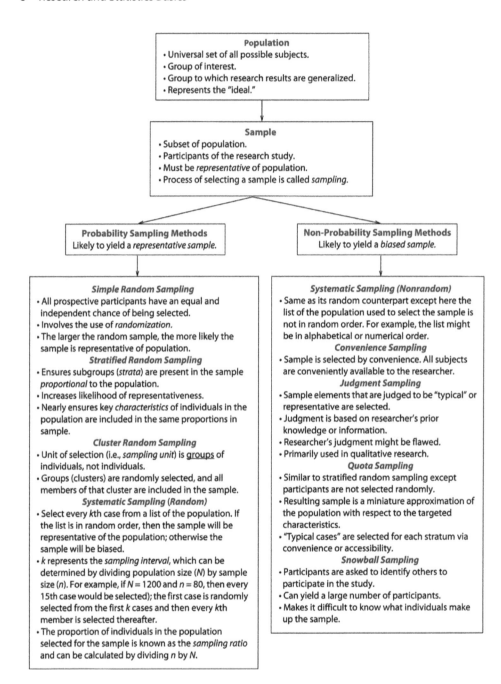

Figure 1.4 A comparison of various sampling strategies.

Non-probability sampling approaches are not randomly based and are more likely to result in a biased sample. Based on our experiences, most students who engage in aviation research as part of a capstone project, master's thesis, or dissertation research generally select their sample by convenience or snowball. Many of these students also have a misconception that if they can acquire a sufficiently large sample, then this would mitigate the inherent bias of a nonrandomly selected sample. This misconception also appears to be deeply rooted for some students even though they were taught that the results of data analysis acquired from a nonrandom sample, regardless of how large it is, will not reflect accurate estimates of the parent population.

To illustrate this point, let's assume we are interested in examining factors related to runway incursions and seek the opinions of all individuals involved in airport operations, including airport managers, ATCs, pilots, ground support crew, and third-party vendors. This would be our population. Let's further assume that we post a questionnaire on the National Association of Airport Executives (NAAE) website and via a snowball approach acquire a very large sample (e.g., 280) of airport managers at U.S. commercial airports who complete the questionnaire. Observe that this sample reflects *some* of the people we have targeted, but this "some" is not representative of "all," and therefore the sample is biased. As observed by Kachigan (1991, p. 99): *"it is better to be approximately correct, than precisely wrong."*

Example 1.2: *Analyzing the Population-Sample Relationship*

Caligan (2012) examined the extent to which "the amount of training a civilian pilot needs to fly an airplane solo is a predictor of future performance" (p. 31). One of his research questions was, "Can the number of hours needed to solo predict the number of hours needed to earn the private pilot license?" Caligan posted information about his study and a hyperlink to his questionnaire on "Internet forums at the Aircraft Owners and Pilots Association, American Bonanza Society, Beechtalk, Cirrus Owners and Pilots Association, Oshkosh 365, and Pilots of America" (p. 39). The total number of respondents was 306, but he discarded 33 incomplete responses. Determine the population, sample, and sampling strategy Caligan used.

Solution
- **Population**. The population is presumed to be all civilian pilots who have at least a PPL.
- **Sample**. The sample consisted of 273 pilots who subscribed to the targeted Internet forums, volunteered to participate, and provided complete responses to the questionnaire.
- **Sampling strategy**. The sampling strategy was convenience. It also could have involved a snowball approach if subscribers to the targeted Internet forums told others about the study who then chose to participate. Although a relatively large sample was acquired, the sample was not representative of the population, and therefore, caution must be exercised when generalizing the results of the sample to the population.

Step 3: Data collection and analysis. After a sample is selected, we next implement the study by collecting and analyzing data. The data we collect and analyze are observations, or measurements, of specific traits or characteristics of people or objects, and the traits or characteristics we measure are specific to the research setting. For example, in our running example of runway incursions, some of the characteristics included pilots' age, biological sex, and total flight hours. Once we have acquired our data, we then use specific statistical procedures to analyze the data, and ultimately help answer the corresponding RQ. Our discussion of these statistical procedures begins in Chapter 2.

Step 4: Interpret results. What emerges from data analysis in Step 3 are the results, or *findings*, of the study. Once we know what the findings are from the sample, we then use them to make *inferences* about the parent population. In other words, we *generalize* the results from the sample back to the population. For example, with respect to our running runway incursions example, we might learn from the sample of 200 records of GA pilots who were involved in a pilot deviation runway incursion that 65% were male with fewer than 1,000 hours of flight time. Assuming we selected these records randomly and therefore have a representative sample, these findings would be reflective of the parent population, which would signal a need to target male GA pilots with fewer than 1,000 flight hours for a safety campaign. If, however, the sample was biased, then we would need to exercise caution in interpreting these results relative to the parent population.

1.3 The Population-Sample Relationship and Sampling Error

As discussed earlier, because it is often prohibitive to study all members of a population, we instead select a representative sample, measure the characteristics of the sample, and then infer what we learned from the sample back to the parent population. Note that this is an *inductive process* because we observe only some instances of the parent population and then generalize our observations to this population. Although it might seem obvious, it is important to recognize that a sample and its parent population are two separate entities. This separation between a sample and its parent population makes it necessary to distinguish between what we measure in the sample and the corresponding measurement in the population. This is because any inferences we make from the sample will only be estimates of the true values in the population.

As an illustration, let's consider the population of all pilots in the United States who are members of the Aircraft Owners and Pilot Association (*AOPA*), and the characteristic we are interested in is members' average flight time. To determine the average flight time of this population, we would have to acquire the total number of flight hours of all U.S. members, add them, and then divide the sum by the total number of members. This can be quite difficult to accomplish. What we would do instead is (a) select a representative sample of the member population, (b) acquire the total flight time from the sample members and calculate this group's average flight time, and (c) use the sample's average flight time to infer the average flight time of the population. Thus, there are two different average flight times: the "true" one for the population, which is unknown, and the one acquired from sample, which is used to *estimate* the population's true average flight time.

To distinguish between the two average flight times, we refer to the "true" but unknown one as the *population parameter* and call the one acquired from the sample the *sample statistic*. Therefore, in the language of research and statistics, population characteristics are referred to as population parameters, and the corresponding sample characteristics are referred to as sample statistics. Population parameters are denoted using Greek letters, and sample statistics are denoted using English letters. For example, the numerical average (called the "mean," which we discuss in Chapter 2) of a population is represented using the Greek letter μ ("mu"), and the corresponding sample statistic is denoted M. Because sample statistics are considered estimates of population parameters, when we make inferences about a population based on sample data, a key question is, "How likely will a sample statistic be truly reflective of the corresponding population parameter?" When examined from the context of measurement, this question reflects the concept of accuracy because we are interested in how close a sample statistic is to the true measure in the population. The presumption is that if the sample represents a good cross-section of the population's members, then the sample statistic will be an accurate estimate of the corresponding population parameter.

To illustrate this relationship between population parameter and sample statistic, let's return to the *AOPA* example. For instructional purposes, and as illustrated in Figure 1.5(a), let's assume that the population consists of 100 members and that the population's true arithmetic average flight time is 3,997 hours. Let's further assume that we randomly select a sample of size $n = 10$ from this population and the average flight time is 5,050 hours. Note that the sample statistic is much larger than the corresponding population parameter (5,050 vs. 3,997 hours). Let's now randomly select a second sample of size $n = 10$ as shown in Figure 1.5(b). Notice that this sample statistic is much closer—and therefore more accurate than that of the first sample statistic—to the true population parameter (4,250 vs. 3,997 hours). In reality, though, we generally do not know the true value of the population parameter and therefore we will not know the accuracy of the sample statistic without analyzing the data from the entire population.

This example also illustrates the concept of sampling error, which is the amount of error that exists between a sample statistic and the corresponding population parameter. To minimize sampling error, we endeavor to select a sample that is representative of the population from which it was selected so that the sample adequately portrays a cross-section of the population relative to the characteristics being studied. Remember: A representative sample makes it possible to generalize the results from the sample back to the parent population with reasonable confidence, whereas an unrepresentative sample, regardless of size, is a biased sample, and findings from biased samples cannot be generalized to the parent population with confidence.

1.4 Quantitative Research Methodologies

As noted earlier, the scope of a research study is given by the study's objective, which is expressed as a purpose statement and corresponding research question. In addition to establishing the framework of a study, a RQ also helps determine the appropriate research methodology, which provides specific information about the way a study is to be conducted. As illustrated in Figure 1.6, quantitative research methodologies may be broadly classified as either *experimental* or *non-experimental*, and non-experimental methodologies—which are also known as *observational* studies—are classified as either *associational* or *descriptive*. Two types of associational studies are *correlational* and *causal-comparative* (also called *ex post facto*), and one type of descriptive study is *survey*, which is not to be confused with a survey instrument.

Experimental Research

An experimental research methodology is what most people think of when they hear the word "research." An experimental study is conducted when trying to demonstrate a *cause-and-effect relationship*. This means that the primary purpose of the inquiry is to

(a)

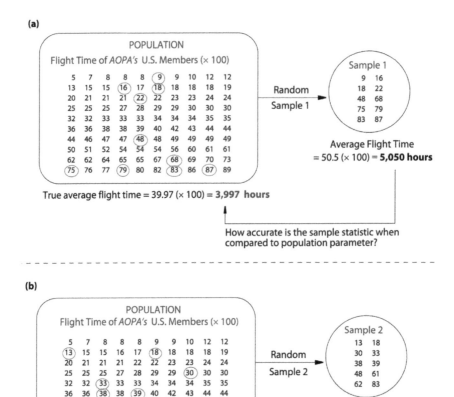

(b)

Figure 1.5 Comparing a sample statistic to the corresponding population parameter. The circled items represent randomly selected flight times.

demonstrate that a person, fact, or condition is indeed responsible for a specific effect. The primary research question asked in experimental studies is expressed generically as "What is the effect of X on Y?" In the context of research, X is the *treatment*, or *treatment condition*, and Y is the *outcome*.

As an example, let's suppose we want to determine if there is any difference in flight students' achievement in an aviation meteorology course between two different instructional approaches: "traditional" (large-group lecture) vs. simulation, where students are presented various weather situations via a simulator. A corresponding research question would be, "What is the effect of two different instructional approaches, traditional vs. simulation, on flight students' achievement in aviation meteorology?" Here, the treatment condition, X, is the two different instructional approaches, and the outcome, Y, is student achievement. In this context, we refer to X as the *independent variable* and Y as the *dependent variable*.

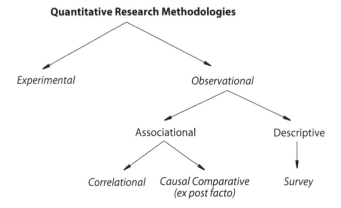

Quantitative Research Methodologies

Figure 1.6   Common types of research methodologies used for answering quantitatively based research questions.

An experimental study also is considered an *intervention* study because it involves two key components: *manipulation* and *control*. The researcher deliberately and systematically manipulates the treatment condition, and controls for the effect of all other factors that might influence the outcome. These other factors are called *extraneous factors* because they are neither pertinent nor relevant to the objective of the study, but they could have an impact on the results of the study if not controlled. For example, if we were examining the effect of sleep deprivation on pilot performance, then possible extraneous factors might be pilots' age, diet, and marital status. An experimental study also involves at least two groups. In the most general sense, one group is the *treatment group*, which receives the experimental treatment, and the second group is called the *control group*, which receives a "pseudo treatment" called a *placebo* designed to look like the actual treatment. It also is possible that instead of a control group, there might be different "levels" of a factor. For example, in the sleep deprivation study, the treatment would be sleep deprivation and the different levels of sleep deprivation might be 0 hours, 6 hours, and 12 hours.

One of the challenges to conducting an experimental study is the control component. It is not always realistically possible, particularly when dealing with people, to ensure that the groups are equivalent in every manner except for the treatment condition. The single best strategy for establishing equivalent groups is through *random assignment*. Random assignment is similar to random selection except now the focus is on *assigning* participants to a group instead of *selecting* participants from a population. This distinction is illustrated in Figure 1.7. Thus, with random assignment, every member of the sample has an equal and independent chance of being assigned to a group. Random assignment helps to account for any possible differences between the groups a priori because chance alone would be the only reason the groups would not be equivalent relative to any possible extraneous factors. Groups formed by random assignment are considered statistically equivalent, which means that any differences between the groups are not related to researcher bias but are a function of chance.

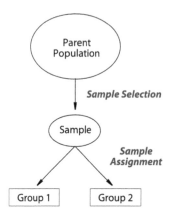

Figure 1.7 Comparison between sample selection vs. sample assignment.

Correlational Research

A correlational research methodology is used when we seek to determine the relationship between two or more research factors. If a relationship exists, then we say the factors are *correlated* (or co-related) with one another, which means the factors share something in common and hence influence each other. The research study we presented in Example 1.1, which examined the relationship between pilots' diet and flight performance, is an example of a correlational study. Similarly, the Caligan (2012) study presented in Example 1.2, which examined the extent to which the number of hours a pilot needs to fly an airplane solo could be used a predictor of the hours needed to earn a PPL, also is a correlational study.

The degree to which two factors are related is determined by measuring the *direction* and *strength* of the relationship. This is done by calculating a *correlation coefficient*, which is a numerical index derived from the raw data. One such index is the *Pearson product moment coefficient of correlation*, denoted r, which is used to measure *linear* relationships and can vary between −1.0 and +1.0. We will discuss this index and the concept of correlation in more detail in Chapter 6. For now, though, note the following:

- If $r > 0$, then the relationship is positive and the two variables tend to move in the same direction. For example, a pilot's age and reaction time are generally positively correlated: As pilots' age increases, their reaction time also increases.
- If $r < 0$, then the relationship is negative and the two variables tend to move in the opposite direction. For example, a plane's altitude and fuel consumption are negatively correlated: As altitude increases fuel consumption decreases.
- If $r = 0$, then there is no relationship between the variables. For example, there is no relationship between the overall average annual salary of U.S. flight attendants and the overall average years of experience of U.S. aircraft maintenance technicians (AMTs).
- If r approaches +1.0 or −1.0, then there is a strong positive relationship or a strong negative relationship, respectively, but if r approaches 0, then the relationship is weak.

Correlational studies generally involve collecting data on multiple variables from a single group. It is important to understand that although correlational research can confirm that two factors are related to each other, it cannot be used to claim a cause-and-effect relationship. For example, we might find a strong, positive relationship between pilots' age and flight time, but in the absence of manipulation and control (the hallmarks of experimental research), we cannot claim an increase in pilot age *causes* an increase in flight time because there might be extraneous factors influencing flight time other than age.

Causal-Comparative Research

In many research endeavors, there often are times when we are unable to directly control or manipulate the factor(s) of interest. For example, suppose we are interested in determining if there is a difference in annual salary among three groups of AMTs based on their highest level of education: high school, 2-year college, and 4-year college. The corresponding research question might be, "What is the effect of AMTs' level of education on annual salary?" In this example, the targeted research factor is "level of education" and the outcome is "annual salary." Recognize that we cannot directly manipulate or control these factors because the effect of "level of education" on "annual salary" has already occurred. What we do instead is ask individuals from the three preexisting groups to specify their annual salary, and then determine if a relationship exists by comparing the salaries among the three groups. As another example, consider the research question, "What is the effect of biological sex on flight students' attitudes toward females in aviation?" In this example, the targeted research factor is a person's biological sex, which involves two preexisting groups: male and female flight students. In this case, though, the groups inherently cannot be manipulated because we cannot assign someone to the male group or the female group.

In each example, you should recognize that we are trying to determine the cause for, or consequences of, differences between preexisting groups. The problem, though, is we cannot examine this using an experimental study because (a) the effect of the research factor on the outcome of interest has already taken place, as demonstrated in the first example, or (b) it is impossible to manipulate the research factor, as demonstrated in the second example. To examine these types of cases, we conduct a causal-comparative study, which also is called ex post facto. Because we cannot manipulate or control the research factors in a causal-comparative study, we cannot claim a cause-and-effect relationship. It also is more hazardous to infer there is a genuine relationship between a treatment and an outcome. Although limited in its interpretation, a causal-comparative study has value because it helps identify possible relationships that were not previously considered. This methodology is commonly used in the medical field where it is referred to as *retrospective* research.

Survey Research

A survey research methodology, or more simply, *survey*, involves collecting information from a group of people to describe various characteristics of the population of which the group is a part. Common characteristics include demographic information such as age, ethnicity, religious preference, and annual earned income, as well as opinions and attitudes about some issues. For example, survey research is used to measure the president's approval rating, high school graduation rates, and the eating habits of children. Similar to any research endeavor, a survey involves collecting data from a sample and then using these sample data to make inferences about the corresponding population. When an entire population is surveyed, it is called a *census*.

There are two major types of surveys: longitudinal and cross-sectional. A *longitudinal survey* involves gathering data at different points in time to study changes over a specified period, whereas a *cross-sectional survey* provides a snapshot of some issue(s) at a specific point in time. An example of the former is Taylor et al.'s (2007) 3-year longitudinal study that annually assessed the flight performance of GA pilots between 40 and 69 years old to determine the influence age and aviation expertise have on flight simulator performance. An example of the latter is the U.S. FAA's (n.d.) annual General Aviation Survey, which gathers information on GA and on-demand Part 135 aircraft activity. Because this is done yearly, the data collected for any given year are considered cross-sectional. Note that if someone were to examine trends in the GA Survey data across multiple years, then this would be considered a longitudinal study.

Accompanying the results of a survey is a corresponding *margin of error*, which describes how well the results from a sample reflect the actual state of affairs in the parent population. For example, if 48% of the people surveyed favor a proposed airport expansion project and the survey has a margin of error of ±5 percentage points, then this means that the percentage of people in the population who favor this project is 48% ± 5%, or between 43% and 53%.

A Note of Caution: Be sensitive to the context in which the term "survey" is used. Survey research is a formal research *methodology* whereas the term "survey" is a generic term that commonly is used to refer to a data collection instrument. To avoid confusion, we suggest using the term *questionnaire* instead of "survey" when referring to a data collection instrument.

Example 1.3: *Research Methodologies*

For each given purpose statement or research objective in a–d, determine the corresponding research methodology you think is most appropriate, and explain why you think it is appropriate. Note that in some cases, depending on how the study is implemented, more than one research methodology might be appropriate.

a. To investigate the effects of three methods for keeping airport managers up-to-date on TSA security protocols. Airport managers may be assigned at random to the different methods.
b. To investigate changes in cognitive functioning by testing a sample of GA pilots older than 50 years every 5 years for the next 20 years.
c. To investigate the extent to which an aviation meteorology prognosis test predicts subsequent aviation meteorology achievement of flight students.
d. To investigate the relationship between GA pilots' flight hours and their level of risk perception.

Solution

a. *Experimental.* We can randomly assign airport managers to one of the three methods, and then after a certain time (e.g., 1 month or 6 months) we would measure managers' level of currency with respect to TSA security protocols to determine which method was most effective.

b. *Survey.* This is an example of a longitudinal study because data are being collected every 5 years for 20 years for the expressed purpose of comparing changes in pilots' cognitive functioning.

c. *Correlational.* This study involves one group (flight students) and we are examining the relationship between two specific measures: scores on the aviation meteorology prognosis test and scores on the aviation meteorology achievement exam.

d. *Correlation or Causal-Comparative.* If flight hours are being reported numerically (e.g., 1,500 hours or 5,617 hours), then this would be a correlational study because we have one group (GA pilots), and we are examining the relationship between participants' flight time and their hazardous attitudes. However, if flight hours are reported as categories (e.g., less than 500 hours, 500 to less than 1,000 hours, 1,000 to less than 2,000 hours, etc.), then we have a *group membership* variable, and we cannot assign participants to these groups because the groups are preexisting based on participants' flight hours. In this case, the research methodology is causal-comparative.

1.5 Variables and Measurement Scales

Variables

To implement the data collection step of the research process, we use well-defined instruments designed to measure the targeted characteristics. In the context of research, we refer to these characteristics as variables, and examples include age, height, number of FAA certifications, number of flight hours, and biological sex. Variables are so named because the characteristics they represent *vary* among individuals. For example, every person does not have the same age or height, every pilot does not have the same number of FAA certifications or number of flight hours, and everyone is not of the same biological sex. In contrast, if a characteristic can assume only a single entity—for example, the number of hours in a day—then it is called a constant. Thus, a constant is a fixed value and is the opposite of a variable.

Depending on the context in which they are applied, variables can be classified in many ways. To illustrate these different classifications, consider the flow chart given in Figure 1.8. One way to classify variables is by the type of research factor they represent. Two types of variables that fall under this classification are independent and dependent variables. An independent variable, or IV, is a variable that affects other variables, and a dependent variable, or DV, is a variable that is affected by other variables. In the context of a research study, the DV represents the object of a study or investigation. It is the consequence of some action that has been administered by an IV, which is the antecedent. This relationship can be represented pictorially as IV → DV. For example, the number of hours of sleep deprivation pilots experience could affect their flight performance. Thus, "hours of sleep deprivation" is the IV and "flight performance" is the DV. Independent variables also are referred to as *manipulated*, *experimental*, or *predictor* variables, and dependent variables also are referred to as *response* or *outcome* variables.

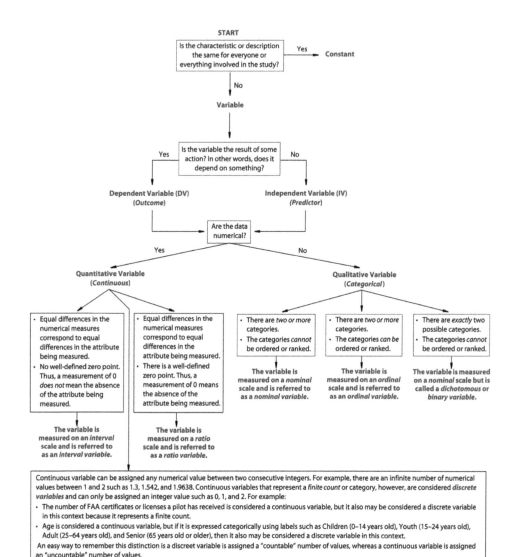

START

| Is the characteristic or description the same for everyone or everything involved in the study? | Yes → Constant |

No
↓
Variable
↓

Yes ← | Is the variable the result of some action? In other words, does it depend on something? | → No

Dependent Variable (DV)
(*Outcome*)

Independent Variable (IV)
(*Predictor*)

Are the data numerical?

Yes   No

Quantitative Variable
(*Continuous*)

Qualitative Variable
(*Categorical*)

- Equal differences in the numerical measures correspond to equal differences in the attribute being measured.
- No well-defined zero point. Thus, a measurement of 0 *does not* mean the absence of the attribute being measured.

- Equal differences in the numerical measures correspond to equal differences in the attribute being measured.
- There is a well-defined zero point. Thus, a measurement of 0 means the absence of the attribute being measured.

- There are *two or more* categories.
- The categories *cannot* be ordered or ranked.

- There are *two or more* categories.
- The categories *can be* ordered or ranked.

- There are *exactly* two possible categories.
- The categories *cannot* be ordered or ranked.

The variable is measured on a *nominal* scale and is referred to as a *nominal variable*.

The variable is measured on an *ordinal* scale and is referred to as an *ordinal variable*.

The variable is measured on a *nominal* scale but is called a *dichotomous* or *binary variable*.

The variable is measured on an *interval* scale and is referred to as an *interval variable*.

The variable is measured on a *ratio* scale and is referred to as a *ratio variable*.

Continuous variable can be assigned any numerical value between two consecutive integers. For example, there are an infinite number of numerical values between 1 and 2 such as 1.3, 1.542, and 1.9638. Continuous variables that represent a *finite count* or category, however, are considered *discrete variables* and can only be assigned an integer value such as 0, 1, and 2. For example:
- The number of FAA certificates or licenses a pilot has received is considered a continuous variable, but it also may be considered a discrete variable in this context because it represents a finite count.
- Age is considered a continuous variable, but if it is expressed categorically using labels such as Children (0–14 years old), Youth (15–24 years old), Adult (25–64 years old), and Senior (65 years old or older), then it also may be considered a discrete variable in this context.
An easy way to remember this distinction is a discreet variable is assigned a "countable" number of values, whereas a continuous variable is assigned an "uncountable" number of values.

Figure 1.8 Flowchart for classifying variables coupled with their corresponding measurement scales.

A second way to classify variables is by the type of data they represent or can be assigned. For example, a quantitative variable represents *numerical* data, whereas a qualitative variable represents *categorical* or non-numerical data. To illustrate the difference between these variable types, consider the type of data we could collect about a person's diet. Quantitative variables might include the number of calories, grams of fat, grams of fiber, and milligrams of sodium a person consumes daily. Qualitative variables, however, would include a *description* or illustration of a person's diet. One description might include the categories "dairy," "vegetables," and "fish." A second description might include the categories "pizza," "pasta," and "meat." Because the type of data that are assigned to quantitative and qualitative variables are numerical and categorical, respectively, we commonly use the term *quantitative or continuous data* to represent any type of numerical data, and *qualitative or categorical data* to represent any type of non-numerical data.

Measurement Scales

Depending on the type of data being collected, variables are measured using specific types of measurement scales, which are referenced in Figure 1.8. A summary of the four main types of measurement scales used in research follows.

Nominal Scale

Variables measured on a nominal scale have a property of *distinctness* (i.e., they differ in kind only), and correspond to "equal to" and "not equal to," which means the characteristic being measured either belongs (=) or does not belong (≠) to a category. A nominal measurement involves placing people, objects, or events into a set of distinct groups that differ on some specific attribute or characteristic. Examples include biological sex (male or female), marital status (e.g., single, married, divorced), religious preference (e.g., Protestant, Catholic, Jewish, Muslim, Other), race/ethnicity (e.g., Caucasian, Black, Asian, Hispanic), employment status (e.g., unemployed, part time, full time), and academic major (e.g., aviation, biology, chemistry, mathematics, physics, psychology). The data are organized into mutually exclusive categories, and numbers used to label the categories cannot be arithmetically manipulated. To analyze nominal data, we assign numbers to the categories, but recognize that these numbers have no quantitative sense. As an illustration, let's assign the numbers 1, 2, 3, 4, and 5, respectively to Protestant, Catholic, Jewish, Muslim, and Other. Note that the magnitude of these numbers is meaningless. For example, just because the number 3 is assigned to Jewish and 1 is assigned to Protestant does not mean that Jews have more religion than Protestants. We also cannot say that Jews are a larger group than Protestants, and it does not make sense to manipulate the numbers using any arithmetic operation. For example, it is meaningless to say that a Protestant plus a Catholic is equal to a Jew (1 + 2 = 3). The numbers are simply labels for the distinct categories of the variable.

Ordinal Scale

Variables measured on an ordinal scale have distinct outcomes (=, ≠), but they also can be placed in a *ranked order* such as first, second, and third. Therefore, they correspond to "less than" (<) and "greater than" (>), and maintain the *transitive property*: If *A* is ranked higher than *B*, and *B* is ranked higher than *C*, then *A* is ranked higher than *C*. Examples include anything that lends itself to a ranked order such as airline rankings based on customer satisfaction or annual gross revenue. The outcomes of a variable are ranked relative to how much of an attribute they possess, and the rankings are independent of any equality of differences. For example, we cannot say that the second item ranked has twice as much of the attribute on which the rankings are based as the fourth ranked item. Furthermore, any numbers we assign to an ordinal measurement are used only to indicate an observation's ranked order. They specify only a relative difference without regard to an absolute difference, and therefore do not necessarily have equal quantitative meaning. Adding, subtracting, multiplying, or dividing rankings also is meaningless.

Interval Scale

Variables measured on an interval scale have properties of distinctness (=, ≠) and order (<, >), and reflect *equal differences,* or *intervals,* between successive categories, which means we can add (+) or subtract (–) numerical values and get meaningful results. For example, suppose two flight students taking an aviation meteorology course scored 90 and 45, respectively, on a 100-point final examination. In this context, it is meaningful to say that one student scored 45 points more (or less) than the other student. However, we *cannot* say that the student who scored 90 has twice as much knowledge of meteorology as the student who scored 45. This is because exam scores do not have a well-defined zero point. For example, if a student scored 0 on the final exam, this does *not* mean the student has zero knowledge of meteorology. Examples of interval scale data include exam scores, IQ scores, calendar years, and temperature.

There is considerable debate in the research literature about whether Likert scores reflect ordinal or interval data. Strictly speaking, Likert scores are *ordinal* data. For example, a score of 5 (Strongly Agree) has more agreement than a score of 4 (Agree), but a score of Agree (4) is *not* three units more than Strongly Disagree (1) because there is no indication that the intervals between successive responses are equal. If we were to assume, however, that the distances between the numerical responses were equal, then a case can be made to treat Likert scores as interval data. In aviation science research involving human factors, it is accepted practice to treat Likert scores as interval data, and in doing so researchers will sometimes refer to Likert scores as **quasi-interval**.

Ratio Scale

Variables measured on a ratio scale have properties of distinctness (=, ≠), order (<, >), and equal intervals (+, –), and they also have a *true zero point,* which means we can multiply (×) and divide (/) numerical values and get meaningful results. For example, if Pilot A has 200 flight hours and Pilot B has 400 flight hours, then it is meaningful to say that "Pilot B has *twice as many* flight hours as Pilot A." Examples of ratio scale data include distance, time, height, weight, and dollars.

When using a statistics program for data analysis, it is important that you recognize the data type being analyzed because this determines what statistical strategy is appropriate. For example, consider the variable "education level." If we ask participants to report the *total number of years* of formal education they completed—where 12 years equals a high school diploma, 14 years equals a 2-year college degree, and 16 years equals a 4-year college degree—then the variable is *continuous* (measured on a ratio scale). However, if we presented participants with categories such as "high school or equivalent," 2-year college degree or equivalent," and "4-year college degree," then the variable is *categorical* (measured on a nominal scale). Observe that for the continuous variable, we would be able to calculate the arithmetic average, but we could not do so for the categorical variable. Although most statistics programs treat variables as either continuous (interval and ratio scales) or categorical (nominal and ordinal scales), it is critical that you familiarize yourself with how your statistics program classifies variables. If you discover that you are unable to perform a certain analysis, it might be because the variable's data type is mislabeled.

Example 1.4: *Variables and Measurement Scales*

Given the purpose statement or research question in a–c, determine the variable being measured, the type of variable, and corresponding measurement scale.

a. Purpose statement: To examine the percentage difference between GA pilots who received their PPL from a Part 141 vs. a Part 61 flight program.
b. RQ: To what extent does an airline's ranking—based on scheduled passenger revenue (millions USD)—relate to the number of passenger bookings?
c. RQ: To what extent does pilot age—measured by categories of years such as younger than 19, 20–29, 30–39, etc.—relate to their total number of flight hours?

Solution

a. The variable is "type of flight program," the variable is categorical, and it is measured on a nominal scale. The variable also is dichotomous.
b. There are two variables: airline's ranking and number of passenger bookings. Because rankings are reported in millions USD, it is quantitative (continuous) and measured on a ratio scale. Because the number of passenger bookings reflects "how many," it is quantitative and measured on an interval scale. From a research perspective, "rankings" is an IV and number of passenger bookings is a DV.
c. There are two variables: pilot age and total flight hours. Because age is reported using categories, it is qualitative and measured on a nominal scale. Flight hours is quantitative (continuous) and measured on a ratio scale. From a research perspective, age is an IV and number of flight hours is a DV.

Chapter Summary

1. Research provides the tools to engage in a formal study, and statistics provides the tools to organize and summarize a study's data so we can make sense of and interpret the findings.
2. A research objective is expressed as a purpose statement and corresponding research question. To avoid ambiguity, key terms and phrases are operationally defined relative to the research setting. The best way to operationally define a psychological or social construct is to cite the instrument that will be used to measure the construct.
3. The research process involves identifying the population of interest, selecting a sample from the population, collecting/analyzing data from the sample related to the characteristic being studied such as average flight hours, and generalizing the results of data analysis to the population. The respective numerical values that describe the characteristic is called sample statistic and population parameter.
4. A sample can be selected from its parent population randomly or nonrandomly. Random samples are more likely to be representative of the parent population than nonrandom samples, which tend to be biased. An outcome of any sampling strategy is sampling error, which is the difference between the sample statistic and its corresponding population parameter.

5. Quantitative research studies involve collecting numerical data, and four methodologies are experimental, correlational, causal-comparative, and survey. Experimental research is designed to establish cause-and-effect relationships, and a key component is manipulation and control. The other three methodologies are considered observational, which are non-experimental. Correlational studies examine the relationship between variables and yield a correlation coefficient, r, which denotes the strength and direction of the relationship. Causal-comparative studies—also known as ex post facto—are like experimental research except there is no direct intervention. This is because the manifestation of the intervention has already occurred, or because it either is not possible or it would be unethical to assign participants to groups. Survey research, which is not to be confused with a survey instrument, is designed to understand peoples' thoughts, opinions, and feelings about a topic. Surveys can be cross-sectional, which reflects a specific time interval, or longitudinal, which involves gathering data across several time intervals to examine trends. Sampling error associated with surveys is called margin of error.

6. A research variable is any entity that is being measured and represents characteristics of people, places, or things that can vary. Variables can be classified as independent (IV), dependent (DV), quantitative, qualitative, continuous, discrete, dichotomous, categorical, predictor, and outcome.

7. Measuring scales on which variables are measured include nominal, which has the property of distinctness; ordinal, which has properties of distinctness and rank; interval, which has properties of distinctness, rank, and addition; and ratio, which has properties of distinctness, rank, addition, and multiplication.

8. The zero point of the interval scale is arbitrary and does not reflect the absence of the attribute being measured (e.g., 0° C does not denote the absence of temperature). The zero point of the ratio scale is well defined—it is an absolute zero—and denotes the absence of the attribute being measured (e.g., a height of 0 m indicates the absence of height).

Vocabulary Check

Biased sample	Longitudinal survey	Ratio variable
Causal-comparative research	Manipulation and control	Representative sample
Census	Margin of error	Research
Cluster random sampling	Nominal measurement scale	Research methodology
Constant	Nominal variable	Research process
Control group	Observational studies	Research question
Convenience sampling	Operational definitions	Sample
Correlation coefficient	Ordinal measurement scale	Sample assignment
Correlational research	Ordinal variable	Sample selection
Cross-sectional survey	Parent population	Sample size
Dependent variable	Population	Sample statistic
Dichotomous variable	Population generalizability	Sampling
Discrete variable	Population parameter	Sampling error
Ex post facto	Power analysis	Simple random sampling
Experimental research	Purpose statement	Snowball sampling
Extraneous factors	Qualitative variable	Statistically equivalent
Generalizability	Quantitative variable	Stratified random sampling
Independent variable	Quota sampling	Survey research
Interval measurement scale	Random assignment	Systematic sampling
Interval variable	Random	Treatment group
Intervention studies	Random sample	Variables
Judgment sampling	Ratio measurement scale	

Review Exercises

A. Check Your Understanding

In 1–10, choose the best answer among the choices provided.

1. Given the RQ: "To what extent are students' perceptions of a flight instructor's ability as a pilot influenced by the flight instructor's personality?" What is the DV?
   a. Pilot
   b. Flight instructor's personality
   c. Flight instructor's ability
   d. Students' perceptions
2. Operational definitions are encouraged in research because they
   a. conform to the requirement of statistical analysis.
   b. make terms used in a study as explicit as possible.
   c. increase the probability that experiments will succeed.
   d. make educational research more easily understood by laypersons.
3. Which of the following represents an operational definition of "test anxiety?"
   a. The inability to do well on an exam.
   b. Failing an exam when the student truly understands the material.
   c. The scores from the self-reported Test Anxiety Inventory (TAI).
   d. The range of emotions students express when taking an exam.
4. A researcher designed a study to investigate CFIs' attitudes in the state of Florida toward gender in aviation. She received permission from 15 flight schools to personally contact the CFIs during their weekly briefings. Which would be an appropriate target population?
   a. All CFIs who have negative attitudes toward gender in aviation.
   b. All CFIs in the world.
   c. All CFIs in Florida.
   d. All CFIs in the 15 flight schools.
5. Although RQs usually concern a _____, the actual research is typically conducted with a _____.
   a. sample, statistic
   b. population, parameter
   c. sample, population
   d. population, sample
6. Based on the responses she received from over 100,000 solicited letters from GA pilots, a representative for the AOPA reports that most GA pilots regret not becoming an ATP. A reasonable evaluation of this finding with respect to the population of GA pilots is that it is
   a. representative because of the large sample size.
   b. representative because respondents were volunteers.
   c. not representative because of the relatively small sample size.
   d. not representative because respondents were self-selected.
7. Choose the research methodology that would be most appropriate to investigate the following RQ: "In what way do married flight attendants deal with the stress of being away from home?"
   a. Survey
   b. Ex post facto
   c. Experimental
   d. Correlational

8. Experimental studies can be differentiated most clearly from observational studies on which of the following characteristics?
   a. The statistical analysis of data.
   b. The use of structured designs.
   c. Collecting numerical data.
   d. A clear intent to establish cause-and-effect relationships.

9. Consider the statement "Jane is preparing for the FAA IRA exam." Which scale should be used to measure this statement?
   a. Nominal
   b. Ordinal
   c. Interval
   d. Ratio

10. A study examined the effect level of education (high school diploma, 2-year college degree, 4-year college degree, and graduate/advanced degree) has on career happiness. The study involved a random selection of people working in the aviation industry, including airline ticket agents, security officers, baggage handlers, and pilots. A person's career happiness was defined as scores on the Career Happiness Rating Scale and could range from 40 to 200; the higher the score, the happier the individual. In this study, "level of education" is considered _____ and "career happiness" is considered _____.
    a. the dependent variable, the independent variable
    b. the dependent variable, a nominal variable
    c. the independent variable, a continuous variable
    d. a discrete variable, a Likert variable

B. Apply Your Knowledge

Following are 12 possible aviation research topics. From this list select *two* topics and then respond to items 1–7 below for each topic.

Flight schools	Aviation safety	Pilot confidence
Alcohol abuse	Accident/incident rate	Flight attendants
Foreign language issues	Airport noise	Age 65 Rule
Spatial ability	Runway incursions	Air traffic controllers

1. Describe what you want to investigate, and explain why you want to investigate it and why you think it merits an investigation.
2. Write an appropriate research purpose statement and corresponding RQ.
3. Identify and operationally define all key terms and phrases.
4. Specify the target population, and *describe* the sampling strategy you propose to use to select the sample.
5. Specify the research methodology you will use to answer the RQ, and explain why this methodology is appropriate.
6. Determine the variables being measured, the type of variables, and corresponding measurement scale.
7. Explain the extent to which you think the results of the study can be confidently generalized to the parent population and why?

References

American Psychological Association (2020). *Publication manual of the American Psychological Association* (7th ed.). https://doi.org/10.1037/0000165-000

Caligan, R. J. Jr. (2012). Time to solo as a predictor of performance in civilian flight training. *The International Journal of Aviation Psychology, 22*(1), 30–40. http://dx.doi.org/10.1080/10508414.2012.635124

Dunbar, V. L. (2015). *Enhancing vigilance in flight instruction: Identifying factors that contribute to flight instructor complacency* (Publication No. 3664585) [Doctoral dissertation, Florida Institute of Technology]. ProQuest Dissertations and Theses Global.

Federal Aviation Administration (n.d.). *General aviation and part 135 activity surveys.* https://www.faa.gov/data_research/aviation_data_statistics/general_aviation/

Hunter, D. R. (2002). *Risk perception and risk tolerance in aircraft pilots* (Rep. No. DOT/FAA/AM–02/17). Washington, DC: Federal Aviation Administration. https://apps.dtic.mil/sti/pdfs/ADA407997.pdf

Kachigan. S. K. (1991). *Multivariate statistical analysis: A conceptual introduction* (2nd ed.). Radius press.

Random.org. (2022). [Online random number generator]. http://www.random.org/integers

Taylor, J. L., Kennedy, Q., Noda. A., & Yesavage, J. A. (2007). Pilot age and expertise predict flight simulator performance. *Neurology, 68*(9), 648–654. https://www.ncbi.nlm.nih.gov/pmc/articles/PMC2907140/

2 Organizing and Summarizing Data

Student Learning Outcomes

After studying this chapter, you will be able to do the following:

1. Determine the shape of the distribution for a set of nominal data and interpret the corresponding frequency distribution table.
2. Determine the shape of the distribution for a set of continuous data, derive and interpret the corresponding grouped frequency distribution table, and interpret the summary statistics.
3. Interpret the mean, median, and mode for a set of continuous data, determine which is the most appropriate measure of central tendency, and determine the shape of the corresponding distribution by examining the relationship among these measures.
4. Interpret the range, standard deviation, Q_1, Q_3, IQR, and the corresponding box plot for a set of continuous data.

2.1 Chapter Overview

The concept of organizing and summarizing data is the foundation of *descriptive statistics*, which is the focus of this chapter. Descriptive statistics include procedures for determining various data characteristics, including how to organize data into tables and graphs, and how to determine key summary statistics. To illustrate this concept, consider the set of data in Figure 2.1, which contains the ages at which a random sample of 20 GA pilots earned their PPL. The data are presented as a nominal variable using age categories, and as a continuous variable using the actual ages.

To organize the nominal data, we present them in a *frequency distribution table*, which provides a summary of the frequency counts associated with each category. For example, of the 20 respondents, 8 earned their PPL between the ages of 21 and 25 years. To organize the continuous data, we report various *single summary statistics*. For example, the *average* age was 24 years, and the ages *ranged* from 18 to 33 years. Thus, organizing and summarizing raw data reduce the data to a form that makes it easier to describe and interpret. Although these results lack specifics, as Kachigan (1991, p. 3) noted: "We sacrifice detail and nuance for parsimony and salience."

DOI: 10.4324/9781003308300-3

Age When Received PPL

GA Pilot #	Raw Age (Continuous)	Age Category (Nominal)
1	29	26–30
2	25	21–25
3	22	21–25
4	18	16–20
5	22	21–25
6	23	21–25
7	19	16–20
8	20	16–20
9	21	16–20
10	25	21–25
11	18	16–20
12	18	16–20
13	30	26–30
14	31	> 30
15	23	21–25
16	31	> 30
17	33	> 30
18	24	21–25
19	21	21–25
20	27	26–30

**Frequency Distribution Table
(Nominal Data)**

Age Group	Frequency
16–20	6
21–25	8
26–30	3
> 30	3

**Single-Summary Statistics
(Continuous Data)**

Report the Numerical Average of the Raw Ages

29 + 25 + 22 + 18 + 22 + 23 + 19 + 20 + 21 + 25
+ 18 + 18 + 30 + 31 + 23 + 31 + 33 + 24 + 21 + 27
= 480
and 480 / 20
= 24
Therefore, average age = 24 years old

Report the Range (Low to High) of the Raw Ages
Ages ranged from 18 to 33 years old.

Figure 2.1 The concept of organizing and summarizing data.

2.2 Examining Distributions

We begin by introducing the concept of a *frequency distribution*, which is a basic statistical strategy that represents a count of how frequently an observation, or score, occurs within a data set as illustrated in Figure 2.1. Although most statistics software packages provide frequency distribution tables and graphs as part of their descriptive statistics output, we will demonstrate for instructional purposes how to prepare them. We begin with how to organize and summarize qualitative data.

Working with Qualitative Data

To illustrate how to organize and summarize qualitative data, we will examine the data in the file Ch_2 AMT Education Level. This data set contains the highest level of education or degree earned as reported by a sample of aircraft maintenance technicians (AMTs) who were part of Uhuegho's (2017) safety climate study involving a national maintenance, re-pair, and overhaul (MRO) facility. The first step in organizing and summarizing these data is to prepare a frequency distribution table as shown in Table 2.1. Note that the first column of Table 2.1 represents the categories of the nominal variable, and the second column re-flects the *frequency*, or count, of the number of AMTs who self-reported their education level relative to the categories. Note also how the table reduces the initial raw data to a smaller, more interpretable form that provides a summary of how the data are distributed: 19 AMTs had at most a high school diploma, 29 had a 2-year college degree, and so forth.

Table 2.1 Frequency Distribution of Participants'
Highest Level of Education

Type of Diploma or Degree	N
High school	19
2-yr college	29
4-yr college	17
Graduate	13
Technical school	12

Table 2.2 Relative Frequency Distribution of Participants' Highest
Level of Education

Type of Diploma or Degree	N	Relative Frequency
High school	19	19/90 = 21.1%
2-yr college	29	29/90 = 32.2%
4-yr college	17	17/90 = 18.9%
Graduate	13	13/90 = 14.4%
Technical school	12	12/90 = 13.3%

A frequency distribution table can be made more informative by including a *relative frequency* column that represents the proportion, or percent, of observations within each category. This is expressed as the ratio of the number of occurrences within a category to the total number of frequencies and is illustrated in Table 2.2. Note how the relative frequency column makes it easy to see the distribution of education level with respect to what proportion of the sample reported having a specific degree. For example, the 2-year college group represented 32.2% of the sample.

To make a distribution table even more informative, we can include two additional columns as shown in Table 2.3: *cumulative frequency* contains a running total of the frequencies, and *cumulative relative frequency* includes a running total of the corresponding percentages for each category. To get these percentages, we divide each cumulative frequency by the total number of frequencies. The presence of these two columns makes it possible to see how many observations occurred above or below a given category. For example, with respect to the cumulative relative frequency column, we can see that more than half of the 90 participants (53.3%) had a high school diploma or a 2-year college degree, and nearly three-fourths (72.2%) had up to a 4-year college degree.

Frequency distributions also can be represented graphically, which often makes it easier to identify patterns or see how various aspects of the data are related. For example, Figure 2.2 contains a bar graph for the data given in Table 2.1. Separate bars represent the categories (Education Level), and the bar lengths correspond to each category's frequency. Note that adjacent bars are separated by spaces. This is done to emphasize that the categories are separate and distinct. Some statistical software packages, however, do not make this distinction when generating bar graphs for nominal data.

Working with Quantitative Data

Unlike qualitative data, quantitative data are numerical and do not contain any predefined categories. This means that to summarize quantitative data in frequency distribution

Table 2.3 Cumulative Relative Frequency Distribution of Participants' Highest Level of Education

Type of Diploma or Degree	N	Relative Frequency	Cumulative Frequency	Cumulative Relative Frequency
High school	19	19/90 = 21.1%	19	19/90 = 20.0%
2-yr college	29	29/90 = 32.2%	19 + 29 = 48	48/90 = 53.3%
4-yr college	17	17/90 = 18.9%	19 + 29 + 17 = 65	65/90 = 72.2%
Graduate	13	13/90 = 14.4%	19 + 29 + 17 + 13 = 78	78/90 = 86.7%
Technical school	12	12/90 = 13.3%	19 + 29 + 17 + 13 + 12 = 90	90/90 = 100.0%

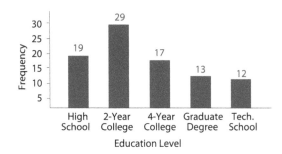

Figure 2.2 Bar graph that corresponds to Table 2.1, which shows the distribution of AMTs' highest level of education.

tables, categories must be created, and the data grouped accordingly. The creation and interpretation of corresponding frequency distribution tables depend on whether the quantitative data are discrete or non-discrete.

Discrete Continuous Data

Recall that discrete data represent a finite count. This characteristic enables the observations to represent categories, and therefore frequency tables involving discrete data are like those involving qualitative data. As an example, let's assume that the number of runway incursions reported at Hartsfield-Jackson Atlanta International Airport (ATL) during a 30-day period was as follows:

5, 7, 4, 6, 5, 1, 0, 1, 3, 2, 0, 4, 4, 1, 3, 0, 0, 1, 0, 0, 2, 2, 2, 0, 2, 0, 0, 1, 1, 0

These data are provided in the file Ch_2 Runway Incursions. Note that these data represent a count of the number of runway incursions that were reported each day for 30 consecutive days. Because the data are discrete, we can use the number of runway incursions (0, 1, 2, 3, 4, 5, 6, 7) as categories, and the frequency would be the number of instances of each category. For example, as shown in Table 2.4, there were 6 days when only one runway incursion was reported, so the category "1 runway incursion" has a frequency of six. Similarly, there were 5 days over this 30-day period where two runway incursions were reported, so the category "2 runway incursions" has a frequency of five. Although we do not show this, we also could extend Table 2.4 by including columns for cumulative frequency and relative cumulative frequency as we did for AMTs' level of education in Table 2.3, and interpretations would be made exactly as we did previously.

To depict the distribution of runway incursions graphically, we use a histogram as shown in Figure 2.3. A histogram looks exactly like a bar graph except it does not have spaces between adjacent bars. In a histogram, all the bars are "touching" because a histogram pictorially represents numerical data, not categorical data. Thus, to emphasize that the data are numerical, the horizontal axis of the graph is ordered numerically just like the way a number line is constructed. One other observation to make is that with discrete data, the rectangular bars are "centered" at the numerical value of each category.

Non-Discrete Continuous Data

In contrast to working with discrete continuous data, non-discrete continuous data do not lend themselves so easily to categories, and therefore we must create them. To illustrate this, let's assume a random sample of 100 people who attended the Oshkosh Air Show

Table 2.4 Frequency Distribution of Number of Runway Incursions Reported at ATL During a 30-Day Period

Number of Runway Incursions	Frequency (f)
0	10
1	6
2	5
3	2
4	3
5	2
6	1
7	1

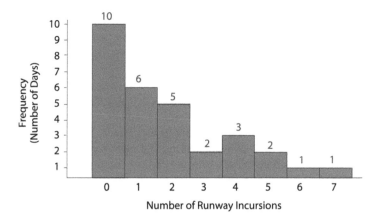

Figure 2.3 Histogram of the number of runway incursions reported at ATL for a 30-day period. The graph corresponds to the frequency distribution given in Table 2.4.

reported their ages. This data set is in the file Ch_2 Oshkosh Air Show Ages, and the ages are presented in numerical order for instructional purposes.

5, 7, 8, 8, 8, 9, 9, 10, 12, 12, 13, 15, 15, 16, 17, 18, 18, 18, 18, 19, 20, 21, 21, 21, 22, 22, 23, 23, 24, 24, 25, 25, 25, 27, 28, 29, 29, 30, 30, 30, 32, 32, 33, 33, 33, 34, 34, 34, 35, 35, 36, 36, 38, 38, 39, 40, 42, 43, 44, 44, 44, 46, 47, 47, 48, 48, 49, 49, 49, 49, 50, 51, 52, 54, 54, 54, 56, 60, 61, 61, 62, 62, 64, 65, 65, 67, 68, 69, 70, 73, 75, 76, 77, 79, 80, 82, 83, 86, 87, 89

Because these data are not discrete, we do not organize them by creating a category for each age and then specifying the corresponding frequency as we did with the runway incursions data. To do so would be extremely inefficient: Of the 100 ages, 62 are *unique* cases, and of these 62 unique cases, 36 have a frequency of 1. For example, only one person reported an age of 5, one person reported an age of 7, and one person reported an age of 10. Thus, such a frequency distribution table would accomplish little with respect to summarizing the data in an informative and concise manner.

Instead, we can create *groups of scores* and record the number of scores, or observations, that fall within each group. For example, in Table 2.5, we created 10 age groups: 5–14 years old, 15–24 years old, etc. These groups are formally called *class intervals*, or simply *classes*, which are just like categories of qualitative data. Note that each class has a *lower limit* and an *upper limit*. For example, the lower limit of the 5–14 class is 5, and the upper limit is 14. The difference between the lower limits of any two consecutive classes is called the *class width*. For example, the lower limit of the first class is 5, the lower limit of the next class is 15, and the difference between them is $15 - 5 = 10$. Thus, the class width for the data in Table 2.5 is 10. Lastly, the classes do not overlap, and therefore, each score falls into only one class. This type of distribution is called a grouped frequency distribution, and once again notice how the table reduces the initial set of raw ages into a more compact and organized form that summarizes the data. A grouped frequency distribution also can be expanded to include cumulative frequencies and cumulative relative frequencies as we did earlier, and interpretations would be made exactly as we did in previous examples.

It is important to recognize that class intervals are not unique and can vary among researchers as well as among different statistics programs. Thus, there is no single correct way to create them. Instead, the following general guidelines are applied: (a) the number of intervals should be around 10; (b) the class width should be a multiple of 2, 5, or 10; (c) the lower limit should be a multiple of the class width, which means that if the width

Table 2.5 Grouped Frequency Distribution of 100 Ages

Age Groups	Frequency	Relative Frequency
5–14	11	11/100 = 11%
15–24	19	19/100 = 19%
25–34	18	18/100 = 18%
35–44	13	13/100 = 13%
45–54	15	15/100 = 15%
55–64	7	7/100 = 7%
65–74	7	7/100 = 7%
75–84	7	7/100 = 7%
85–94	3	3/100 = 3%

is 5, then the lower limit of each interval should be a multiple of 5 (e.g., 5, 10, 15, ...); and (d) all intervals should be the same width. Examining Table 2.5 relative to these general guidelines:

(a) There are exactly 10 intervals.
(b) The class width is 10.
(c) The lower limit is a multiple of the class width (10).
(d) All intervals are the same width.

As a result, the grouped frequency table is compliant with the general guidelines. Depending on the data, though, there might be an exception to the last guideline. For example, sometimes when working with age data, it might be informative to create a class interval such as younger than 20 years old (< 20) or older than 65 years old (> 65). With respect to the former, the class width does not have a lower limit, and in the latter case, the class width does not have an upper limit.

The corresponding histogram for the grouped frequency distribution of the ages example is shown in Figure 2.4. Note that unlike a histogram for discrete continuous data, we label the horizontal axis using the lower limits of the class intervals for each bar. For example, working from left-to-right, the first bar represents the frequency of ages from 5 to *younger than 15*, the second bar represents the frequency of ages starting at 15 but *younger than 25*, and so forth. This is because age data are non-discrete continuous. So, if a person's self-reported age is 44 years, 11 months, 29 days, this age would be reflected in the fourth bar of the histogram because it is younger than 45 years.

Most statistical software packages provide the histogram of a grouped frequency distribution, but not the corresponding table, which means we must prepare the table manually from the histogram. It also is important to recognize that the number of classes and class width generated by a statistics program might be different than what we want, and hence we might have to modify these parameters. Most packages provide users with this option.

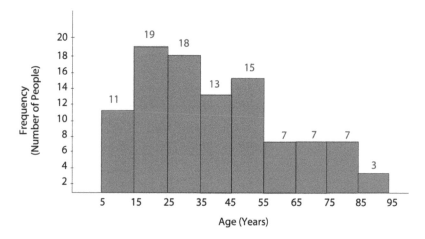

Figure 2.4 Histogram of ages of 100 people who attended the Oshkosh Air Show. The graph corresponds to the grouped frequency distribution given in Table 2.5.

Shapes of Distributions

One characteristic that describes a distribution is its shape. Although the shape of a distribution is described by a formal mathematical equation, we nevertheless can describe the shape of a distribution in a less technical manner using general terms. Figure 2.5 presents common shapes of distributions, and a brief discussion of each shape follows.

Uniform Distribution

A *uniform*, or *rectangular*, distribution is one in which each numerical value of a variable occurs equally often. Such a distribution is infrequent, but still is possible. For example, if we were to record the annual salaries of all first-year pilots flying for a regional airline, the distribution of these salaries most likely will resemble a rectangular shape.

U-Shaped Distribution

A *U-shaped* distribution is parabolic in form where the observed values of a variable are either very high or very low with little middle ground. *U*-shaped distributions can be either a regular *U* or an inverted *U*. An example of the former would be the number of aviation events (incidents or accidents) relative to pilot age. The number of aviation events tends to be high with younger pilots, declines as pilots become older, but then at a certain point, the number of events begins to rise again as pilots become even older.

Bell-Shaped Distribution

A *bell-shaped* distribution is one in which the highest frequencies of a variable occur in the center, but then become lower to the left and right of the center. Bell-shaped distributions will vary in profile—some may be elongated whereas others might be more spread out.

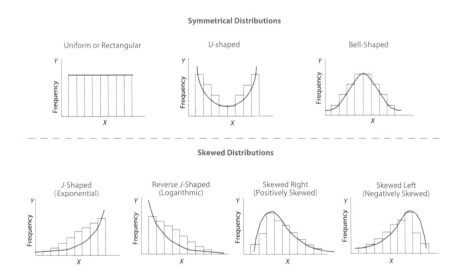

Figure 2.5 Common shapes of distributions partitioned into two main categories: symmetrical and skewed. (Adapted from Kachigan, 1991, p. 31.)

One very important bell-shaped distribution is the *standard normal distribution*, which has a specific shape and an exact mathematical definition. We will discuss normal distributions in Chapter 3.

J-Shaped Distribution

A *J-shaped* distribution resembles the letter *J* and is similar to an exponential curve, where most of the observations accumulate to the right, and a backward *J* distribution is similar to a logarithmic curve, where most of the observations accumulate to the left. An example of a regular shaped *J* is the frequency of people receiving social security benefits with respect to their ages. Very few people receive social security benefits at an early age, but as age increases, so does the frequency of recipients.

Skewed Distribution

A *skewed* distribution is one in which the frequency of a variable is relatively high at one end, but then decreases considerably and gradually tapers off at the other end. The part of the graph where the frequency tapers off is called the *tail* of the distribution, and the distribution is named with respect to the tail. Therefore, a *positively skewed distribution* is *skewed right*, and a *negatively skewed distribution* is *skewed left*. For example, if we were to consider the frequency of test scores on an extremely difficult examination, the corresponding distribution most likely would be skewed right, with fewer students scoring high. Conversely, the distribution of test scores on an extremely easy examination most likely would be skewed left, with few people scoring low.

A skewed distribution signals the presence of *outliers*, which are extreme scores relative to the other scores in a distribution. Outliers generally consist of *contaminated data*, which include recording errors or calculation errors, or *rare cases*, which reflect a valid but extremely rare observation. An entry of 3,000 dual hours given in the past 90 days by a CFI is most likely contaminated data. An entry of 45,000 aggregate flight hours by a Part 121 pilot could be considered a rare case, particularly if the pilot has 40 years of service and has flown mostly international routes.

The first three shapes shown in Figure 2.5—uniform, *U*, and bell—are more generally classified as symmetrical distributions. This is because if we were to split these distributions in half by placing an imaginary vertical line down the center, the line would split the distribution into two equal halves such that the left and right sides would be mirror images of each other. On the other hand, nonsymmetrical distributions would taper off either to the right or left. These types of distributions are more generally referred to as skewed distributions and include those that are *J*-shaped. Thus, the shape of any distribution may be considered in the most general sense as being either symmetrical or skewed. One final observation: Although it is rare for a set of data to match one of the shapes perfectly, the shape of the data's distribution will have an appearance that approximates those shown.

2.3 Measures of Central Tendency

In addition to describing the shape of a distribution, we also can describe a distribution's central point where most of the observations are located or clustered. The central point of a distribution is formally called the *measure of central tendency*. This measure describes an entire data set by identifying its most typical, or most representative, observation. It is a single score that identifies the center of a distribution. By specifying the measure of central tendency in conjunction with the shape of the distribution, we can provide a very detailed summary of an entire set of raw data in a very concise manner.

When considered informally, the concept of central tendency may be best understood as an "average." For example, if a car averages 30 miles per gallon, although the car's mileage will not be exactly 30 miles per gallon every time we calculate its mileage, it does give us a good idea of what we can typically expect. The concept of average also is useful for making comparisons. For example, according to the FAA (2021; Table 13), the average age of an ATP was 46.0 years in 2001, but 51.3 years in 2021. As you can see from these examples, the concept of central tendency enables us to use a single metric to describe what is typical, or most representative, for a sample or population.

Although the concept of central tendency is intuitive and simple to understand, it is not always easy to agree on where the center of a distribution is located. Unlike symmetrical distributions such as those depicted in Figure 2.5, where the center is in the "middle," the center of a skewed distribution is problematic. For example, consider the distribution of Figure 2.3, which represents the number of runway incursions reported at ATL in a 30-day period. Where do you think the center of this distribution is located? Some might say it is at 0 because this is where most of the scores are located. However, 0 is not in the "middle." There are more scores above 0 than at 0, which infers that the center should be above 0.

Because not all data sets will form a single, perfectly symmetrical distribution that will enable everyone to agree on its center, there is no single procedure for determining a distribution's measure of central tendency. To reconcile this problem, statisticians have developed three measures of central tendency, *mode*, *median*, and *mean*, and each measure is determined using different procedures and has its own advantages and disadvantages. As we discuss these measures, it is important for you to keep in mind the fundamental purpose of central tendency: It is to identify a single score that is most representative of the distribution. This should help determine which measure is most appropriate for a given distribution.

Mode

One measure of central tendency is called the mode and represents the numerical value around which most of the data cluster. For example, in the data set 2, 4, 4, 4, 7, 8, 62, the mode is 4 because it is the most frequently occurring score (it occurs three times). Similarly, for the runway incursion data given earlier, the mode is 0 because it is the most frequently occurring observation. In some cases, it is possible for a distribution to have more than one mode. A distribution with two modes is called *bimodal*, and a distribution with more than two modes is called *multimodal*. It also is possible for distributions to have no mode.

As a measure of central tendency, the mode's major advantage is that it is easy to determine, and its interpretation makes sense, namely, the single observation that is most representative of a distribution is the one that occurs most frequently. Another advantage is that outliers have no impact on the mode. The mode is limited, however, with respect to its mathematical properties and therefore is not used in advanced statistical analyses involving quantitative data, but it is quite appropriate for qualitative data.

Median

Because the concept of central tendency refers to the center of a distribution, a second logical and sensible measure of central tendency would be the score that is literally in the center. This truly central value is called the median, denoted *Mdn*, and it divides a distribution in half so that 50% of the scores are below this measure and 50% of the scores are above this measure. The median also represents a *measure of position* and is equivalent to the 50th *percentile*. For example, if you score at the 50th percentile on a standardized test, then 50% of the examinees scored lower than your score, and 50% of the examinees scored higher than your score. We will discuss measures of position later in this chapter.

To determine the median, we arrange the data in numerical order and then select the middle number. For example, using the data set given for the mode, the median is 4 because there are three scores below it (2, 4, 4) and three scores above it (7, 8, 62).

$$2 \quad 4 \quad 4 \quad \mathbf{4} \quad 7 \quad 8 \quad 62$$
$$\text{Median}$$

If a distribution contains an even number of pieces of data, then the median is the score that lies between the middle two numbers. For example, given the data set

$$3 \quad 3 \quad \mathbf{4} \quad \mathbf{5} \quad 7 \quad 8$$

the median lies midway between 4 and 5. Thus, the median is $(4 + 5)/2 = 9/2 = 4.5$. Note that the median is not necessarily an actual score in the distribution. It is simply the point where 50% of the data lie below it, and 50% of the data lie above it.

In addition to it making the most sense, the median also is insensitive to outliers. For example, in the first data set above, the last entry (62) is an extreme score relative to the rest of the data set. However, because the median is a measure of position, outliers have no impact on the median. Thus, the median is influenced by the position of the observations and not their magnitude. It is this feature of being unaffected by outliers that makes the median the most appropriate measure of central tendency for skewed distributions. However, as was the case with the mode, the median is limited with respect to its mathematical properties and therefore is not used in advanced statistical analyses.

Mean

The third measure of central tendency is the mean, and it is what most people think of with respect to the concept of average. The mean is the arithmetic average and is calculated by dividing the sum of the scores by the total number of scores. For example, the mean of our working data set is

$$\text{Mean} = \frac{2+4+4+4+7+8+62}{7} = \frac{91}{7} = 13$$

The mean has a distinct advantage over the other measures of central tendency because it involves every score of a distribution. This advantage, though, also becomes the mean's primary disadvantage. Because it is based on every score in a distribution, the mean is very sensitive to distributions that contain outliers. For example, note that the mean of 13 is not as representative of this data set as are the mode and median. This is because the outlier, 62, is "pulling" the mean toward it. Thus, the mean gets dragged toward the tail of skewed distributions and therefore is not a representative measure of central tendency for skewed distributions. This is illustrated in Figure 2.6 where the relationship among the three measures of central tendency is examined with respect to symmetrical and skewed distributions. The mean is the most important measure and serves as the foundation of all statistical analyses. We denote the mean of a sample as M, and for the corresponding population parameter, we use the Greek letter *mu* (pronounced "mew"), which is denoted by the symbol μ.

Of the three measures of central tendency, the mean is the preferred measure because it uses every score in the distribution. However, there are situations when it is either not possible to calculate the mean or the mean is not particularly representative of the data set. For example, if a distribution is skewed, contains an unknown or undetermined score, includes data from a survey item that has no upper or lower limit (e.g., a category like "65 years or older"), or scores are measured on an ordinal scale, then the median is more appropriate. Similarly, if data are measured on a nominal scale or you are measuring discrete variables, then the mode is more appropriate. Independent of these guidelines, though, when reporting descriptive statistics for a research study, it is beneficial to report and discuss all three measures of central tendency.

2.4 Measures of Dispersion

The Concept of Variation and Why We Need to Consider It

At this stage of our discussion, we have shown that we can reduce an entire set of raw data by describing the shape of its distribution and its central tendency. Although a

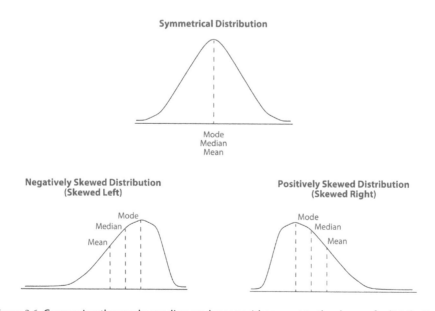

Figure 2.6 Comparing the mode, median, and mean with respect to the shape of a distribution.

distribution's shape and central tendency are quite informative, they do not always pro-vide a complete summary of a given set of data. To describe a distribution more com-pletely, we need to consider a third characteristic that represents the amount of spread or dispersion there is in the data. This third characteristic is called *variation*, and it de-scribes the extent to which the numerical values of the scores in a distribution *vary* among themselves.

To understand why we need to consider variation in addition to shape and central ten-dency when describing a distribution, consider the three data sets given in Figure 2.7(a), which also are provided in the file Ch_2 Need for Variation. Each data set consists of 10 randomly selected scores from a research study that asked 90 participants to self-report the number of years of experience they had working in the aviation profession. As illus-trated in Figure 2.7(a), all three distributions have the same shape (symmetrical), and all three samples have nearly the same mean, $M = 17$.

In Figure 2.7(b), we superimposed the corresponding continuous curves for these dis-tributions and centered them at $M = 17$. Note that data set A's spread is very narrow, with scores ranging from 16 to 19, a difference of 3; data set B's spread is a little more moder-ate, with scores ranging from 14 to 20, a difference of 6; and data set C's spread is almost flat, with scores ranging from 2 to 33, a difference of 31. As a result, it is not sufficient to describe a distribution solely by its shape and mean. We also need to know how much spread there is in the data.

As a result, to summarize a distribution more completely and accurately, we need to include a description of its measure of variation in addition to its shape and measure of central tendency. A measure of variation helps answer questions such as: "How much do the scores differ among the members of a sample?" "How widely dispersed are the scores in a data set?" and "How far do individual scores tend to be from the mean of the distribu-tion?" Common approaches to describing variability include the *range, interquartile range, five-number summary*, and *standard deviation*.

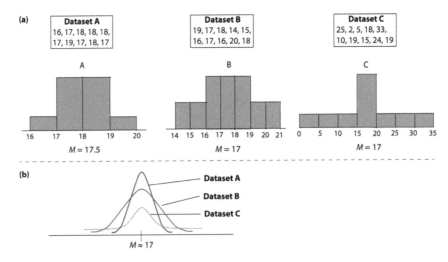

Figure 2.7 Demonstration of why it is not sufficient to describe a distribution solely by its shape and mean. Information about the amount of spread there is in the data also is needed.

Range

The most intuitive measure of variation is the range, which is simply the difference between the highest and lowest observations. For example, as shown in Figure 2.7(a), the respective ranges for data sets A, B, and C are $19 - 16 = 3$, $20 - 14 = 6$, and $33 - 2 = 31$, respectively. Although the range represents a single metric that describes the amount of variation in a distribution—for example, the higher the number, the more spread out the data—it is often more informative to report the highest and lowest scores than the actual range. The range is straightforward and easy to determine, but it suffers from the same problem as the median in that it does not involve every observation. The range is based solely on two scores, the highest and lowest scores. As a result, it does not always provide an accurate description of a distribution's variation. The range also is highly sensitive to outliers.

Interquartile Range

To help mitigate the shortcomings of the range, we can report the interquartile range (IQR), which ignores the lowest and highest scores and instead covers the middle 50% of a distribution. The *IQR* is sometimes called "the middle fifty." To find the *IQR*, we first partition a distribution into quarters, or *quartiles*, as shown in Figure 2.8. Note the following:

- The first quartile, Q_1, is the boundary value that separates the lower 25% of a distribution from the upper 75%. This is also known as the *lower quartile*.
- The second quartile, Q_2, is the boundary value that separates the distribution into two halves so that 50% of the data lie below it and 50% of the data lie above it. Note that Q_2 is equal to the median.
- The third quartile, Q_3, is the boundary value that separates the upper 25% of a distribution from the lower 75%. This is also known as the *upper quartile*.

The interquartile range is the distance between Q_1 and Q_3, and is equal to $Q_3 - Q_1$, which represents the middle 50% of the distribution. Although the concept of quartiles

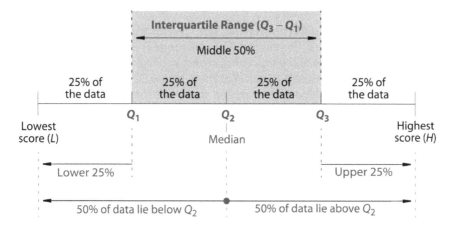

Figure 2.8 The interquartile range is the distance between the first quartile (Q_1) and the third quartile (Q_3) and represents the middle 50% of a distribution.

is easy to understand, there is no single, unified approach for determining the boundary values for Q_1 and Q_3. Thus, it is common for textbooks, statistics software packages, graphing calculators, and online statistics resources to use different approaches to determine quartiles. For our purposes, we will simply recognize that the interquartile range is the difference between Q_1 and Q_3 and use what is reported from our statistics program.

To see the improvement the *IQR* makes to the range, consider the following set of sample data, which reflects the number of miles CFIs and flight students respectively commute one-way to the airport for flight instruction. These data are provided in the file Ch_2 IQR Example.

CFIs	2	6	10	19	28	30	33	40
Flight Students	2	3	4	6	7	8	10	40

Note that the range for each data set is 40 – 2 = 38. This implies that the variation is the same for both CFIs and flight students. This is not the case, though, because for flight students, 40 is an outlier relative to the rest of the data than it is for CFIs. Based on our statistics program, the *IQR* for the CFIs' data is 25.25, but the *IQR* for the flight students' data is 6.25. Remember: Your statistics software might yield different interquartile results. Because outliers do not influence the *IQR*, it is apparent that the CFIs' data set has more variation than the flight students' data set. As a result, the *IQR* provides a more stable measure of variation than the range because it is less likely to be influenced by outliers.

The Five-Number Summary and Box Plots

When we combine Q_1, the median, and Q_3 with the lowest and highest scores of a distribution, we get the five-number summary, which provides a better picture of the variation of a distribution. The five-number summary for CFIs' and flight students' one-way commute distances to the airport based on our statistics program is given in Table 2.6.

We also can graphically represent a five-number summary using a box plot. As illustrated in Figure 2.9, a box plot consists of a "box" placed around the *IQR* and includes a vertical line within the box at the median (Q_2). Horizontal lines (called "whiskers") extend from the left and right sides of the box to the lowest and highest scores, respectively. The whiskers, however, never extend $1.5 \times IQR$ or more below Q_1, or $1.5 \times IQR$ or more above Q_3. This is because observations that exceed these thresholds are considered outliers, which are represented using a dot. Most statistics software programs provide box plots as part of their descriptive statistics output.

Table 2.6 Five-Number Summary for One-Way Commute Distances to the Airport

	Lowest	Q_1	Mdn	Q_3	Highest
CFIs	2	7	23.5	32.25	40
Flight Students	2	3.25	6.5	9.5	40

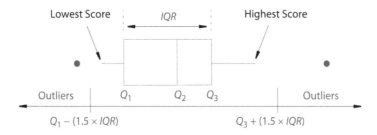

Figure 2.9 General illustration of a boxplot.

Standard Deviation

Although the *IQR* is more informative than the range, one of its drawbacks is that it only considers the middle 50% of the scores in a distribution and ignores the other 50%. Similarly, although the five-number summary and corresponding box plot give us a good feel for the variation of a distribution, their main drawback is that they involve five numbers. What we seek is a single index of variation that involves all the observations of a distribution, not just a few selected ones. Because variation measures the amount of dispersion there is within a data set, it follows logically that the more dispersion there is among the scores in a distribution, the more scattered the scores will be about the mean. Thus, a logical alternative to the range, *IQR*, and five-number summary would be to measure how much the scores in a distribution deviate from the mean. This logical alternative is called standard deviation and is graphically illustrated in Figure 2.10.

Given this focus, the best way to determine standard deviation would be to measure the amount of deviation between each score and the mean, add these deviation scores, and then find the *average deviation*. Although this sounds reasonable, we encounter a small problem mathematically. To understand why, consider the simple data set of 4, 6, 8, 12, 15, which has a mean of $M = 45/5 = 9$. As shown in Table 2.7, the overall sum of the deviations from the mean is 0, and hence the corresponding average deviation from the mean also is 0. This is because the mean is the center of a distribution and therefore the sum of the deviation scores above the mean will always equal the sum of deviation scores below the mean. To fix this problem, we square each deviation score and then add the squared deviations. This sum of the squared deviations is denoted *SS*. We then divide *SS* by the sample size, *N*, to get the *mean squared deviations*. Thus, for the running example, the mean of the squared deviations is 80/5 = 16.

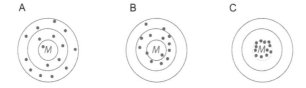

Figure 2.10 Visualizing the concept of standard deviation using different targets where each target is represented by concentric circles with the mean (*M*) in the center, and scores are represented by "dots." As scores become more clustered about the mean (and each other), the average distance scores are from the mean, which is standard deviation, becomes smaller.

Table 2.7 Calculating Standard Deviation

Score (X)	Deviations from Mean	Squared Deviations	Calculations
4	$4 - 9 = -5$	$(-5)^2 = 25$	**Population**
6	$6 - 9 = -3$	$(-3)^2 = 9$	• Variance = Mean of $SS = \dfrac{SS}{N} = \dfrac{80}{5} = 16$
8	$8 - 9 = -1$	$(-1)^2 = 1$	
12	$12 - 9 = 3$	$(3)^2 = 9$	• Standard Deviation = $\sqrt{Variance} = \sqrt{16} = 4$
15	$15 - 9 = 6$	$(6)^2 = 36$	**Sample**
Sums 45	0	80	• Variance = Mean of $SS = \dfrac{SS}{N-1} = \dfrac{80}{4} = 20$
			• Standard Deviation = $\sqrt{Variance} = \sqrt{20} = 4.47$

Note. The sum of the squared deviations, denoted *SS*, is 80.

This resulting measure is formally called variance, which also is a measure of dispersion. However, because variance is in squared units, it is not very meaningful to use it as a descriptive statistic because most variables are not measured in squared units (e.g., distance, flight hours, IQ, heart rate). To reconcile this, we perform one more operation, namely, we take the principal square root of the variance. Thus, for our running example, the principal square root of 16 is 4.0. This result is the standard deviation of the distribution, denoted *SD*, and represents the *average distance the scores in a distribution are from the mean*. For our running example, *SD* = 4.0 indicates that the average distance the scores are from the mean is 4 units. As with any measure of dispersion, the smaller the standard deviation, the less spread there is in the data.

We conclude this section on measures of dispersion with the following observations:

1. When discussing variance, we make a distinction between population variance and sample variance, which is illustrated in Table 2.7.
   • **Population variance** is calculated by dividing the sum of the squared deviations by the total sample size, *N*, and is denoted by the square of the lowercase Greek letter *sigma*, σ^2 (read "sigma squared"). Thus, the corresponding formula is $\sigma^2 = \dfrac{SS}{N}$.
   • **Sample variance** is calculated by dividing the sum of the squared deviations by $(N - 1)$ and is denoted by SD^2. Thus, the corresponding formula is $SD^2 = \dfrac{SS}{N-1}$.
   We divide by $(N - 1)$ because when we estimate the population variance using sample data, the result underestimates the population variance. Dividing by $(N - 1)$ yields an unbiased estimate of the population variance based on sample observations. (*Note:* Statistics software packages report sample variances.)
2. When discussing standard deviation, we also make a distinction between population standard deviation and sample standard deviation.
   • **Population standard deviation** is denoted by the lowercase Greek letter *sigma*, σ, and the corresponding formula is $\sigma = \sqrt{\dfrac{SS}{N}}$.

- **Sample standard deviation** is denoted as *SD*, and the corresponding formula is

$$SD = \sqrt{\frac{SS}{N-1}}.$$

3. Because standard deviation is equal to the "square root of the mean of the squared deviations," it is commonly referred to as the root mean square, or more simply, *RMS*.
4. The greater the standard deviation, the greater the *spread* the distribution has, the more *scattered* the scores are about the mean, the more *heterogeneous* the group is, the more *dispersion* there is among the scores, and the more *volatile* the scores.
5. Standard deviation always should be reported with the mean because the combination of these two measures provides a very good description of a distribution.

2.5 Measures of Position: Percentiles and Quartiles

Recall from Chapter 1 that the data we acquire from any research endeavor is first collected in raw form, which means that the data are obtained without any manipulations, and then converted to derived form to make it more meaningful. For example, if we answer 45 questions correctly on the FAA's Instrument Rating–Airplane (IRA) written examination, then our raw score is 45. To make this score more meaningful, we convert it to a percentage by dividing 45 by 60, which is the total number of items on the exam: 45/60 is 0.75 = 75%.

In addition to converting a raw score to a percentage, we also can convert raw scores to percentiles. Percentiles divide sets of data into 100 equal parts, and therefore, 100% is the basis of measure. Anyone who has ever taken a standardized test such as the FAA IRA exam, IQ tests, or Graduate Record Examinations (GREs) has encountered percentiles. A percentile always is interpreted as the *percentage of scores that was lower than your score*. To determine a person's percentile score for an exam, we need to know the number of people who scored below the person's score and then divide that number by the total number of people who took the exam.

For example, with respect to the FAA IRA examination, let's assume that 200 flight students took the exam and 80 students scored lower than 45. Given this information, our percentile is 80/200 = 0.40 = the 40th percentile, which means that 40% of the students who took the exam scored lower than us, or alternatively, we scored higher than 40% of the people who took that exam. This also means that 60% of the examinees scored higher than us.

Percentiles often are partitioned into quarters or quartiles, which we presented earlier in our discussion of the *IQR*. For example:

- $Q_1 = P_{25}$, which is the 25th percentile, and represents the position where 25% of the data in a distribution lie below it and 75% lie above it.
- $Q_2 = P_{50}$, which is the 50th percentile, and represents the position where 50% of the data lie below it and 50% lie above it. (This also is the median.)
- $Q_3 = P_{75}$, which is the 75th percentile, and represents the position where 75% of the data lie below it and 25% lie above it.

A percentile is more informative than a percent because it represents a measure of *relative position*. In other words, a percentile provides us with information about a score's position relative to the rest of the scores in the distribution.

Chapter Summary

1. A frequency is a simple count of the number of times an observation occurs. Frequencies are organized into a simple frequency distribution table with two columns: one for each observation and one for the number of times the observation occurs. Additional columns may be included for relative frequencies, cumulative frequencies, and cumulative relative frequencies.
2. A grouped frequency distribution table is used to organize continuous data. The categories are called class intervals and consist of lower and upper limits, which signify the range of data values that will be placed within an interval, and the class width is the difference between the lower limits of any two consecutive classes.
3. A bar graph is a graphical representation of a frequency distribution involving qualitative data. Spaces are placed between bars to signify the categories are distinct.
4. A histogram is a graphical representation of a frequency distribution involving quantitative data. A histogram resembles a bar graph except adjacent bars "touch" to signify the data are numerical. If data are discrete, then the bars are "centered" at the numerical value of each category. If data are continuous, then the width of the bars is equal to the width of the class intervals from the accompanying grouped frequency distribution table.
5. A distribution is commonly described by its shape. Two general shapes are symmetrical and skewed. A symmetrical distribution is one in which the left and right sides are mirror images of each other with respect to an imaginary vertical line that splits the distribution into two equal halves. A skewed distribution is one that tails off either to the right or left.
6. A measure of central tendency represents the central point of a distribution and includes the mean, median, and mode. The mean is the arithmetic average and is denoted M for a sample and μ for a population. The mean is the most stable measure of central tendency, but it is sensitive to outliers and therefore not representative of skewed distributions. The median (*Mdn*) is the middle score in a set of ranked data where 50% of the scores lie above it and 50% of the scores lie below it. The median is not sensitive to outliers, and therefore is appropriate for skewed distributions. The median also is equivalent to Q_2, which is the 50th percentile. The mode represents the most frequently occurring score and is appropriate for distributions involving qualitative data.
7. A measure of dispersion, or variation, is a numerical value that describes how much spread there is in the data. If the scores are widely spread out then the variation is large; if the scores are clustered together, then the variation is small. Two common measures of dispersion are range and standard deviation. The range is the difference between the highest and lowest observations, and standard deviation is the average distance the scores of a distribution deviate from the mean. The standard

deviation of a sample is denoted *SD*, and the standard deviation of a population is denoted σ.

8. Two measures of position are percentiles, which separate a distribution into 100 equal parts, and quartiles, which separate a distribution into four equal parts: Q_1 separates the lower 25% of a distribution from the upper 75%, Q_2 separates the distribution in half and is equal to the median, and Q_3 separates the lower 75% of a distribution from the upper 25%.

10. The interquartile range is the middle 50% of a distribution and is defined as $IQR = Q_3 - Q_1$.

11. The five-number summary provides an overall summary of the amount of variation in a distribution and consists of the lowest score, Q_1, *Mdn*, Q_3, and the highest score.

12. A box plot graphically depicts the five-number summary and consists of a box that borders the interquartile region with a vertical line inside the box at the median. "Whiskers" extend from the box to the lowest and highest scores, respectively. If outliers are present, whiskers stop prior to the boundaries that separate outliers from the rest of the data. The left boundary is $Q_1 - (1.5 \times IQR)$, and the right boundary is $Q_3 + (1.5 \times IQR)$.

Vocabulary Check

Bar graph	Histogram	Relative frequency
Bell-shaped distribution	Interquartile range (*IQR*)	Root mean square (*RMS*)
Bimodal	J-shaped distribution	Skewed distribution
Box plot	Mean	Skewed left
Class interval	Measure of central tendency	Skewed right
Class width	Measure of dispersion	Standard deviation
Cumulative frequency	Median	Sum of squared deviations (*SS*)
Cumulative relative frequency	Mode	Symmetrical distribution
Five-number summary	Outliers	U-shaped distribution
Frequency distribution	Percentile	Uniform distribution
Frequency distribution table	Quartile	Variance
Grouped frequency distribution	Range	Relative frequency

Review Exercises

A. Check Your Understanding

In 1–10, choose the best answer among the choices provided.

1. Given the following two distributions, determine which statement is correct.

    A: 8 6 5 5 4 3 3 3 1
    B: 27 6 5 5 4 3 3 3 1

    a. The two distributions have equal means.
    b. The two distributions have unequal medians.

    c.   The standard deviation of distribution A is greater than the standard deviation of distribution B.

    d.   The median is the better measure of central tendency in distribution B than the mean.

2. Given the set of scores, 5, 5, 5, 10, 7, 7, 7, what would be the shape of the distribution?

    a.   Uniform

    b.   Skewed left

    c.   Skewed right

    d.   Symmetrical

3. To describe the distribution of average daily teaching load of flight students for CFIs at a local flight school, a spokesperson reports that a few CFIs have a very heavy teaching load, whereas many CFIs have a very light load. This implies that among the staff of CFIs:

    a.   The mean teaching load is higher than the median.

    b.   The median teaching load is higher than the mean.

    c.   The median teaching load and the mean teaching load are the same.

    d.   None of the above is correct.

4. Which of the following statements is correct?

    a.   The mean is a point in the distribution below and above which 50% of the cases lie.

    b.   The mode is the least stable measure of central tendency.

    c.   The median is equal to the sum of the scores divided by the number of scores.

    d.   The median is the same value as the P_{25}.

5. If the standard deviation of the scores on a test is $SD = 0$, we should conclude:

    a.   Nobody got even one question correct.

    b.   Half the scores were above the mean, and half the scores were below the mean.

    c.   A computational error was made.

    d.   Everybody got the same score.

6. The table below contains the number of days retired U.S. naval aviators spent in the Intensive Care Unit (ICU) at a local veteran's hospital who were admitted during the 5-day period December 23–27, 2022. The total sample size is _____ and the median number of days spent in the ICU is _____.

Number of Days	1	2	3	4	5
Number of Patients	20	30	12	9	9

    a. $N = 15, Mdn = 1$

    b. $N = 197, Mdn = 3$

    c. $N = 80, Mdn = 2$

    d. $N = 95, Mdn = 4$

7. ICAO's Test of English for Aviation (TEA) was administered to 100 nonnative English speaking flight students. Which statement CANNOT be an accurate description of the distribution of the TEA scores?

    a.   The majority of students had scores above the mean.

    b.   The majority of students had scores above the median.

    c.   The majority of students had scores above the mode.

    d.   All the students got the same score.

8. A random sample (N = 200) of the Florida Young Aviators Club, which consists of pre-teenage children, yielded the following distribution of heights: M = 46 in. (117 cm), Mdn = 45 in. (114 cm), SD = 3 in. (7.6 cm), Q_1 = 43 in. (109 cm), and Q_3 = 48 in. (122 cm). About 100 children in the sample have heights that are
   a.   less than 43 in. (109 cm)
   b.   less than 48 in. (122 cm)
   c.   between 43 in. (109 cm) and 48 in. (122 cm)
   d.   more than 46 in. (117 cm)
9. When would a score that is 15 points above the mean be considered an outlier?
   a.   When the population mean is much larger than 15.
   b.   When the population standard deviation is much larger than 15.
   c.   When the population mean is much smaller than 15.
   d.   When the population standard deviation is much smaller than 15.
10. Given the box plots in Figure 2.11, which of the following statements CANNOT be justified?
   a.   Data set A and data set B have the same number of data points.
   b.   The interquartile range of data set A is equal to the interquartile range of data set B.
   c.   The median of data set A is less than the median of data set B.
   d.   The range of data set A is equal to the range of data set B.

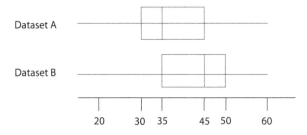

Figure 2.11 **Chapter 2/Exercise A-10.**

B. Apply Your Knowledge

The data file labeled Ch_2 Exercises Part B contains research data collected from a random sample of N = 340 certified flight instructors (CFIs). The data set consists of 10 independent variables (IVs) and one dependent variable (DV). A description of each variable follows.

X_1 = Participants' biological sex (male or female)
X_2 = Participants' age in years
X_3 = Participants' race/ethnicity
X_4 = Participants' marital status
X_5 = Participants' highest level of education
X_6 = Total number of years participants held a CFI certificate
X_7 = Participants' total hours dual given
X_8 = Participants' total hours dual given in previous 90 days
X_9 = Participants' total flight time (in hours)
X_{10} = Types of certificates participants currently hold

$Y =$ Complacency was assessed using a 7-item instrument measured on a traditional 5-point Likert response scale. Thus, scores could range from 7 to 35, with higher scores indicating a greater likelihood toward complacency as a flight instructor.

1. Based on the IVs and DV, what would be an appropriate overall RQ for this study?
2. What would be an appropriate research methodology/design for this study? Explain.
3. Using your statistics software, find the appropriate measures of central tendency and variability for Y and $X_1 - X_9$. Complete a chart like the one below. The first row is completed for you as a guide.

Factor	Appropriate Measure of Central Tendency	Appropriate Measure of Variability	Reason	Actual Measure of Central Tendency
X_1	Mode	Not applicable	Data are nominal	Male

4. Explain how you would determine the measure of central tendency for X_{10}?
5. Prepare a descriptive statistics summary table for all the continuous variables. Use the table below as a guide. (Round to two decimal places.)
6. Consider the outliers associated with X_8. Determine which outliers you think are rare

Factor	N	M	SD	Range (L–H)	Shape of Distribution

cases and which you think might be contaminants. Explain.
7. Based on the results for Y, how would you assess CFIs' overall level of complacency?
8. Based on the results for X_1 and X_5, to what population do you think the results of this study would be generalizable? Why?
9. Interpret in the context of the given research setting Q_1, Mdn, and Q_3 for Y.
10. What type of distribution do the complacency (Y) scores form, and what is the shape of this distribution if the outliers are removed? Justify your response relative to the three measures of central tendency.

References

Federal Aviation Administration (2021/Table 13). *U.S. civil airmen statistics.* https://www.faa.gov/data_research/aviation_data_statistics/civil_airmen_statistics
Kachigan, S. K. (1991). *Multivariate statistical analysis: A conceptual introduction* (2nd ed.). Radius press.
Uhuegho, K. O. (2017). *Examining the safety climate of U. S. based aviation maintenance, repair, and overhaul (MRO) organizations.* [Doctoral dissertation, Florida Institute of Technology]. https://repository.lib.fit.edu/handle/11141/1371

Part B

Making Reasonable Decisions about a Population

.

3 *Z* Scores, the Standard Normal Distribution, and Sampling Distributions

Student Learning Outcomes

After studying this chapter, you will be able to do the following:

1. Transform scores from a raw score distribution to *z* scores using the *z* score formula and interpret the corresponding results.
2. Use the standard normal curve to solve an application problem involving a normal distribution and express the solution as a proportion, probability, or percentile.
3. Use a statistics program to determine if the distribution of a given set of continuous data is normal or near normal in form, and standardize the distribution using *z* scores.
4. Use a statistics program to construct a sampling distribution of sample means for a set of continuous data, and apply the central limit theorem.

3.1 Chapter Overview

In Chapter 2, we demonstrated how a raw score of 45 on the 60-item FAA IRA exam can be made more meaningful by converting it to a percentage: $45/60 = 75\%$. In Chapter 2, we also presented two additional raw score manipulations, percentiles and quartiles, which provide information about a score's position relative to the rest of the scores in the distribution. Neither derived score, though, is interpretable with respect to the mean. For example, if we were to assume that the mean raw score for the IRA exam is 48, which is equivalent to 80%, all we would know is that a score of 45 is below the mean, but we do not know anything about the score's position relative to the mean.

To know how far a raw score is above or below the mean, we also need to know the standard deviation of the distribution. In other words, we need to know how the scores in the distribution vary about the mean. This information is critical because the position of a raw score with respect to the mean can vary. For example, as illustrated in Figure 3.1, given a mean score of 48, if the standard deviation is 8, then a raw score of 45 would be approximately one-half standard deviation below the mean. If, however, the standard deviation is 1, then a score of 45 is three standard deviations below the mean. Thus, the mean and standard deviation of a distribution are used to determine various measures of *relative position with respect to the mean*.

Note also from Figure 3.1 that because the location of a score within a distribution is with respect to the distribution's mean, each score's position would change for different distributions because they would have different means and standard deviations. One way

DOI: 10.4324/9781003308300-5

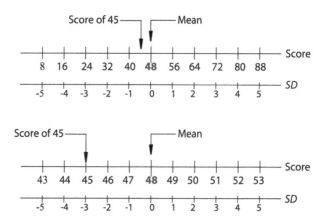

Figure 3.1 Demonstration of how the standard deviation of a distribution can affect how far a score is from the mean.

to overcome this problem is to convert different distributions of raw scores to distributions of *standard scores* that are based on the same mean and standard deviation. In this chapter, we introduce the concept of standard scores, apply this concept to the normal curve, which leads to the *standard normal distribution*, and then present a very important concept known as the *sampling distribution of sample means*.

3.2 *z* Scores

Consider the situation where two brothers, Roy and Steve, are comparing their respective scores on the FAA IRA exam. Although the FAA issues a simple Pass/Fail, let's assume that Roy's score is 52 and Steve's score is 50. Let's further assume that Roy took the exam in 2018 and Steve took it in 2021. Would it be fair for the two brothers to compare their scores? Let's consider another situation and assume that in addition to the IRA exam, Steve also took the FAA Flight Instructor Instrument–Airplane (FII) exam in 2021 and his score was 45. Can we claim that Steve's performance on the IRA exam was better than his score on the FII exam?

The answer to both questions is "no." This is because the distributions are different in each situation. Because each distribution has its own mean and standard deviation, we cannot directly compare Roy's 2018 IRA exam score to Steve's 2021 IRA exam score, and we cannot compare Steve's IRA and FII scores to each other. To be able to make direct comparisons, we need to standardize these different distributions by converting the raw scores to standard scores. The most widely used standard score is the *z* score. A *z* score describes the exact location of a raw score within a distribution by standardizing raw scores. The resulting scores indicate the direction (+ or –) and distance from the mean the scores are located. To transform a raw score to a *z* score, we use the formula

$$z = \frac{x - \mu}{\sigma}$$

where *x* is the raw score, μ is the corresponding distribution's mean, and σ is the distribution's standard deviation. Note that the numerator of the *z* score formula is a *deviation score*, which measures the distance between a raw score and the distribution's mean. We divide the deviation score by standard deviation because we want to measure distance with respect to standard deviation units.

One of the benefits of a *z* score is it enables us to interpret raw scores from different distributions. This is because a *z* distribution is a *standardized* distribution with mean $\mu = 0$ and standard deviation $\sigma = 1$. When every raw score of a distribution is converted into *z* scores, the corresponding *z* distribution will have the same shape as the original raw score distribution except now the raw scores are with respect to a mean of 0 and a standard deviation of 1.

Returning to the first situation involving Roy and Steve, let's assume that the mean and standard deviation for the IRA exam in 2018 was $\mu = 48$ and $\sigma = 8$, and in 2021, $\mu = 48$ and $\sigma = 4$. Given this information, we can now standardize Roy and Steve's respective IRA exam scores and then compare them:

- Roy: If $x = 52$, then $z = (52 - 48)/8 = 4/8 = 0.5$
- Steve: If $x = 50$, then $z = (50 - 48)/4 = 2/4 = 0.5$

So, although Roy had a higher raw score, when standardized both brothers' scores were one-half standard deviation above the mean, and hence, relatively speaking, neither score was "better" than the other one. Let's now compare Steve's 2021 IRA and FII scores. We will assume that the mean and standard deviation for the FII exam in 2021 was $\mu = 42$ and $\sigma = 2$.

- IRA Exam: If $x = 50$, then $z = (50 - 48)/4 = 2/4 = 0.5$
- FII Exam: If $x = 45$, then $z = (45 - 42)/2 = 3/2 = 1.5$

Although Steve's raw IRA exam score was 5 points higher than his FII exam score, when standardized, his FII score actually was "better," relatively speaking, than his IRA score because the former was 1.5 standard deviations above the mean vs. one-half standard deviation above the mean. Most statistics software have a feature that will automatically transform raw scores to *z* scores.

3.3 The Standard Normal Distribution

The Concept of a Normal Distribution

Recall from Chapter 2 that we demonstrated that nearly all distributions can be classified as being symmetrical or skewed (see Figure 2.5). Our primary focus in this section is on one type of symmetrical distribution, namely, bell-shaped. This is because many human characteristics often yield frequency distributions that are bell-shaped. As an example, consider the simple task of flipping a fair coin.

- When we flip a coin one time, there are two possible outcomes: heads (H) or tails (T).
- When we flip a coin two consecutive times, the number of possible outcomes is $2 \times 2 = 2^2 = 4$, namely: HH, HT, TH, TT.
- When we flip a coin three consecutive times, the number of possible outcomes is $2 \times 2 \times 2 = 2^3 = 8$, namely: HHH, HHT, HTH, HTT, THH, THT, TTH, TTT.

- If we extend this to flipping a coin five consecutive times, the number of possible outcomes is $2^5 = 32$, and on 10 flips, the number of possible outcomes is $2^{10} = 1024$.

This is illustrated in Figure 3.2, which shows frequency distribution tables of getting "heads" on three flips, five flips, and 10 flips of a coin, respectively, and their corresponding histograms. We also superimposed a line graph, called a *frequency polygon*, to get a better idea of each distribution's shape. Note that the shapes are symmetrical and bell-shaped. Further note that as the number of coin flips increases, the corresponding

(a) Results from Flipping a Coin Three Consecutive Times

Number of "Heads"	Number of Possible Outcomes (*f*)	Actual Outcomes
0	1	TTT
1	3	HTT, THT, TTH
2	3	HHT, HTH, THH
3	1	HHH

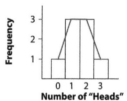

(b) Results from Flipping a Coin Five Consecutive Times

Number of "Heads"	Number of Possible Outcomes (*f*)	Actual Outcomes
0	1	TTTTT
1	5	HTTTT, THTTT, TTHTT, TTTHT, TTTTH
2	10	HHTTT, HTHTT, HTTHT, HTTTH, THHTT, THTHT, THTTH, TTHHT, TTHTH, TTTHH
3	10	HHHTT, HHTHT, HHTTH, HTHHT, HTHTH, HTTHH, THHHT, THHTH, THTHH, TTHHH
4	5	HHHHT, HHHTH, HHTHH, HTHHH, THHHH
5	1	HHHHH

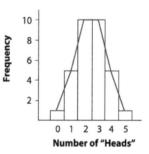

(c) Results from Flipping a Coin 10 Consecutive Times

Number of "Heads"	Number of Possible Outcomes (*f*)
0	1
1	10
2	45
3	120
4	210
5	252
6	210
7	120
8	45
9	10
10	1

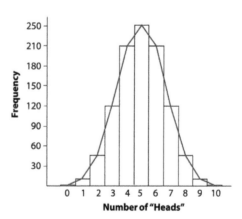

Figure 3.2 Frequency distribution tables and corresponding histograms and frequency polygons of the distributions for the number of "heads" in (a) 3 flips, (b) 5 flips, and (c) 10 flips of a fair coin.

frequency polygons begin to approximate a bell-shaped curve. This becomes more apparent in Figure 3.3, which contains the results for 20 coin flips.

An empirical distribution that is bell-shaped is known as a normal distribution, and the corresponding continuous curve that results from an infinite number of observations is called a normal curve. As presented in Chapter 2, Figure 2.5, bell-shaped distributions are symmetrical with the highest frequencies occurring in the middle, and the remaining frequencies gradually tapering off in either direction. With respect to the corresponding curve, this tapering shows that the tails of the distribution are asymptotic—that is, they approach the horizontal axis but never touch it. Furthermore, as standard deviation increases, the shape becomes flatter and more spread out. Although we will not work with it, the following mathematical equation determines the exact shape of a normal distribution:

$$Y = \frac{e^k}{\sigma\sqrt{2\pi}}$$

where Y = the height of the curve directly above any given X value in the frequency distribution
e = the base of the natural logarithm (≈ 2.7183)

$$k = \frac{-(X - \mu)^2}{2\sigma^2}$$

π = the ratio of the circumference of a circle to its diameter (≈ 3.1416)
μ = the mean of the distribution
σ = the standard deviation of the distribution

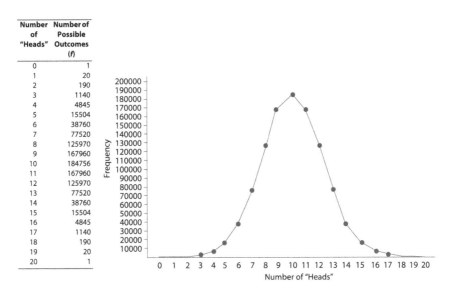

Number of "Heads"	Number of Possible Outcomes (f)
0	1
1	20
2	190
3	1140
4	4845
5	15504
6	38760
7	77520
8	125970
9	167960
10	184756
11	167960
12	125970
13	77520
14	38760
15	15504
16	4845
17	1140
18	190
19	20
20	1

Figure 3.3 As the number of flips of a coin increases, the distribution approaches a bell-shaped curve as shown by the given frequency polygon, which represents the number of "heads" in 20 flips of a coin.

For any given *X* value, a *Y* value is calculated, and the corresponding (*X, Y*) point is plotted. Because we are dealing with an infinite number of observations (i.e., *X* values), the result is a smooth curve when the points are connected.

A key mathematical property of any normal distribution is that the area under the normal curve is exactly equal to 1. The implication of this property is that the total area under a normal curve represents the total number of observations in the distribution. In other words, *we can interpret the area under a normal curve as representing 100% of the observations of a distribution.* Furthermore, because the total area under the curve is equal to 1, proportions of this area also can be interpreted as probabilities.

To illustrate this concept, let's revisit the runway incursion data from Chapter 2, which was provided in the data file Ch_2b Runway Incursions and depicted as a histogram in Figure 2.3. Recall that the data set reflected the number of runway incursions at ATL over a given 30-day period. Instead of using a histogram, though, we prepared a *frequency distribution graph* as shown in Figure 3.4, which partitions the bars of the histogram into 1-unit squares. For example, because 10 of the 30 days had 0 runway incursions, the corresponding "bar" has 10 squares, which correspond to its frequency.

Observe from Figure 3.4 that the entire distribution consists of 30 squares, which is 100% of the distribution, and each outcome represents a proportion of the entire distribution. For example, focusing on the shaded area, an outcome of "4 runway incursions" represents 3 of the 30 squares, which is 3/30 = 1/10 of the distribution and thus 10% of the distribution consists of four runway incursions.

Let's now consider this from a probability perspective. In very simple terms, the probability of an event *A*, denoted *P(A)*, is given by the ratio

$$P(A) = \frac{\text{The number of successful outcomes } (S)}{\text{The total number of possible outcomes } (T)} = \frac{\text{Success } (S)}{\text{Total } (T)} = \frac{S}{T}$$

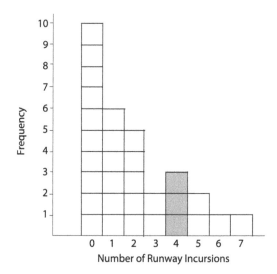

Figure 3.4 A *frequency distribution graph* for the number of runway incursions reported at ATL for a 30-day period. The data correspond to the frequency distribution in Table 2.4.

Applying this to our runway incursion example, consider the question, "What is the probability that if we were to randomly select 1 day from the 30-day period, there would be four runway incursions on that day?" Because there are 3 days in which four runway incursions occurred ("success") out of 30 days ("total"), the probability is

$$P(4 \text{ runaway incursions}) = \frac{3}{30} = \frac{1}{10} = 0.10$$

In other words, there is a 0.10 probability—or a 10% chance—that the day we select will have four runway incursions. Note how we have equated proportion, percentage, and probability by representing a distribution as a frequency distribution graph. Doing so enables us to visualize how each outcome reflects a proportion of area relative to the overall area of the graph, which is 100% of the distribution.

Also recognize that the terms percentage, proportion, and probability all relate to the same concept, but their usage is contextually oriented. Percentages and proportions are commonly used in the context of comparing a fractional part to a whole. For example, in 2020, 58,541 of the 691,691 pilots in the United States were women (Women in Aviation, International, n.d.). This ratio would be commonly reported as a percentage, 8.4%, and the corresponding proportion would be approximately 1/12. Probability, on the other hand, is more commonly used in the context of "chance." When applied to the women in aviation statistic, if we were to randomly select one U.S. woman pilot from all U.S. pilots, then the chance that this pilot is a woman is .084. Alternatively, we also could say there is less than a 10% chance that we would select a woman.

The Standard Normal Distribution

Recall from our earlier discussion that we can standardize any distribution by converting the raw scores of the distribution to standard scores, and that the most widely used standard score is the z score. Let's now consider the situation where the distribution we seek to standardize is a normal distribution, and we standardize this distribution using z scores. When we transform the raw scores of a normal distribution into standard scores, the corresponding distribution results in a standard normal distribution. Furthermore, because the distribution has been standardized using z scores, the standard normal distribution will have a mean of 0 and a standard deviation of 1. This standard normal distribution also has the same shape as the raw data distribution, and it maintains all the properties of a normal distribution. The only difference is that all the raw observations are now with respect to a mean of 0 and a standard deviation of 1.

To demonstrate this, consider the data set Ch_2 Exercises Part B, which involved a study involving factors associated with CFI complacency (Dunbar, 2015). Although it is not statistically prudent to do so because data integrity could be comprised, we modified this data set by deleting the missing data and outliers for instructional purposes. The modified data set is provided in the file Ch_3 Standard Normal Distribution. As illustrated in Figure 3.5(a), the distribution of the raw complacency scores approximates a normal distribution. To emphasize this, we superimposed a smooth curve over the histogram. Note that the mean is 15.7 and the standard deviation is 3.3. In Figure 3.5(b), we converted the raw complacency scores to z scores and then plotted these standard scores. Note that the shape of the distribution is the same as that of the raw scores, except now the mean is 0 and the standard deviation is

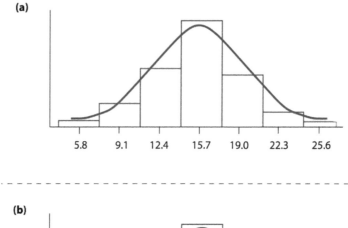

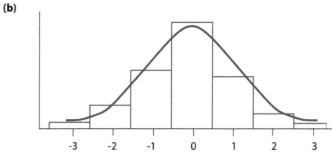

Figure 3.5 Demonstration of how standardizing a distribution with z scores does not affect the shape of the distribution relative to the raw data. The figure is with respect to the complacency scores from the data set Ch_3 Standard Normal Distribution.

1. (*Note:* Your statistics program might prepare these histograms using a different grouped frequency distribution.)

When we partition a normal curve into separate intervals that are exactly 1, 2, and 3 standard deviation units from the mean in either direction, approximately 68% of the area under the curve lies within one standard deviation of the mean, approximately 95% of the area under the curve lies within two standard deviations of the mean, and approximately 99.7% of the area under the curve lies within three standard deviations of the mean. This partitioning, which is illustrated in Figure 3.6(a), is known as the empirical rule, or more directly as the 68–95–99.7 rule. One implication of this rule is that any score in a data set that approximates a normal distribution would be considered an outlier if it is more than three standard deviations below or above the mean.

Because each interval represents a portion of the total area under the curve, we can determine the proportion of observations, which is the relative frequency, of an interval by measuring the area under the curve in that interval. For example, as shown in Figure 3.6(b), when applied to the complacency scores data, approximately 68% of the 255 scores are between 12.4 and 19.0. In other words, $0.68 \times 255 \approx 174$ scores are within this range. Similarly, approximately 2.14% of the scores, which is about six scores, are between 22.3 and 25.6. Checking the raw data from the data file, there are 187 scores between 13 and 19, and seven scores between 23 and 26. We are slightly "off" in our results because our data set is empirically based with

N = 255, and the corresponding distribution *approximates* a normal distribution. Nevertheless, by knowing the area under the curve for an interval, we can determine the proportion of observations that fall within that interval.

Also recognize that we can examine these results from a probability perspective. For example, the probability of a complacency score between 13 and 19 is approximately .68, and the probability of a complacency score between 23 and 26 is approximately .214. Another way to look at this is if we were to randomly select a complacency score from the data set, there is approximately a 68% chance that this score will be between 13 and 19, and there is approximately a 2% chance that the score will be between 23 and 26. Keep in mind that these results are based on a *theoretical* distribution. Empirically, though, based on our sample data of *N* = 255, the probability of randomly selecting a score between 13 and 19 would be 187 successful outcomes divided by 255 total possible outcomes, which is 187/255 = 0.733, so there is a 73.3% chance that this would occur.

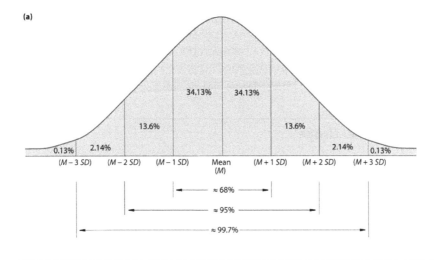

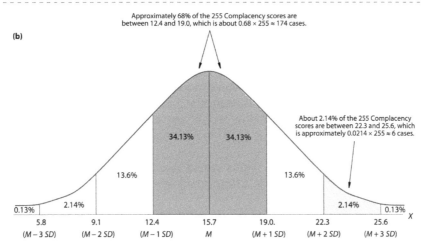

Figure 3.6 In (a), are the approximate percentages of scores that lie within 1, 2, and 3 standard deviations of the mean in a normal distribution. In (b), the normal distribution is applied to the complacency scores.

To understand the distinction between theoretical vs. empirical probability, consider the experiment of flipping a coin. In theory, we expect the probability of getting "heads" on a single flip of a coin to be one-half or 50% of the time. What this means is that if we were to perform this experiment an unlimited number of times, then in the long run we will get "heads" about half the time. If we now perform this experiment a specific number of times, say exactly 20 times, it is possible that we could get "heads" 18 times, which means the probability of getting "heads" is 90%. This is empirical probability, and it refers to the *relative frequency* of the occurrence of an event. So, when we say that the probability of getting "heads" when tossing a coin is one-half, this means that when a coin is tossed *many* times, about half of those tosses will turn up heads and half will turn up tails. It does *not* mean, for example, that if a coin fails to turn up heads on the first toss, then it must turn up heads on the second toss.

To determine the proportion of observations that fall above or below any given z value in a standard normal distribution, all we need to do is consult a single table that provides the proportion of area under the normal curve with respect to specific z values. This table, known as the unit normal table, or z table, is given in Appendix A, Table 1, and lists the area under the standard normal curve between the mean and the indicated z value. Examples of how to use the z table are provided in Figure 3.7. We conclude this discussion by presenting two aviation-based applications involving the standard normal curve.

Example 3.1: *Fuel Usage*

Let's assume a commercial airline recorded the fuel burn from takeoff to landing for all the routes it flew over the past 2 years and the fuel usage was normally distributed. Let's further assume that the Chicago-to-Newark route averaged 90 minutes of fuel usage with a standard deviation of 10 minutes, and the flight plan called for 1.5 hours of fuel. (a) Apply the empirical rule to this scenario. (b) Do you think the airline would approve a pilot's request for 30 minutes of extra fuel (above what was called for in the flight plan)? Why/Why not?

Solution

(a)   Applying the empirical rule:

- Approximately 68% of flights would require between 80 and 100 minutes of fuel, which means that 34% would require at most 10 minutes of extra fuel.
- Approximately 95% of flights would require between 70 and 110 minutes of fuel, which means that 13.6% would require at most an extra 20 minutes of fuel.
- Approximately 99.7% of flights would require between 60 and 120 minutes of fuel, which means that 2.14% would require at most an extra 30 minutes of fuel.

(b)   To give pilots an extra 30 minutes of fuel would provide them with a total of 2 hours of fuel, which is 3 deviations above the mean. As noted in Part (a), only 2.14% of flights would require at most 30 minutes of extra fuel. Therefore, it would not be prudent (statistically speaking) to grant this request. Carrying this extra fuel also would increase the aircraft's weight, which would cause the aircraft engines to burn additional fuel.

A. Finding the Area to the Right of a z Score

1. To the <u>right</u> of z = 1.15:
The area between z = 0 and z = 1.15 = 0.3749.
Therefore, the area to the *right* of z = 1.15 is
0.5000 − 0.3749 = 0.1251.

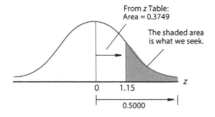

2. To the <u>right</u> of z = − 0.75:
The area between z = 0 and z = − 0.75 = 0.2734.
Therefore, the area to the *right* of z = − 0.75 is
0.5000 + 0.2734 = 0.7734.

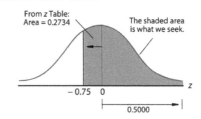

B. Finding the Area to the Left of a z Score

1. To the <u>left</u> of z = 1.15:
The area between z = 0 and z = 1.15 = 0.3749.
Therefore, the area to the *left* of z = 1.15 is
0.5000 + 0.3749 = 0.8749.

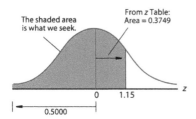

2. To the <u>left</u> of z = − 0.75:
The area between z = 0 and z = − 0.75 = 0.2734.
Therefore, the area to the *left* of z = − 0.75 is
0.5000 − 0.2734 = 0.2266.

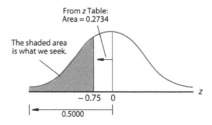

C. Finding the Area Between Two z Scores

1. Between z = 1.15 and z = 2.37:
• The area between z = 0 and z = 1.15 = 0.3749.
• The area between z = 0 and z = 2.37 = 0.4911.
Therefore, the area between z = 1.15 and z = 2.37
is 0.4911 − 0.3749 = 0.1162.

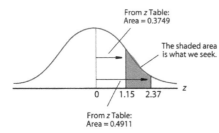

2. Between z = − 0.75 and z = − 1.95:
• The area between z = 0 and z = − 0.75 = 0.2734.
• The area between z = 0 and z = − 1.95 = 0.4744.
Therefore, the area between z = − 0.75 and z = − 1.95
is 0.4744 - 0.2734 = 0.201.

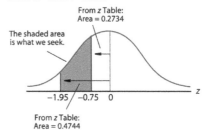

Figure 3.7 Examples of how to find area under the standard normal curve using the z table (Table 1/Appendix A).

Example 3.2: *Aircraft Seat Width*

Let's assume that a private jet manufacturer in North America wants to design seats in its aircraft so they are wide enough to fit most men. Let's further assume that the design engineers use anthropometric data from Fullenkamp et al. (2008), who reported that the sitting hip breadth for men is normally distributed with a mean of $\mu = 38.2$ cm (15 in.) and a standard deviation of $\sigma = 3.6$ cm (1.4 in.). (a) Determine the sitting hip breadth for men that would separate the smallest 75% from the largest 25%. (b) Determine the sitting hip breadth for men that would separate the smallest 99% from the largest 1%. (c) Do you think it would be "better" to design seats based on (a) or (b) and why? (d) Why would it be necessary to make a distinction between the smallest and largest possible sitting hip breadths as in (a) and (b). In other words, why not simply design seats to accommodate 100% of men?

Solution

(a) To determine the sitting hip breadth for men that would separate the smallest 75% from the largest 25%, we draw a normal curve and insert a line that separates the curve so that 75% of the area lies below this point and 25% lies above this line (see Figure 3.8).

We next consult Table 1 from Appendix A to determine which z score is associated with an area that is approximately 25% above the mean. Note that for $z = 0.67$, the proportion of area is 0.2486, and for $z = 0.68$, the proportion of area is 0.2517. We choose the former. We now calculate the corresponding x value using the z formula:

$$z = \frac{x - \mu}{\sigma}$$
$$0.67 = \frac{x - 38.2}{3.6}$$
$$(0.67)(3.6) = (x - 38.2)$$
$$(0.67)(3.6) + 38.2 = x$$
$$x = 2.412 + 38.2$$
$$x = 40.6$$

Therefore, if the jet manufacturer designs its seats to be 40.6 cm (16 in.) wide, then approximately 75% of men in North America will be able to fit in them.

(b) The solution is done exactly as was done in (a) except now we are looking for a z score that corresponds to an area equal to approximately 49% of the curve between the mean and the z score as shown in Figure 3.9. From Table 1 note that for $z = 2.33$, the area is 0.4901. When we solve for x in the z score equation, we get 46.6 cm (18.3 in.).

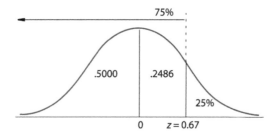

Figure 3.8 For Example 3.2(a).

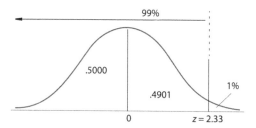

Figure 3.9 For Example 3.2(b).

Therefore, if the jet manufacturer designs its seats to be 46.6 cm wide, then approximately 99% of North American men will be able to fit in them.

(c) The answer is "it depends" on who will benefit. For example, it would be better for the jet manufacturer to use the smaller width because it would save them money, but it would be better for North American male passengers if the jet manufacturer used the larger width because 99% of the population would be able to fit in them.

(d) To design seats to accommodate 100% of males would be cost-prohibitive for the jet manufacturer.

3.4 The Sampling Distribution of Sample Means

The Concept of a Sampling Distribution

Recall from Chapter 1 that research studies generally involve selecting a sample from a population, working with this sample, and then using the results from the sample to make inferences about the parent population. Further recall that the results we acquire from a sample are not always identical to the corresponding population parameters because of sampling error. We demonstrated this in Chapter 1/Figure 1.5 by (a) presenting a hypothetical population of AOPA's U.S. members' flight time (×100), (b) randomly selecting two samples of size 10, and (c) recording the means of each sample and comparing them to each other and to the population mean. Because of sampling error, the sample means were not only different from each other, but they also were different from the population mean. Because we rely on sample statistics to estimate population parameters, the concept of sampling error is important: It provides us with a metric of how a sample statistic is expected to vary from the corresponding population parameter.

To get a better idea of the amount of sampling error, instead of focusing on one or two samples as we did in the AOPA example, a better approach would be to estimate the variability that could be expected from *many different random samples of the same size drawn from the same population*. To illustrate this, we used the random number generator from Random.org (2022) to select 25 random samples of size $n = 5$ from the hypothetical AOPA population. We then calculated the mean for each sample, how far each sample mean deviated from the population mean of 39.97, and examined the distribution of the 25 sample means. This is illustrated in Figure 3.10. Although not shown in Figure 3.10, the population distribution approximates a normal distribution.

(a)

POPULATION

Flight Time of *AOPA's* U.S. Members (× 100)

5	7	8	8	8	9	9	10	12	12
13	15	15	16	17	18	18	18	18	19
20	21	21	21	22	22	23	23	24	24
25	25	25	27	28	29	29	30	30	30
32	32	33	33	33	34	34	34	35	35
36	36	38	38	39	40	42	43	44	44
44	46	47	47	48	48	49	49	49	49
50	51	52	54	54	54	56	60	61	61
62	62	64	65	65	67	68	69	70	73
75	76	77	79	80	82	83	86	87	89

$N = 100$
$\mu = 39.97$
$\sigma = 21.93$

(b)   **Randomly select 25 samples each of size $n = 5$ from the same population and calculate the mean of each sample**

Sample No.	Sample[a]	M	Deviation from μ[b]	Sample No.	Sample[a]	M	Deviation from μ[b]
1	83, 49, 56, 25, 86	59.8	19.83	14	69, 33, 10, 77, 8	39.4	-0.57
2	8, 61, 52, 49, 61	46.2	6.23	15	44, 44, 34, 38, 19	35.8	-4.17
3	22, 9, 30, 42, 49	30.4	-9.57	16	65, 20, 9, 8, 54	31.2	-8.77
4	36, 54, 16, 46, 54	41.2	1.23	17	38, 36, 30, 56, 28	37.6	-2.37
5	65, 79, 82, 36, 34	59.2	19.23	18	48, 23, 24, 54, 61	42.0	2.03
6	47, 30, 39, 46, 47	41.8	1.83	19	34, 68, 89, 34, 38	52.6	12.63
7	18, 73, 52, 29, 50	44.4	4.43	20	47, 21, 48, 16, 16	29.6	-10.37
8	34, 5, 8, 18, 52	23.4	-16.57	21	86, 76, 60, 54, 47	64.6	24.63
9	13, 5, 18, 33, 29	19.6	-20.37	22	30, 70, 15, 25, 7	29.4	-10.57
10	62, 28, 30, 27, 50	39.4	-0.57	23	46, 33, 82, 73, 33	53.4	13.43
11	61, 24, 34, 40, 8	33.4	-6.57	24	30, 33, 44, 79, 54	48.0	8.03
12	33, 38, 13, 80, 67	46.2	6.23	25	42, 19, 15, 67, 49	38.4	-1.57
13	61, 67, 29, 51, 62	54.0	14.03				

Note. [a]All sample data represent flight hours (× 100). [b]The population mean is $\mu = 39.97$ (3,997 hours).

(c)   **Examine the distribution of the 25 sample means**

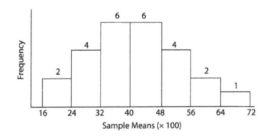

Class	Sample Means	Freq.
16 to < 24	23.4, 19.6	2
24 to < 32	30.4, 31.2, 29.6, 29.4	4
32 to < 40	39.4, 33.4,39.4, 35.8, 37.6, 38.4	6
40 to < 48	46.2, 41.2, 41.8, 44.4, 46.2, 42.0,	6
48 to < 56	54.0, 52.6, 53.4, 48.0	4
56 to < 64	59.8, 59.2	2
64 to < 72	64.6	1

Mean of Sample Means: 41.64
Std. Dev. of Sample Means: 11.5

Figure 3.10 Concept of sampling distribution of sample means.

Observe from Figure 3.10(c) that when we collected the means of each sample, we effectively created a new sample that consists of the 25 sample means. In other words, we now have a *sample of sample means*. In a similar fashion, if we had targeted the median, range, or standard deviation instead of the mean, then we would have a sample of sample medians, sample ranges, and sample standard deviations, respectively. Thus, every sample statistic acquired from a collection of repeated samples yields a sample comprised of that sample statistic.

Every sample statistic also has its own sampling distribution. For example, with respect to the mean, the distribution of the sample means is called the *sampling distribution of sample means*. With respect to the 25 AOPA samples, the corresponding sampling distribution of the sample means is illustrated in Figure 3.10(c) where the mean of the distribution is 41.64, the standard deviation is 11.5, and the corresponding histogram approximates a normal distribution. Thus, a sampling distribution is simply a set of data relative to a specific sample statistic such as the mean that has been acquired from many samples of the same size randomly selected from the same population. Sampling distributions are also just like the raw score distributions we discussed in Chapter 2 except instead of working with a single sample we are working with multiple samples. Furthermore, in the case of a sampling distribution of sample means, instead of focusing on the mean from one sample, we are focusing on the mean of many samples, that is, a sample of sample means.

Notice that this approach to using 25 different random samples of the same size randomly selected from the same population provides us with a much better understanding of how much a sample mean (M_i) varies from the true population mean (μ) than relying on a single sample. For example, in Figure 3.10(b), we can see that the mean for Sample 21 was nearly 25 units away from the population mean, whereas the mean for Sample 10 deviated approximately one-half unit from the population mean. Because each sample mean represents an estimate of the same population mean, any variation among these sample means must be attributed to sampling error. Thus, Sample 21 had a greater amount of sampling error than Sample 10.

The Sampling Distribution of the Mean

Given the mean's importance and usefulness in statistics, we now focus exclusively on the sampling distribution of the mean. Although we could extend our AOPA example from Figure 3.10 by selecting many more random samples—for example, why limit it to 25 samples, why not select 1,000, 10,000, or even 1 million random samples?—besides being time consuming, we would still be dealing with empirical sampling distributions. What we need is a *theoretical sampling distribution* comprised of an infinite number of sample means. To generate such a distribution, though, we would need to select *all possible samples* of the same size from the same population.

To make the concept of a theoretical sampling distribution more concrete, let's assume we have the first four odd integers written separately on four pieces of paper as shown here.

1   3   5   7

We will now place these numbered papers in a hat, mix them up, and draw and record the number. We will then place this numbered paper back into the hat and repeat the procedure one more time. By placing the paper back into the hat after its initial draw, each subsequent drawing represents an *independent event.*

As an example of what we mean by this, consider the probability of drawing two red cards in succession—one after the other—from a standard deck of 52 playing cards. (Recall that the probability of an event is "success" divided by "total.") Observe that the P(1st card is red) = 26/52 but the P(2nd card is red) depends on what we do with the first card drawn. If we *replace the first card* before we draw the second card, then the P(2nd card is red) = 26/52 and the two events are independent. However, if we initially draw a red card and do not replace it before we draw the second card, then the P(2nd card is also red) = 25/51, and the second event is *dependent* on the first. Thus, the concept of independent events from a probability perspective is that the probabilities must stay constant from one selection to the next if there is more than one trial. This means we must *sample with replacement.*

Returning to our example, we will consider the four numbered papers as our popula-tion, and we will randomly select "all possible samples" of size $n = 2$ from this population. Because we will be sampling with replacement, the probability of drawing 1, 3, 5, or 7 is the same, namely, 1/4, and as shown in Figure 3.11(a), the corresponding probability distri-bution is rectangular with $\mu = 4.0$.

Because the population is of size $N = 4$, the number of "all possible samples" of size $n = 2$ is 16. We could get 1 on the first selection followed by 1 on the second selection, {1, 1}; we could get 1 on the first selection followed by 3 on the second selection, {1, 3}; and continuing in this manner, we would get samples of {1, 5}, {1, 7}, {2, 1}, {2, 3}, and so forth. As depicted in Figure 3.11(b), all possible samples of size $n = 2$ are listed using a *tree diagram*. We also show in Figure 3.11(b) the respective means for these 16 samples, and the corre-sponding probability distribution of the 16 sample means.

Note from Figure 3.11(b) that the *shape* of the probability distribution of the sample means is *not* rectangular as was the distribution of the parent population. Instead, it is symmetrical. Also note that the *mean* of the distribution of sample means—that is, the "mean of the sample means," which we will call the *grand* mean and denote μ_M—is equal to the mean of the parent population, namely, 4.0. This is reasonable

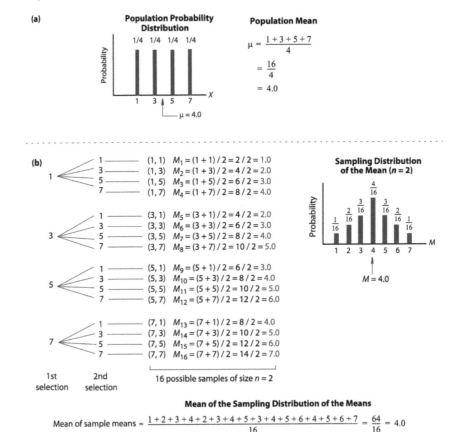

Figure 3.11 Illustration of the concept of a sampling distribution of sample means.

because the sample means of randomly selected samples are expected to be a close approximation of the population mean. When considered from the concept of the sampling distribution of sample means, this also makes sense because we are randomly selecting *all possible samples* of the same size from the same population. As a result, the sampling error should be zero. This is because the positive errors are expected to balance the negative errors, and therefore the average sampling error of the distribution of sample means should be zero. This implies that on average the "mean of the sample means" (the grand mean) will equal the corresponding population mean. Furthermore, because positive sampling errors equal negative sampling errors, a single sample mean is as likely to underestimate a population mean as to overestimate it. Therefore, we can claim that a sample mean is an *unbiased statistic*, which implies that it is both an unbiased estimate as well as a reasonable estimate of the population mean.

Applying this relationship between the grand mean and the corresponding population mean to the AOPA example from Figure 3.10, the grand mean is 41.64, but the population mean is 39.97, a difference of 1.67 units (which is 167 hours). The reason this relationship does not hold is because we are dealing with an *empirical* sampling distribution of sample means based on 25 randomly selected samples of size $n = 5$. We did not randomly select "all possible samples" of size $n = 5$. Still, though, the grand mean of the 25 sample means is relatively "close" to the population mean.

The Variability of a Sampling Distribution of the Mean

We now turn our attention to the variability of the sampling distribution of sample means. First recall that variability refers to how much spread there is in the data, and one measure of variability is standard deviation. When applied to a single sample of raw scores, standard deviation represents the average distance any single score is from the sample's mean: small standard deviations indicate little spread among the data, which indicates the scores are bunched up around the mean, whereas large standard deviations indicate greater dispersion.

When considered from the context of the sampling distribution of the mean, the concept of standard deviation is the same. The only difference is in the interpretation. Instead of examining the average distance a single score is from the mean of a sample, the focus is on the average distance the mean of any sample is from the grand mean. Because the context of standard deviation is now with respect to the sampling distribution of sample means, we refer to the standard deviation of a sampling distribution of sample means as the standard error of the mean *(SEM)*, or more simply, standard error *(SE)*. We also denote standard error as σ_M to distinguish it from the standard deviation of a population of individual scores (σ), where the subscript M reflects a sampling distribution of sample means. Because of its similarity to the standard deviation of a distribution of raw scores, standard error describes the distribution of sample means by providing a measure of how much difference is expected from one sample to another: If standard error is small, the sample means are close together and have similar numerical values. If standard error is large, the sample means are scattered over a wide range and there are large differences from one sample mean to another. For example, in the empirical AOPA example of Figure 3.10(c), the standard deviation of 11.5 indicates that the means from the 25 samples deviate, on average, 11.5 units from the grand mean. This is relatively large and suggests there is considerable dispersion among the sample means.

Let's now return to the sampling distribution of Figure 3.11, which emulated a theoretical sampling distribution of sample means where we randomly selected all possible samples of size $n = 2$ from the same population. Recall that two observations of this theoretical sampling distribution were (a) the probability distribution of the population was uniform, but the probability distribution of the sample means was symmetrical (bell-shaped), and (b) the grand mean was equal to the population mean. What about the differences in variability between the population and the sample of sample means?

To answer this question, we calculated the respective standard deviations, as demonstrated in Table 3.1 (see also Chapter 2's presentation of standard deviation). Observe from Table 3.1 that the standard deviation of the population is approximately $\sigma = 2.236$ whereas the standard deviation of the sampling distribution of means is approximately $\sigma_M = 1.58$. The relationship between these measures of variability is

$$\sigma_M = \frac{\sigma}{\sqrt{n}}$$

Table 3.1 Standard Deviation Calculations for the Population {1, 3, 5, 7} and the Distribution of Sample Means Based on All Possible Samples of Size $n = 2$ Randomly Selected from the Same Population

		Population		
	X	$X - M$	$(X - M)^2$	• Sum of squared deviations (SS) = 20.
	1	$1 - 4 = -3$	$(-3)^2 = 9.0$	• Population variance (σ^2) = SS/N = 20/4 = 5.0
	3	$3 - 4 = -1$	$(-1)^2 = 1.0$	• Population standard deviation $(\sigma) = \sqrt{5}$
	5	$5 - 4 = 1$	$(1)^2 = 1.0$	≈ 2.236
	7	$7 - 4 = 3$	$(3)^2 = 9.0$	
Sums	16	0	20.0	

Distribution of Sample Means ($M_1, M_2, M_3, \ldots, M_{16}$)
where M_G = the grand mean ("mean of the sample means") as illustrated in Figure 3.11.

	M_i	$X_M - M_G$	$(M_i - M_G)^2$	• Sum of squared deviations (SS) = 40.
	1.0	$1 - 4 = -3$	$(-3)^2 = 9$	• Variance $(\sigma^2{}_M)$ = SS/N = 40/16 ≈ 2.5
	2.0	$2 - 4 = -2$	$(-2)^2 = 4$	
	3.0	$3 - 4 = -1$	$(-1)^2 = 1$	• Standard deviation $(\sigma_M) = \sqrt{2.5} \approx 1.58$
	4.0	$4 - 4 = 0$	$(0)^2 = 0$	
	2.0	$2 - 4 = -2$	$(-2)^2 = 4$	
	3.0	$3 - 4 = -1$	$(-1)^2 = 1$	
	4.0	$4 - 4 = 0$	$(0)^2 = 0$	
	5.0	$5 - 4 = 1$	$(1)^2 = 1$	
	3.0	$3 - 4 = -1$	$(-1)^2 = 1$	
	4.0	$4 - 4 = 0$	$(0)^2 = 0$	
	5.0	$5 - 4 = 1$	$(1)^2 = 1$	
	6.0	$6 - 4 = 2$	$(2)^2 = 4$	
	4.0	$4 - 4 = 0$	$(0)^2 = 0$	
	5.0	$5 - 4 = 1$	$(1)^2 = 1$	
	6.0	$6 - 4 = 2$	$(2)^2 = 4$	
	7.0	$7 - 4 = 3$	$(3)^2 = 9$	
Sums	64	0	40	

In other words, the standard deviation of the sampling distribution of sample means (i.e., standard error) is equal to the standard deviation of the population divided by the square root of the sample size. Applying this to our theoretical example in Figure 3.11, if we divide the standard deviation of the population (σ = 2.236) by the square root of 2, which is approximately 1.414, we get 1.58 (rounded to two decimal places). To explain why this relationship holds would require a mathematical proof that is beyond the scope of this textbook. In the absence of a mathematical proof, though, we could confirm this relationship empirically. For example, let's apply this relationship to the AOPA example in Figure 3.10, where the standard deviation of the population is σ = 21.93, the sample size is n = 5, and the square root of 5 is approximately 2.24:

$$\sigma_M = \frac{\sigma}{\sqrt{n}} = \frac{21.93}{\sqrt{5}} = \frac{21.93}{2.24} = 9.8$$

Comparing this result to the actual standard deviation of the sample of sample means, 11.5, we are relatively "close" given that we selected only 25 random samples of size n = 5 as opposed to *all possible samples* of size n = 5.

When applied to a theoretical sampling distribution of sample means as illustrated in Figure 3.11, standard error also measures how well an *individual* sample mean represents the entire distribution of sample means: It provides a measure of how much distance is reasonable to expect between a sample mean and the overall "mean of the sample means" (i.e., the grand mean) of the distribution of sample means. Furthermore, and perhaps more importantly, because the grand mean is equal to the population mean, the standard error provides a measure of how much error to expect on *average* between a sample mean M and the population mean, μ. Thus, the standard error represents the average distance between a sample mean and the population mean.

As expressed by the standard error equation, you also should recognize that the magnitude of standard error primarily is a function of sample size (n). For example, if we were to randomly select samples of size n = 16 from the same population in which the standard deviation is σ = 8, then the standard error is σ_M = 2. Using this same population, if we were to now randomly select samples of size n = 100, then the standard error is σ_M = 0.8. Thus, as sample size increases, standard error becomes smaller, and the more probable it is that the sample mean will be close to the unknown true population mean. Based on this discussion of sampling distributions, we now can derive a conceptual understanding of how accurately a sample statistic such as the mean estimates the corresponding population parameter. Information culled from a sampling distribution enables us to surmise how accurately a sample statistic estimates the corresponding population parameter. This will be discussed in greater detail in Chapter 4.

Sampling from Normal vs. Non-Normal Distributions

In the theoretical sampling distribution of sample means presented in Figure 3.11, we noted that the shape of the distribution of the population was uniform, but the shape of the sampling distribution of sample means was symmetrical. Furthermore, we also indicated that the shape of the parent population for the empirical AOPA example of Figure 3.10 approximated a normal distribution, and the corresponding sampling distribution of sample means also had the characteristics of a normal distribution. Based on these two observations, it appears that regardless of the shape of the parent population's distribution, the sampling distribution of sample means will be normally distributed. Although this is indeed true, it is important to distinguish whether we are sampling from a normal or non-normal distribution. Following are two theorems, presented without proof, that formalize our discussion of sampling distributions of the mean.

Sampling from a Normal Distribution

Given a parent population that is normal in form with a mean μ and standard deviation σ, if all possible samples of size n are randomly selected from this population, then the corresponding sampling distribution of sample means

(a) will have an overall mean that is equal to the population mean—that is, the mean of the means (μ_M) will equal μ;

(b) will have a standard deviation that is equal to the standard deviation of the population divided by the square root of the sample size: $\dfrac{\sigma}{\sqrt{n}}$; and

(c) will be normal in form.

If the parent population is not normally distributed, then the central limit theorem (CLT) is applied.

Sampling from a Non-Normal Distribution: The Central Limit Theorem

Given a parent population not normal in form with a mean μ and standard deviation σ, if all possible samples of size n are randomly selected from this population, then the corresponding sampling distribution of sample means

(a) will have an overall mean that is equal to the population mean—that is, the mean of the means (μ_M) will equal μ;

(b) will have a standard deviation that is equal to the standard deviation of the population divided by the square root of the sample size: $\dfrac{\sigma}{\sqrt{n}}$; and

(c) will approximate a normal distribution as the sample size increases.

With respect to part (c) of the CLT, a common question is, "How large of a sample is needed for the sampling distribution to approximate a normal distribution?" The answer is "it depends." Remarkably, a sample size as small as $n = 30$ often results in a sampling distribution of sample means that is very near normal in form, even when the original population is nowhere near a normal distribution. However, if the parent population is extremely skewed, a relatively large sample size (e.g., at least 50) might be required.

   We end this chapter with a review of why we gave so much attention to discussing the concept of normal distributions. First, when analyzing population data involving quantitative variables, the results generally lead to a normal or near-normal distribution.

Second, even if the population distribution is not normal, the CLT assures us that given a sufficient sample size, the sampling distribution will approximate a normal distribution. This latter point is critical because it facilitates statistical inference, which involves making decisions about a population based on sample data. We are now ready to discuss the concept of statistical inference, which we do beginning in Chapter 4.

Chapter Summary

1. A z score is a standard score that maps a corresponding raw score to an exact location within the distribution. To transform raw scores to standard scores, we apply the z-score formula

$$z = \frac{(\text{Raw score}) - (\text{Mean})}{\text{Standard deviation}}$$

Converting raw scores into z scores makes it possible to compare scores from different distributions once they have been standardized.

2. The empirical rule—also known as the 68–95–99.7 rule—refers to the partitioning of a normal curve such that approximately 68% of the observations lies within one standard deviation of the mean, approximately 95% of the observations lies within two standard deviations of the mean, and approximately 99.7% of the observations lies within three standard deviations of the mean.

3. A normal distribution is symmetric (bell-shaped) in form with most scores located at the center, and the remaining scores taper off in either direction. The actual shape of a normal distribution is defined by a precise mathematical equation.

4. A standard normal distribution is a normal distribution in which all its raw scores are transformed into z scores with a mean of $\mu = 0$ and a standard deviation of $\sigma = 1$ The total area under a standard normal curve is 1, the mean divides the area in half with 50% on either side, and nearly all the area under the curve is between $z = -3.00$ and $z = 3.00$.

5. When a normal or near-normal distribution of raw scores is transformed into a standard normal distribution, the area under the curve and above the interval between the mean and a specific z score may be found using Table 1, Appendix A. Depending on the given context, this area may be interpreted in terms of percentage, proportion, or probability.

6. A sampling distribution consists of a set of scores associated with a specific sample statistic derived from repeatedly drawing random samples of the same size from the same population. If the sample statistic is the mean, then the distribution consists of a "sample of sample means" and is called the sampling distribution of sample means.

7. For a sampling distribution of sample means, if all possible samples of the same size n are randomly selected from the same population with mean μ and standard deviation σ, then (a) the mean of the sample of sample means (the grand mean) will equal μ, and (b) the standard deviation of the sample of sample means will equal $\frac{\sigma}{\sqrt{n}}$, which is called standard error of the mean (*SEM*), or more simply, standard error (*SE*).

8. The standard error of the mean, which applies to a sampling distribution of sample means, denotes the average distance a sample mean is from the grand mean. Because the grand mean equals μ, if the standard error is small, then the sample means are clustered around the population mean, which implies that any single sample randomly selected from the population will have a mean that is a good estimate of the population mean.

9. If the samples of a sampling distribution of the mean are drawn from a parent population that is normally distributed, then the corresponding distribution of sample means also will be normally distributed. If the samples of a sampling distribution of the mean are drawn from a parent population that is not normal in form, then the central limit theorem indicates that the corresponding distribution of sample means will approximate a normal distribution as the sample size increases.

Vocabulary Check

Area under a normal curve	Sample of sample means
Central limit theorem (CLT)	Sample with replacement
Deviation score	Sampling distribution
Empirical vs. theoretical probability	Sampling distribution of sample means
Empirical rule	Standard error (SE)
Grand mean	Standard error of the mean (SEM)
Independent events	Standard normal curve
Mean of the sample means	Standard normal distribution
Normal curve	Standard score
Normal distribution	Standardized distribution
Probability distribution	Unit normal table
Probability of an event A	z score
Proportion vs. Probability vs. Percentile	

Review Exercises

A. Check Your Understanding

In 1–10, choose the best answer among the choices provided.

1. If the scores on an exam were approximately normally distributed and your score was one standard deviation above the mean, then you probably scored better than _____ of those taking the exam.
   a. 50%
   b. 68%
   c. 84%
   d. 98%

2. Which of the following is true for a normal distribution?
   a. Around 30% of the scores will be located within one standard deviation of the mean.
   b. Around 50% of the scores will be located within one standard deviation of the mean.

c. Around 70% of the scores will be located within one standard deviation of the mean.

d. Around 90% of the scores will be located within one standard deviation of the mean.

3. Which of the following is NOT true of the normal curve?

a. It is useful for estimating what percent of scores will fall above and below certain points.

b. The mode is equal to the median.

c. The median is equal to the mean.

d. Large z scores occur more often than small z scores.

4. Lauren is enrolled in a very large Aviation Psychology class. On the first exam, the class mean was $M_1 = 75$ ($SD_1 = 10$), and on the second exam, the class mean was $M_2 = 70$ ($SD_2 = 15$). Lauren scored 85 on both exams. Assuming the scores on each exam were approximately normally distributed, which of the following statements accurately depicts Lauren's performance on the two exams relative to the rest of the class?

a. She scored much better on the first exam.

b. She scored much better on the second exam.

c. She scored about equally well on both exams.

d. It is impossible to tell because the class size is not given.

5. Consider the table below, which reports the number of U.S. airports and the number of meters of runways that are either paved (The World Factbook, 2021a) or unpaved (The World Factbook, 2021b). If one of these airports is selected at random, what is the approximate probability that it will have a paved runway?

	Number of Airports	
Total Runway (meters)	Paved	Unpaved
> 3,047	189	1
2438–3047	235	6
1524–2437	1478	140
914–1523	2249	1552
< 914	903	6760
Total	5054	8459

a. 0.35

b. 0.37

c. 0.53

d. 0.66

6. The population {2, 3, 5, 7} has mean $\mu = 4.25$ and standard deviation $\sigma = 1.92$. When sampling with replacement, there are 16 different possible ordered samples of size $n = 2$ that can be selected from this population. The mean of each of these 16 samples is computed. For example, the mean of the sample {2, 5} is 3.5. The sampling distribution of the means of the 16 samples has its own mean, μ_M (the mean of the means) and its own standard deviation σ_M (the standard error). Which of the following statements is true?

a. $\mu_M = 4.25$ and $\sigma_M = 1.92$
b. $\mu_M = 4.25$ and $\sigma_M > 1.92$
c. $\mu_M = 4.25$ and $\sigma_M < 1.92$
d. $\mu_M > 4.25$

7. The researcher of a human aviation factors study involving "round dial" equipped aircraft reported that the distribution of the diameters of the round dials was approximately normal with a standard deviation of 7.62 mm (0.3 in.). How does the diameter of a round dial at the 67th percentile compare with the mean diameter?
   a.   3.35 mm (0.132 in.) above the mean
   b.   5.1 mm (0.20 in.) below the mean
   c.   3.35 mm (0.132 in.) below the mean
   d.   11.17 mm (0.44 in.) above the mean

8. The standard error of the mean serves as a rough indicator of the average amount by which sample means deviate from
   a.   the population mean.
   b.   the one observed sample mean.
   c.   some population characteristic.
   d.   each other.

9. In a hypothetical study, let's assume that an aviation safety officer conducted a nationwide survey of flight schools and reported that approximately 4% of the flight schools' training fleet had a defective magneto. Let's further assume that the safety officer randomly selected two samples of aircraft across the United States and tested for the defective component. Sample A had 150 planes and Sample B had 250 planes. Which of the following statements is true?
   a. $M_A < M_B$
   b. $M_A > M_B$
   c. $SE_A < SE_B$
   d. $SE_A > SE_B$

10. A negative z score signifies that the original score is
   a.   less than the mean
   b.   negative
   c.   the opposite of the mean
   d.   a small number

B. Apply Your Knowledge

In Exercises 1–5, use the empirical rule, Table 1 from Appendix A, and the concept of the standardized normal distribution to respond to the given questions.

1. Let's assume the manager of a fixed-base operator (FBO) asked the pilots who used the FBO's services to specify their age. After collecting data for 6 months, the results approximated a normal distribution with a mean of 43 years and a standard deviation of 6 years. Based on these data, the manager decided to play music inside the FBO facility that catered to the 37- to 49-year-old age group. On what basis did the manager make this decision?

2. Let's assume that the College of Aeronautics at a local university administers a college placement test of mathematical ability (CPT-MA) at the beginning of each fall semester to incoming first-year flight students, and students who score 13 or lower are placed in a remedial mathematics course. If the scores from the CPT-MA are normally

distributed with a mean of 18 and standard deviation of 3, and if there are 100 incoming first-year flight students, approximately how many students will be placed in a remedial mathematics course?

3. According to the U.S. Department of Health and Human Services (2021), the poverty rate for a four-person household in 2021 was $26,500. To bring attention to how low salaries are for first officers at regional airlines, let's assume that an aviation student researcher at a local university acquired a convenience sample of $n = 75$ first officers working at regional airlines throughout the United States. Let's further assume that the salaries approximated a normal distribution with a mean of $40,500 and a standard deviation of $5,045. If 20 of the 75 participants live in a four-person household, are any of these first officers' salaries at the poverty rate? Why?

4. An aircraft maintenance aptitude test has mean $\mu = 300$ and standard deviation $\sigma = 25$. If the scores approximate a normal distribution, what score would be at the upper quartile?

5. A common design requirement for aircraft seats is they must fit the range of people who fall between the 5th percentile for women and the 95th percentile for men. If this requirement is adopted, what are the minimum and maximum sitting distances (i.e., legroom)? For the sitting distance, use the buttock-to-knee length. Assume men's buttock-to-knee lengths are normally distributed with $\mu = 59.7$ cm (23.5 in.) and $\sigma = 2.8$ cm (1.1 in.), and women's buttock-to-knee lengths are normally distributed with $\mu = 57.6$ cm (22.7 in.) and $\sigma = 2.54$ cm (1.0 in.).

In 6–10, use the following research description.

> In response to an increase in the number of pilot deviation (PD) runway incursions (RIs) among general aviation (GA) pilots 65 years or older, an aviation human factors researcher hypothesizes that a plausible explanation is memory related: She believes that older GA pilots' short-term memory is failing, and they are not remembering ATC instructions. After seeing a television ad for a supplement that claims to improve memory, she decides to test the effect of the supplement on older pilots. Her rationale is that if pilots' memory capacity could be improved, then this might result in a decline in PD RIs. She randomly selects a sample of GA pilots who are 65 years or older and has them take one capsule of the supplement daily for 6 months. Using a simple memory test that requires memorizing a sequence of 12 digits in the order in which they are presented, she found that the time it takes to memorize these digits is normally distributed with a mean of 72 seconds and standard deviation of 18 seconds (the longer the time, the less the memory capacity).

6. What is the research question for this study?
7. What are the independent and dependent variables?
8. Let's assume the researcher randomly selects a 73-year-old GA pilot, has him take the supplement daily for 6 months, and then administers the memory test. If the pilot memorizes all 12 digits in the correct sequence in 47 seconds, what is the probability that someone 65 years or older who does NOT take the supplement will be able to complete the test in fewer than 47 seconds?
9. Does it appear that the supplement influences memory capacity? Explain.
10. If the researcher wants to identify the 10% of the 65-year-old or older population that takes the longest time to complete the memory test, what minimum time defines this group?

11. As part of a study involving the international travel and tourism industry, Abdullah (2019) examined the relationship between the 14 factors of the travel and tourism competitiveness index (TTCI) and airline passenger seat capacity, which was measured as the per capita annual average of weekly available seat kilometers (PCAAWASK). A copy of Abdullah's PCAAWASK data for the 90 countries that were part of his study is given in the file Ch_3 Exercises Part B-11. (a) Report the descriptive statistics of this data set and identify the shape of the corresponding distribution. (b) Randomly select 60 samples, each of size $n = 5$, and prepare a table of the respective sample means. (c) Apply the central limit theorem to this sampling distribution of sample means and comment on the results.

References

Abdullah, H. (2019). *The travel and tourism competitiveness index: Examining the reciprocal relationship among the TTCI factors relative to Porter's (1998) diamond model and airline passenger seat capacity for the countries of the world.* [Doctoral dissertation, Florida Institute of Technology]. https://repository.lib.fit.edu/handle/11141/3015

Dunbar, V. L. (2015). *Enhancing vigilance in flight instruction: Identifying factors that contribute to flight instructor complacency* (Publication No. 3664585) [Doctoral dissertation, Florida Institute of Technology]. ProQuest Dissertations and Theses Global.

Fullenkamp, A. M., Robinette, K., & Daanen, H. A. M. (2008). Gender differences in NATO anthropometry and the implication for protective equipment. Air Force Research Laboratory [AFRL-RH-WP-JA-2008–0014]. https://apps.dtic.mil/sti/pdfs/ADA491083.pdf

Random.org. (2022). [Online random number generator]. http://www.random.org/integers

The World Factbook (2021a). *Airports–with paved runways.* https://www.cia.gov/the-world-factbook/field/airports-with-paved-runways/

The World Factbook (2021b). *Airports–with unpaved runways.* https://www.cia.gov/the-world-factbook/field/airports-with-unpaved-runways/

U.S. Department of Health and Human Services (2021). *2021 poverty guidelines.* https://aspe.hhs.gov/topics/poverty-economic-mobility/poverty-guidelines/prior-hhs-poverty-guidelines-federal-register-references/2021-poverty-guidelines

Women in Aviation, International (n.d.). *Current statistics of women in aviation careers in U.S.* https://www.wai.org/industry-stats##

4 The Concept of Statistical Inference

Student Learning Outcomes

After studying this chapter, you will be able to do the following:

1. Construct and/or interpret a confidence interval, including its precision and accuracy in parameter estimation.
2. Write research, null, and alternative hypotheses for a research study.
3. Conduct and interpret the results of a hypothesis test involving the mean (σ known).
4. Engage in sample size planning relative to both precision and power analysis perspectives.

4.1 Chapter Overview

In the previous chapters, our focus was on descriptive statistics. In research, though, we must go beyond describing the observations we make. We also need to know how likely we can *infer* the observations we make from a sample to the parent population. In other words, "To what extent can we apply the results acquired from a sample back to the population from which the sample was selected?" To answer this question, we use *inferential statistics,* which are a set of statistical procedures that help us make informed decisions or conclusions about a population based on sample data. When considered in this context, inferential statistics is *the science of making reasonable decisions based on limited information*. Because research studies often employ inductive reasoning, any inferences we make based on sample data will not be absolute, which leads to the question, "How will we know if the inferences we are making about the population using sample data are accurate or reasonable?" In this chapter, we present two approaches to statistical inference—parameter estimation and hypothesis testing—that will help answer this question.

4.2 Parameter Estimation

Point and Interval Estimates

One approach to statistical inference is called parameter estimation, which is considered a direct approach because it estimates population parameters directly by using sample observations as approximations of the true population characteristics. For example, in our discussion on the sampling distribution of sample means from Chapter 3/Figure 3.10, we randomly selected 25 samples of size $n = 5$ from the hypothetical

DOI: 10.4324/9781003308300-6

population of $N = 100$ AOPA members' flight hours. Although we know that the population true mean is $\mu = 3{,}397$ hours, in reality, the population mean will be unknown to us. As a result, we could estimate the population means by using the mean from any of the samples. This approach of using a sample statistic to estimate the corresponding parameter is called a *point estimate*. Thus, if we used Sample 11's mean of $M = 3{,}340$ hours to estimate the population mean, then the point estimate would be $\mu = 3{,}340$ hours.

Although point estimates provide an approximation of the true population parameter, they do not indicate how "close" they are to the population parameter. For example, focusing on Sample 11 again, the point estimate of 3,340 hours is "off" by 657 hours when compared to the true population mean of 3,997 hours, whereas Sample 20's point estimate of 2,960 hours is "off" by 1,037 hours. Although we can calculate how far off these point estimates are, the true population mean will be unknown to us, and therefore, we will not be able to determine how much a point estimate deviates from the true population mean. A better approach to parameter estimation would be to use an *interval estimate*, which is a range of probable values within which the parameter falls. For example, instead of using the sample mean as a point estimate of the population mean, we could report that the population mean lies within an interval bounded by two values, *a* and *b*. An interval estimate is more desirable because it gives us some idea of how close the estimate is to the true parameter.

To illustrate this concept, first recall from Chapter 3 that if the parent population's distribution is approximately normal, then the sampling distribution of the sample means will be normal as well. Furthermore, even if the parent population is not normal in form, the central limit theorem (CLT) indicates that the sampling distribution of sample means will approximate a normal distribution if the samples are sufficiently large. Obviously, with our current AOPA example, a sample size of $n = 5$ would not be considered sufficiently large, but the parent population did approximate a normal distribution so we will proceed.

Next, as illustrated in Figure 4.1(a), recall from the empirical rule that approximately 95% of a normal distribution lies within 2 standard deviations of the mean. Further recall, as illustrated in Figure 4.1(b), that the standard deviation of a sampling distribution of sample means is equal to the standard error of the mean (σ_M), which is the standard deviation divided by the square root of the sample size, $\dfrac{\sigma}{\sqrt{n}}$. Applying this to our AOPA example and as depicted in Figure 4.1(c), we should expect that approximately 95% of all randomly selected samples of size $n = 5$ will have means within 2 standard errors of the population mean—that is, between $\mu - 2\sigma_M$ and $\mu + 2\sigma_M$. The calculation of the lower and upper bounds of this interval follows.

Lower Bound	Upper Bound
$\mu - 2\sigma_M$	$\mu + 2\sigma_M$
$\mu - (2)\left(\sigma/\sqrt{n}\right)$	$\mu + (2)\left(\sigma/\sqrt{n}\right)$
$\mu - (2)\left(2193/\sqrt{5}\right)$	$\mu + (2)\left(2193/\sqrt{5}\right)$
$\mu - (2)\,(2193/2.24)$	$\mu + (2)\,(2193/2.24)$
$\mu - (2)\,(979.02)$	$\mu + (2)\,(979.02)$
$\mu - 1958$	$\mu + 1958$

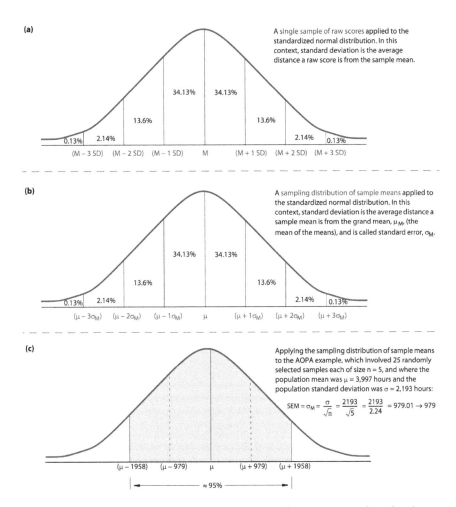

Figure 4.1 Comparison of a distribution for a single sample (a) vs. a sampling distribution of sample means (b), and the application of the sampling distribution of sample means to the AOPA example where 25 random samples each of size *n* = 5 were selected (c).

Thus, if we were to randomly select all possible samples of size *n* = 5 from the same hypothetical AOPA population, then we would expect approximately 95% of these samples to have a mean within 1,958 hours of the true population mean. Note from Chapter 3/Figure 3.10 that 22 samples (88%) have sample means within this interval. (Samples 1, 9, and 21 have sample means outside this interval.) Although 88% is 7% off from 95%, remember that we did not select *all possible samples of size n* = 5. We only selected 25 samples.

Confidence Intervals

Conceptual Foundation

An interval estimate is an improvement over a point estimate because it gives us an idea of how "close" a sample is to the true population mean. If we now were to associate a probability factor expressed as a percentage to an interval estimate, then we could indicate the degree of confidence we have that a sample mean will fall within this interval. In the AOPA example, we reported that approximately 95% of all random samples of size $n = 5$ selected from the same hypothetical AOPA population will have a sample mean within 1,958 hours of the true population mean. This infers that if we were to randomly select *one* sample of size $n = 5$ from this population, then there is approximately a 95% chance that this sample mean will be within 1,958 hours of the true population mean.

What is missing from this interval estimate, though, are the specific boundaries of the interval. Instead of saying there is a 95% chance that the mean of a randomly selected sample is expected to fall within a certain distance of the population mean, we would like to provide a *range of probable values* within which the parameter falls. In other words, we want to state with a certain level of confidence that the population mean lies within an interval bounded by *a* and *b*. To construct these boundaries, we use the corresponding sample's point estimate, which reflects the center of the interval. For example, if we used Sample 11 from Chapter 3/Figure 3.10, then the point estimate is Sample 11's mean, $M = 3,340$, and the corresponding boundaries would be between 1,958 hours below and 1,958 hours above 3,340:

$$\left[3340-1958, 3340+1958\right]$$
$$=\left[1382, 5298\right]$$

In other words, based on the results from Sample 11, we expect with an approximate 95% level of confidence that the mean number of flight hours for the hypothetical population of AOPA members will be between 1,382 hours and 5,298 hours.

This result is called a confidence interval (CI), and it represents the range of probable values that can be assigned to the population mean along with a specific degree of certainty. In other words, based on the results of one sample (in this case, Sample 11), we would conclude that the mean number of flight hours among all AOPA members is between 1,382 hours and 5,298 hours, and there is an approximately 95% level of confidence associated with this interval. Thus, a CI includes two key components: the specification of the limits or boundary values of the interval, and the associated probability that the population parameter is contained within that interval of probable values. Statistics programs report the boundaries of a CI separately as lower and upper bounds. When writing CIs in text, though, we follow the American Psychological Association (APA) style guidelines and use square brackets:

$$95\% \, \text{CI}: \mu = \left[a, b\right]$$

Thus, for the AOPA example, 95% CI: $\mu = [1,382, 5,298]$. Note that we also include the targeted parameter to avoid any ambiguity or misinterpretation.

Focusing on the level of confidence for a moment, we established a 95% level of confidence in the AOPA example. You should recognize that a 95% CI means that 47.5% of the area under the normal curve lies to the left and to the right of the mean. Consulting Table 1/Appendix A, this corresponds to a z score of $z = \pm 1.96$. In the AOPA example, though, we based our calculations on 2 standard errors of the mean as illustrated in Figure 4.1(c). Consulting the z table again, this corresponds to an area of 0.4772, which equates to a total area under the normal curve of 95.44%. Thus, for the AOPA example, we really have a 95.44% CI. Furthermore, although we focused on the 95% CI, we also could have established a different degree of confidence. The most widely used confidence levels are 90%,

95%, and 99%. Per the z table, if the level of confidence is 90%, then 45% of the area under the normal curve is between the mean and 1.65 standard deviations, which corresponds to $z = \pm 1.65$. Similarly, if the level of confidence is 99%, then 49.5% of the area under the normal curve is between the mean and 2.58 standard deviations, which corresponds to $z = \pm 2.58$.

The General Form of a Confidence Interval

Although we will rely on our statistics program to report CIs, it is instructional to understand the underlying foundation of CIs. First, recognize that in the most general sense, a sample statistic, which is the point estimate, and its corresponding parameter are related as follows:

$$Point\ Estimate = Population\ Parameter \pm Error$$

Using the mean as an illustration, this would be expressed as:

$$M = \mu \pm Error$$

This relationship indicates that when using a sample statistic such as the mean to estimate the corresponding population parameter, the estimate will not be accurate because we are relying on sample data, which inherently involves sampling error. Algebraically manipulating this relationship, we get the following equivalent statement:

$$Population\ Parameter = Point\ Estimate \pm Error$$

Once again, using the mean as an illustration, we have:

$$\mu = M \pm Error$$

As we did with the AOPA example earlier, using our knowledge of sampling distributions and the CLT, we can determine a range of values around μ that will probably contain most of the sample means if we were to select hundreds of thousands of samples (or *all possible samples*) of the same size from the same population. More specifically:

- The standard error of the mean, σ_M, which represents the average distance between a sample mean M and μ, provides information about the typical magnitude of the estimation error.
- The CLT tells us that as n increases, the distribution of sample means approaches a normal distribution regardless of the shape of the parent population. We also know that in a normal distribution, approximately 95% of the observations will be within 1.96 standard errors of the mean.
- Because the standard deviation of the distribution of sampling errors is equal to the standard error of the mean, and because $\mu = M \pm Error$, it follows that with 95% certainty

$$\mu = M \pm (1.96)(\sigma_M)$$

Alternatively, we could write the 95% CI for the population mean μ, when *M* is based on a random sample from a normal population as

$$M - (1.96)(\sigma_M) < \mu < M + (1.96)(\sigma_M)$$

As noted earlier, though, we will use square brackets to express CIs:

$$95\% \; CI : \mu = [a, b]$$

where $a = M - (1.96)(\sigma_M)$ and $b = M + (1.96)(\sigma_M)$

- This CI indicates that we are 95% certain that the true population mean μ will be within 1.96 standard errors of the mean. Note that we do not know exactly where the true population mean will fall within this interval, but we are 95% certain that it will be within this interval. This CI also implies there is a 5% chance that the mean will fall outside this interval.

In summary, the general form of a CI regardless of the parameter under discussion and involving a standard normal distribution is

$$Population \; Parameter = Point \; Estimate \pm (SE)(Corresponding \; z \; score)$$

where:

- *Population Parameter* is the targeted population parameter (e.g., μ, σ, σ^2).
- *Point Estimate* is the corresponding sample statistic (e.g., *M*, *SD*, SD^2).
- *SE* is the standard error associated with the targeted parameter.
- Corresponding *z* score is the *z* score for the level of confidence. (*Note:* In Chapter 5, we will replace *z* scores with *t* scores.)

Example 4.1: *Confidence Intervals*

Open the data file Ch_4 CFI Complacency Scores, which contains the complacency scores from the file Ch_3 Standard Normal Distribution from Chapter 3. Use your statistics program to produce a set of summary descriptive statistics that includes *M*, *SD*, *SE*, and the corresponding 95% CI. Calculate the 95% CI for CFIs' level of complacency and compare this CI with what your statistics program generated (assume σ = 5).

Solution

- First recognize that the target population is all CFIs in the United States, and our objective is to estimate the mean complacency score of this population from the given sample, which was randomly selected from all possible samples of size *n* = 255 from the same population. Therefore, the context is the sampling distribution of sample means.

- The descriptive statistics are $M = 15.7$, $SD = 3.3$, $SE = 0.21$, and the 95% CI for the mean is [15.1, 16.3]. Because we are assuming a population standard deviation of $\sigma = 5$, the corresponding standard error of the mean is

$$\sigma_M = \frac{\sigma}{\sqrt{n}} = \frac{5}{\sqrt{255}} = \frac{5}{15.9687} = 0.313$$

- We now construct the lower and upper bounds of the 95% CI.

  Lower Limit: $M - (\sigma_M)(1.96) = 15.7 - (0.313)(1.96) = 15.7 - 0.61348 = \mathbf{15.0865}$
  Upper Limit: $M + (\sigma_M)(1.96) = 15.7 + (0.313)(1.96) = 15.7 + 0.61348 = \mathbf{16.31348}$

  Therefore, the 95% CI for $\mu = [15.1, 16.3]$. Based on the sample data, we believe with 95% certainty that the mean complacency score of all CFIs in the United States will fall within the interval [15.1, 16.3].
- The calculated 95% CI is nearly identical to what our statistics program provided. The difference is because we used the assumed population standard deviation, $\sigma = 5$, when calculating SE, whereas our statistics program used SD.

Interpreting Confidence Intervals

Many students will interpret a 95% CI as we did in Example 4.1: "We are 95% certain that the population mean will fall within the given interval." Although, technically, this is an accurate interpretation, it does not reflect the true meaning of the CI. Although we use the mean of one sample to construct a 95% CI that is applied to the parent population, it is *incorrect* to say there is a 95% chance that the true population mean will fall within the boundaries of the confidence interval. A more appropriate interpretation is:

> *In the long run, approximately 95% of the CIs that are constructed by using hundreds of random samples of the same size selected from the same normally distributed population with mean μ will include the true population mean between the lower and upper boundaries.*

This is illustrated in Figure 4.2 where we constructed the 95% CIs for the 25 randomly selected samples each of size $n = 5$ from the hypothetical AOPA population data of flight hours given in Chapter 3/Figure 3.10(a). Note that 23 samples (92%) include the population mean $\mu = 3,997$ hours within their intervals. The exceptions are Samples 9 and 21 whose CIs do not include the population mean.

Because this technical interpretation can become a bit tiresome, two other ways to interpret the 95% CI involving the mean are as follows:

- 95% of the time we can expect the mean to fall within the given interval.
- If we were to randomly select 100 samples of the same size from the same population, then 95 of these samples would have a mean within the given interval.

Using these alternative interpretations for Example 4.1:

- 95% of the time we can expect the mean complacency score of CFIs in the United States to be between 15.1 and 16.3.

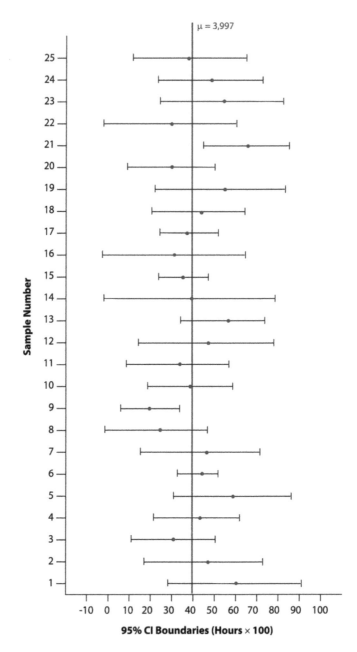

Figure 4.2 Shown are the 95% confidence intervals for the 25 randomly selected samples from the hypothetical AOPA population data of flight hours given in Figure 3.10(a) in Chapter 3. The CIs are represented by horizontal lines, the lower and upper bounds are presented as "end" bars, and the corresponding sample means are pictured as closed circles. (Adapted from Kachigan, 1991, p. 97.)

- If we were to randomly select 100 samples of size $n = 255$ from the same population of CFIs in the United States, then 95 of these samples will have a mean complacency score between 15.1 and 16.3.

Accuracy in Parameter Estimation (AIPE) and Sample Size Planning

When reporting a CI, a key question is, "How accurate is the CI in estimating the population parameter?" A CI's accuracy in parameter estimation, or AIPE, is a function of the CI's width, which is referred to as *precision*. Narrow CIs are more precise than broad CIs and therefore have higher AIPE. For example, if we were to predict that tomorrow's temperature will be between −20° C (−4° F) and 35° C (95° F), we would have high AIPE because the temperature is likely to fall within this interval. The precision, though, would be low because the interval's width is so broad.

As another example, look at the 95% CIs from Samples 1 and 6 as depicted in Figure 4.2 from our AOPA discussion. These CIs were from separate random samples of size $n = 5$ drawn from the same hypothetical AOPA population of members' flight hours. Note that both CIs include the true population parameter, but the width of the two samples is 63 units and 18.4 units, respectively. Because Sample 6's CI has a narrower width than that of Sample 1, it is a more precise interval and therefore has higher AIPE.

The key to achieving high AIPE is to control the CI's width (i.e., its precision), and the best way to do this is to focus on sample size. This is because a CI is a function of standard error: the higher the standard error, the wider the interval. Recall that we defined standard error by the fraction $\dfrac{\sigma}{\sqrt{n}}$. Further recall that if a fraction has a constant numerator but its denominator gets increasingly smaller, the numerical value of the fraction decreases (e.g., 1/10 vs. 1/100 vs. 1/1000). Therefore, if we can acquire a sufficiently large sample size, then we will be able to increase the precision (narrower width) of the CI, which will lead to higher AIPE. To acquire a sample size that is "sufficiently large" introduces the concept of sample size planning, which is done prior to conducting a study. In the absence of such systematic planning, the extent to which a CI has high precision and high AIPE would be left to chance, which is the case for Samples 1 and 6 of the AOPA example, because both had the same sample size of $n = 5$.

To determine a sufficiently large sample size for parameter estimation, we use the equation

$$n^* = \left(\frac{z \times \sigma}{E} \right)^2$$

where:
- n^* is the estimated size of the sample needed relative to how precise we want the CI. We denote this n^* to distinguish it from the actual sample size, n.
- z is the corresponding z value for the targeted level of confidence. Recall $z = 1.65$ for 90%, $z = 1.96$ for 95%, and $z = 2.58$ for 99% confidence. Note that in this context, we only use half of the area under the curve, and we are assuming a normal distribution.
- σ is the standard deviation of the population.
- E is the maximum amount of error and equates to one-half the width of the CI.

Example 4.2: *Precision, AIPE, and Sample Size*

Use the CFI complacency scores from Example 4.1 to determine the sample size needed to estimate the mean complacency score of the population of CFIs in the United States if we want to be accurate within 2 points and with 95% certainty. Assume the complacency scores form a normal distribution and $\sigma = 5$.

Solution

- $z = 1.96$ for a 95% confidence level.
- $\sigma = 5$.
- $E = 2$ because we want to be within 2 points (i.e., ±2) of the true mean score. Given that E is one-half the width of the CI, this means that the width of the corresponding CI will equal 4.
- Applying the sample size equation:

$$n^* = \frac{z \times \sigma}{E}^2$$

$$= \frac{1.96 \times 5}{2}^2 = \frac{9.8}{2}^2 = (4.9)^2 = 24.01 \rightarrow 25$$

Therefore, we will need *at least* 25 participants to be able to estimate within 2 points with 95% confidence the true mean complacency score of CFIs in the United States.

Putting this result in context, observe from Example 4.1 that: (a) the actual sample size was $n = 255$; (b) the corresponding 95% CI was [15.1, 16.3]; (c) the width of this CI is 16.3–15.1 = 1.2; and (d) one-half of this width is 0.6, which is E in the sample size equation. As a result, the 95% CI from Example 4.1 estimates the true mean complacency score of CFIs in the United States within six-tenths of a point. When we compare this to Example 4.2, where a sample size of $n = 25$ produced a 95% CI that was within 2 points of the true mean score (a width of 4), the 95% CI from Example 4.1 is more precise and has higher AIPE. Observe how an increase in sample size leads to a reduction in standard error, which in turn leads to a narrower width and higher AIPE.

The lesson to be learned from this presentation is that as researchers, we need to engage in sample size planning *before* we begin a study. If the focus of the study is to directly estimate a population parameter using CIs, then we need to decide how much error we are willing to accept and gauge the resulting estimated sample size against the likelihood of being able to select as large of a sample. This interactive "play" among maximum error (E), confidence level, and sample size (n) frequently results in adjusting these parameters to the point where the result will be both feasible and acceptable. Thus, in the context of estimating population parameters via confidence intervals, sample size planning involves determining acceptable probability and error levels that will lead to a sufficiently narrow CI and high AIPE.

4.3 The Nature of Hypothesis Testing

In the previous section, we used sample statistics to develop point or interval estimates to approximate population parameters. An alternative to this direct approach is to use a hypothesis test, which is a formal procedure that uses an *indirect* approach to statistical inference. This indirect approach involves first making a *hypothesis* about a population parameter such as the mean, and then using sample statistics to either support or re-fute this hypothesis. The basis for the hypothesis could come from several sources such as a preliminary study, a literature review of past studies, theory, or possibly anecdotal evidence.

To illustrate the nature of hypothesis testing, let's consider a study that examines the efficacy of a FAA test preparation software package. Most flight students are familiar with the various test preparation packages offered by companies such as Gleim, Shep-pard Air, Jeppesen, and ASA. For our example, let's assume that a new company, BIM, enters the market and claims that flight students who use its software to prepare for the 60-item FAA IRA exam will score above the national average. As depicted in Figure 4.3, we begin with a population with a known mean. In the BIM study, we will assume that the population mean IRA exam score is $\mu = 48$. This population is considered an *un-treated population* because it has not been exposed to using BIM's software to prepare for the exam. We randomly select a sample from this untreated population and admin-ister the treatment by having participants prepare for the exam using BIM's software. Prior to treatment, the sample is considered an *untreated sample*, but after treatment, the sample becomes the *treated sample*. Because we cannot treat and test the entire population, we rely on the sample statistics of the treated sample to assess the state of affairs in the corresponding *treated population*. In other words: *If the entire popula-tion of FAA IRA exam test-takers were to prepare for the exam using BIM's software, to what extent would the mean of the treated sample reflect the mean of the treated population?* By answering this question, we are indirectly making an inference about the treated population's mean by using the treated sample's mean to either support or refute the hypothesis.

What Is a Hypothesis?

Because a hypothesis test involves hypotheses, a common question students ask is, "What is a hypothesis?" In the most general sense, a hypothesis is an opinion or belief a person offers—based on limited evidence or without any assumption of its truth—to explain some phenomenon. For example, an aviation accident investigator, based on eyewitness accounts of the sound a plane made before crashing, might initially hypothesize that the

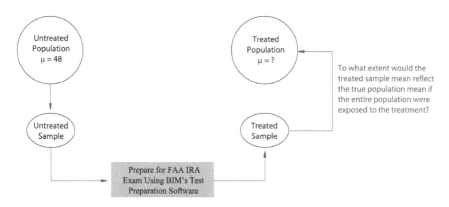

Figure 4.3 Conceptual view of hypothesis testing involving the mean.

aircraft experienced a mechanical malfunction prior to crashing. This hypothesis helps guide investigators with respect to the direction the investigation should take and the type of data to collect. The results of this investigation will then either support or refute the hypothesis, and adjustments to the investigation are then made relative to these results.

When considered from the perspective of a research study, there are two types of hypotheses: research and statistical. A research hypothesis is a statement researchers make to express the expected relationship or difference between variables and is derived directly from the research question. As an example, let's revisit the BIM study where we assumed that the national mean raw score for the IRA exam is $\mu = 48$. Let's now test BIM's claim via a research study. Following are the corresponding purpose statement and research question:

Purpose Statement
The purpose of this study is to test BIM's claim that students who use its test preparation software will score above the national average.

Research Question
What is the effect of BIM's software with respect to its claim that using it will enable students to score above the national average?

Let's now answer this RQ by re-expressing it as a *prediction* about the expected result. Based on BIM's claim, we predict the mean IRA exam score of flight students who prepare for the exam using BIM's software will be higher than 48. This prediction statement is the research hypothesis and is expressed as follows:

Research Hypothesis
Students who use BIM's software will score higher than the national average of $\mu = 48$.

Thus, a research hypothesis is a statement of the researcher's expectations relative to a relationship between variables. In other words, it states the expected relationship or difference between variables in such a manner that it will answer the RQ.

To determine whether the study's data will provide sufficient evidence to support or refute a research hypothesis, we must first re-express the research hypothesis in statistical form so that it lends itself to being tested via a hypothesis test. This re-expression yields two different, but complementary statistical hypotheses: null and alternative.

Null Hypothesis

A null hypothesis, denoted H_0, is a statistical statement about a population parameter and represents a "chance explanation" about whatever treatment effect was observed in the sample. It states that whatever effect or relationship we observed occurred by chance and therefore any similar effect or relationship in the parent population also will be random. Thus, the null hypothesis maintains that the current state of affairs in the population holds and effectively says: "Unless there is compelling evidence to the contrary, there is no change, no treatment effect, no difference, no relationship." Because the null hypothesis is a statistical statement, we include the words *statistically significant*, or more simply, *significant*, when we express it. So, with respect to our running example, the parameter of interest is the mean BIM IRA exam score, and the null hypothesis is expressed in symbols and words as

H_0: $\mu_{BIM} = 48$. *Students' mean IRA exam score, after preparing for the exam using BIM's package, will not be significantly different than the national average of $\mu = 48$.*

Alternative Hypothesis

The complement to the null hypothesis is the alternative hypothesis, denoted H_1 (or H_A), which is a statement about the same population parameter used in the null hypothesis. It specifies that the population parameter will be different from that given in the null hypothesis. It is a prediction of what we believe will happen and essentially is the same as the research hypothesis except it is expressed as a statistical statement. Based on the research hypothesis of our running example, the corresponding alternative hypothesis is

H_1: $\mu_{BIM} > 48$. *Students' mean IRA exam score, after preparing for the exam using BIM's package, will be significantly higher than the national average of $\mu = 48$.*

Using the BIM study as an example, and with respect to the corresponding research hypothesis, an alternative hypothesis may be expressed in three ways:

Research Hypothesis	Alternative Hypothesis
1. BIM users' mean score will be *different*.	$\mu_{BIM} \neq 48$
2. BIM users' mean score will be *greater than* 48.	$\mu_{BIM} > 48$
3. BIM users' mean score will be *less than* 48.	$\mu_{BIM} < 48$

Furthermore, because the null hypothesis always contains the equal sign, which generically reflects "no significant difference or effect," the alternative hypothesis complements the null hypothesis. In those instances where the alternative hypothesis is greater than (>) or less than (<), some textbooks will emphasize this complementary relationship between the null and alternative hypotheses by placing the *full* mathematical complement in parentheses besides the null hypothesis as follows:

Null Hypothesis	Alternative Hypothesis
1. $\mu_{BIM} = 48$	$\mu_{BIM} \neq 48$
2. $\mu_{BIM} = 48$ ($\leq$)	$\mu_{BIM} > 48$
3. $\mu_{BIM} = 48$ ($\geq$)	$\mu_{BIM} < 48$

Although we choose not to do so, the reader is to assume that the signs of all three relationships (=, >, <) have been accounted for with respect to the null and alternative hypotheses.

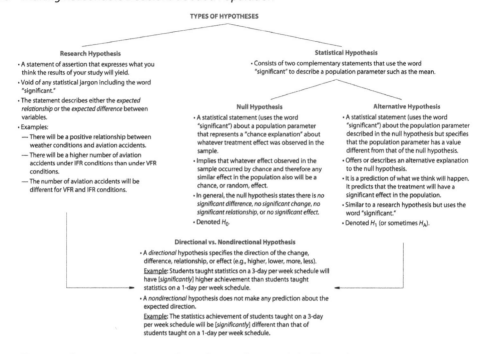

Figure 4.4 Summary and comparison of research vs. statistical hypotheses.

An overall summary of the different types of hypotheses we presented is provided in Figure 4.4, and a summary of our running example involving BIM's claim is provided in Figure 4.5.

Example 4.3: *Writing Research and Statistical Hypotheses*

Given the descriptions in a–c, write the research, null, and alternative hypotheses.

a. Refer to Example 4.1. Let's assume that the mean complacency score in the population of CFIs is $\mu = 20$, which we believe is too high. (*Note:* Complacency scores as measured by Dunbar, 2015, could range from 7 to 35, with higher scores reflecting a higher level of complacency in flight instruction.) As a result, we administer a 15-hour flight instruction safety workshop to a random sample of CFIs designed to reduce complacency in flight instruction.

b. Kettley et al. (2008) reported that women worry more about their finances, whereas men have a more complacent attitude. As a result, and relative to Dunbar (2015), we use Kettley et al.'s findings as a guide for a study that examines complacency in flight instruction involving male and female CFIs.

c. According to Zippia (n.d.), the mean age of CFIs is $\mu = 45.5$ years. Based on anecdotal information, we do not believe that the reported mean age is correct.

Solution

a. *Research Hypothesis:* CFIs who take this workshop will have a lower level of complacency relative to the population mean.

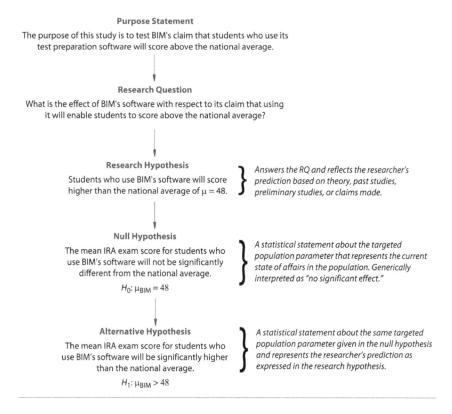

Purpose Statement

The purpose of this study is to test BIM's claim that students who use its test preparation software will score above the national average.

Research Question

What is the effect of BIM's software with respect to its claim that using it will enable students to score above the national average?

Research Hypothesis

Students who use BIM's software will score higher than the national average of $\mu = 48$.

} *Answers the RQ and reflects the researcher's prediction based on theory, past studies, preliminary studies, or claims made.*

Null Hypothesis

The mean IRA exam score for students who use BIM's software will not be significantly different from the national average.

H_0: $\mu_{BIM} = 48$

} *A statistical statement about the targeted population parameter that represents the current state of affairs in the population. Generically interpreted as "no significant effect."*

Alternative Hypothesis

The mean IRA exam score for students who use BIM's software will be significantly higher than the national average.

H_1: $\mu_{BIM} > 48$

} *A statistical statement about the same targeted population parameter given in the null hypothesis and represents the researcher's prediction as expressed in the research hypothesis.*

Figure 4.5 A summary of the progression from purpose statement to statistical hypotheses of a research study relative to the BIM IRA exam scores example.

Null Hypothesis: H_0: $\mu_{Workshop}$ = 20. The mean complacency score of CFIs who take the workshop will not be significantly different than the untreated population mean.

Alternative Hypothesis: H_1: $\mu_{Workshop}$ < 20. The mean complacency score of CFIs who take the workshop will be significantly lower than the untreated population mean.

b. *Research Hypothesis:* Male CFIs will have higher complacency scores than female CFIs.

Null Hypothesis: H_0: μ_{Male} − μ_{Female} = 0 (or alternatively, μ_{Male} = μ_{Female}). There is no significant difference in mean complacency scores between male and female CFIs.

Alternative Hypothesis: H_1: μ_{Male} > μ_{Female}. Male CFIs' mean complacency scores will be significantly higher than female CFIs' mean complacency scores.

c. *Research Hypothesis:* The mean age of CFIs is not 45.5 years.

Null Hypothesis: H_0: μ_{Age} = 45.5. The mean age of CFIs is not significantly different than the population mean age.

Alternative Hypothesis: H_1: μ_{Age} ≠ 45.5. The mean age of CFIs is significantly different than the population mean age.

What Is Statistical Significance?

The reason the word "significant" is included in statistical hypotheses is to empha-size the distinction between an "effect" vs. a *statistical* effect, which infers statistical significance. To illustrate the concept of statistical significance, consider the data file Ch_4 IRA Test Scores for BIM Users, which contains hypothetical IRA exam scores for 81 flight students who prepared for the exam using BIM's software. The mean score of this data set is $M = 52$, which is 4 points higher than the assumed untreated population mean of $\mu = 48$. So, based on this sample, it appears that BIM's claim is true because using its software resulted in a 4-point increase above the national average. When considered from a statistical perspective, though, is this 4-point difference *statistically significant*?

When we conduct a hypothesis test, the inferences we make about a population pa-rameter are made relative to a decision involving the null hypothesis. Based on the results from analyzing the sample data, we will make one of two decisions:

- *Reject H_0* (and accept H_1), which indicates that the effect we observed in the sample *probably exists* in the parent population.
- *Fail to reject H_0*, which indicates that the effect we observed in the sample *probably does not exist* in the parent population.

To determine if we should reject or fail to reject H_0, we focus on either (a) a *probability threshold* called the significance, or alpha-level approach; or (b) a *probability factor* called the *p*-value approach. To understand the difference between these two probabilities, we will apply the BIM example to each approach separately.

The Alpha-Level Approach to Significance

If we were to assume that the true population mean for the FAA IRA exam is indeed $\mu = 48$, then how likely would we get a treated sample mean of $M = 52$, which is so differ-ent from the true population mean? If it is *unlikely* to occur, then this infers that the true population mean cannot be $\mu = 48$ as stated in the null hypothesis. In other words: If the true population is indeed $\mu = 48$, then it would be *unlikely* for us to get a treated sample mean of $M = 52$.

The challenge now is, how do we define "unlikely?" From a probability perspective, we commonly define an unlikely event as one that occurs less than 5% of the time (1 time out of 20), or one that occurs less than 1% of the time (1 time out of 100). Depending on the situation, an unlikely event also could be a little less restrictive, such as 10%, which is sometimes acceptable for dissertation research, or more stringent, such as one-tenth of a percent (.001). These respective probabilities, which we denote using the Greek letter alpha (α) and establish prior to conducting the study, represent *significance levels*. When examined from the perspective of the normal distribution, they split the normal curve into two parts: the *critical region*, which also is referred to as the *rejection region*, and the *noncritical region*. This is illustrated in Figure 4.6, which is based on a probability of 5% (i.e., $\alpha = .05$). Note that the boundary value that separates the two regions is the *z* score, which corresponds to the area under the curve relative to the significance level and is obtained from Table 1/Appendix A.

For example, as shown in the lower part of Figure 4.6(a), if the significance level is α = .05, and the entire 5% area is in the left tail, then the boundary is $z = -1.65$ and we have a *one-tailed test to the left.* On the other hand, if the entire 5% area is in the right tail, then the boundary is $z = 1.65$ and we have a *one-tailed test to the right.* If, however, as shown in Figure 4.6(b), the full probability associated with the alpha level is split between both tails, then it is considered a *two-tailed test* and there are two critical regions. A one-tailed test corresponds to a directional, alternative hypothesis that expresses either a "less than" (<) or "greater than" (>) relationship, and a two-tailed test reflects a non-directional, alternative hypothesis that expresses a "not equal" ($\neq$) relationship as illustrated earlier in Example 4.3(c). Figure 4.7 depicts the z score boundaries for the critical regions relative to four significance levels, α = .10, .05, .01, and .001

Let's now apply this to our BIM example where H_0: $\mu = 48$ and H_1: $\mu > 48$, which means we have a one-tailed test to the right. Let's assume that when we calculate the z score relative to the treated sample mean, we get $z^* = 3.6$, so denoted to distinguish it from the table z score. When z^* is compared to z, observe from Figure 4.8(a) that $z^* = 3.6$ falls within the critical region because it is greater than $z = 1.65$. Therefore, we have statistical significance,

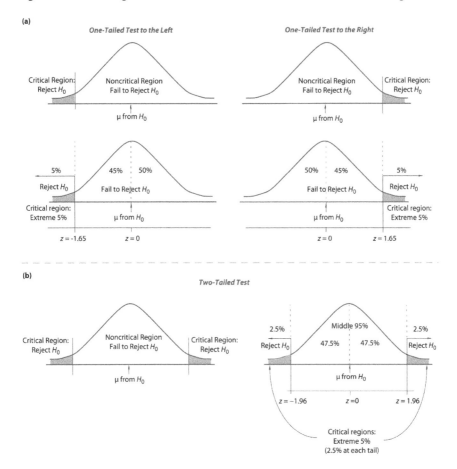

Figure 4.6   Critical z boundary values for a one-tailed test to the left and right, and for a two-tailed test with respect to a significance level of α = .05.

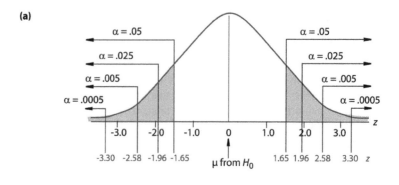

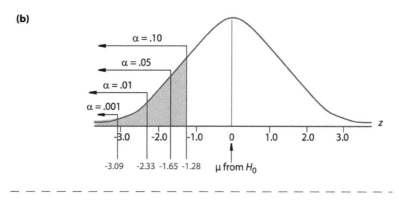

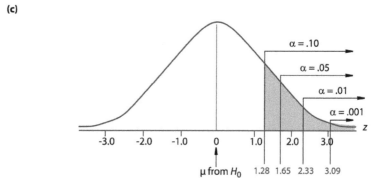

Figure 4.7 The *z* boundary values for the critical regions relative to four commonly used significance levels: $\alpha = .10, .05, .01,$ and $.001$.

and our decision is to *reject the null hypothesis* because the treated sample mean is an *unlikely outcome* based on the contention that the null hypothesis is true. As a result, not only do the sample data support BIM's claim, but the difference between the treated sample mean and the hypothesized population mean is statistically significant. If, however, z^* fell outside the critical region—for example, a treated sample mean of $M = 49.5$ corresponds to $z^* = 1.25$—then we would not have statistical significance and our decision would be to *fail to reject the null hypothesis*

The p-Value Approach to Significance

Unlike the alpha level, which we establish a priori, the *p* value is a post hoc probability, which means it is determined after-the-fact. To illustrate what we mean by this, let's return to our BIM example, which resulted in a calculated *z* value ("after-the-fact") of $z^* = 3.6$. What we now want to do is determine the probability associated with z^*. As shown in Figure 4.8(b) and based on Table 1/Appendix A, the area under the curve between $z = 0$ and $z = 3.6$ is .4998, which means that the area to the right of $z = 3.6$ is .0002. This is the *p* value, and it answers the question

> *If the null hypothesis is presumed to be true—that is, the true mean IRA exam score in the population is indeed $\mu = 48$—then what is the likelihood of getting a z score of 3.6?*

Based on the results of data analysis, this probability is .0002, which is extremely small (i.e., very unlikely) if the null hypothesis were indeed true. This result shows, from a statistical perspective, how incompatible the sample data are with respect to the null hypothesis based on the presumption that the null hypothesis is true. Thus, small *p* values (close to 0)

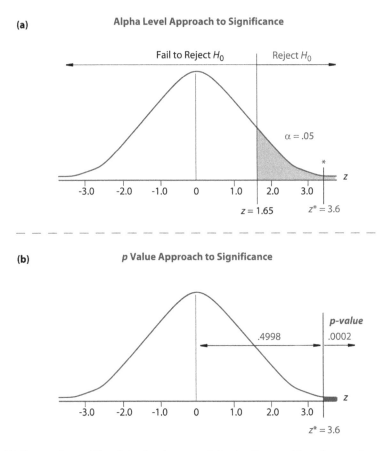

Figure 4.8 Comparison of the alpha-level approach to significance (a) vs. the *p*-value approach to significance (b).

provide evidence of distrust of the null hypothesis and make H_0 suspect, but large p values (close to 1) suggest that the sample data are consistent with what we would expect if the null hypothesis were indeed true.

Because most statistics programs report p values, many students simply compare p to alpha to determine if the results are statistically significant. For example, if $p < \alpha$, then reject H_0 and report significance, but if $p > \alpha$, then fail to reject H_0 and report no significance. Although this comparison is common practice, it is not appropriate to compare p to α. The reader is encouraged to review Greenland et al. (2016) and Wasserstein and Lazar (2016) to understand why this is the case. For our purposes, we will use the alpha-level approach to significance. We also will report p values because they are beneficial—for example, they provide us with a metric of how good the data fit the overall hypothesized model—but we will not use them to determine statistical significance.

Example 4.4: *Examining α and p*

Continuing with the BIM example, let's assume that 100 flight students prepared for the IRA exam using BIM's software. The data file Ch_4 Example 4.4 contains the corresponding hypothetical exam scores. Let's further assume that the mean exam score for the population of IRA test-takers is $\mu = 48$, the standard deviation is $\sigma = 12$, the scores are normally distributed in the population, and the null and alternative hypotheses relative to BIM's claim are H_0: $\mu_{BIM} = 48$ and H_1: $\mu_{BIM} > 48$. **(a)** Determine if the mean of the sample data is statistically significant for $\alpha = .05$ and $z^* = 1.25$. **(b)** Calculate and interpret the corresponding p value.

Solution

(a) The treated sample mean is $M = 49.5$, which is 1.5 points higher than the assumed population mean, but is this 1.5-point difference statistically significant? Referencing Figure 4.7(c), we observe that for a one-tailed test to the right for $\alpha = .05$, the boundary z score is $z = 1.65$. As shown in Figure 4.9(a), $z^* = 1.25$ is to the left of z and in the noncritical region. Therefore, we fail to reject the null hypothesis and conclude that the sample data do not provide sufficiently strong evidence to claim statistical significance. So, although preparing for the IRA exam using BIM's software yielded a higher mean score than that of the untreated population, the difference is not statistically significant. However, this result could still be *practically significant*.

(b) If $z^* = 1.25$, then based on the z table and as shown in Figure 4.9(b), the area under the curve is .3944, which means that $p = .1056$. Therefore, if the null hypothesis is presumed to be true, then the likelihood of getting a z score of $z = 1.25$ is $p = .1056$, which is probable. Therefore, the treated sample mean $M = 49.5$ is consistent with what we should expect given a true population mean of $\mu = 48$.

Statistical Significance vs. Practical Significance

When we read that the results of a study are "statistically significant," it is easy to want to place a value judgment on the results because in English the word "significant" means

(a)

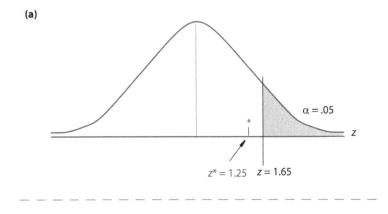

(b)

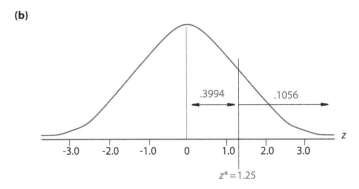

Figure 4.9 Contrasting views of the alpha-level approach (a) vs. the *p*-value approach (b) to significance based on the BIM example.

something meaningful or important. Do *not* be misled into assuming importance, however. In the context of statistics, statistical significance simply says that the sample data and corresponding results of data analysis provide sufficiently strong evidence to reject the null hypothesis. Statistical significance does not say anything about whether the results are meaningful or important. Only we can make that determination.

This notion of determining whether a result is meaningful or important is grounded in the concept of *practical significance*. Practical significance is not a statistical question, but instead is a judgment call that we make with respect to the context in which a conclusion is made. In Example 4.4, we reported that the 1.5-point difference between the treated sample and the untreated population mean IRA scores was not statistically significant. From a practical perspective, though, this 1.5-point difference still might be meaningful or important. Thus, regardless of the statistical significance of a research finding, the result might or might not have practical significance, which becomes a judgment call by researchers, policyholders, FAA personnel, education officials, and others.

4.4 Hypothesis Tests Involving the Mean (σ Known)

The z Test

In our introduction to the concept of hypothesis testing in the previous section, we re-stricted our discussion to making inferences about the population mean relative to z scores. Therefore, the corresponding hypothesis test is called a z test. (In later chapters, we will extend this to t and F tests.) Also recall that when we evaluated H_0 using z scores, we used the z table (Table 1/Appendix A), which is based on the standard normal dis-tribution, to determine the boundary values for the critical region, and the test statistic was relative to the sampling distribution of sample means. As a result, a z test is based on the assumptions that (a) the sample is randomly selected; (b) the population mean and standard deviation are known; and (c) the population distribution is normal in form, or the sample size is sufficiently large to be compliant with the CLT.

We now present a detailed example that highlights the procedure for performing a hy-pothesis test. To demonstrate this procedure, we will focus on the information given in Ex-ample 4.3(a), which involved assessing the effectiveness of a workshop to reduce CFIs' level of complacency to flight instruction and assume that $\mu = 20$ and $\sigma = 5$. We also will use the hy-pothetical data given in the file Ch_4 Complacency Scores for Hypothesis Testing Example.

Step 1. Identify the targeted parameter, and formulate null/alternative hypothe-ses. The targeted parameter is the mean, and as given in Example 4.3(a), the null and alternative hypotheses are:

$$H_0: \mu_{Workshop} = 20$$
$$H_1: \mu_{Workshop} < 20$$

Step 2. Determine the test criteria. The test criteria involve three components:

Level of Significance (α). We will set alpha to $\alpha = .05$.

Test Statistic. The *test statistic* refers to the statistical test we will use, and each test has its own distribution. Because we are using the sampling distribution of sample means and z scores, the test statistic is given as follows:

$$z* = \frac{M - \mu}{\dfrac{\sigma}{\sqrt{n}}}$$

The Critical Boundary. The critical t value is acquired from Table 1/Appendix A. Because we have a one-tailed test to the left, the corresponding z boundary is $z = -1.65$ (see Figure 4.7).

Step 3. Collect data, check assumptions, calculate the test statistic, and determine the corresponding p value.

Collect Data. The data are provided in the file Ch_4 Complacency Scores for Hy-pothesis Testing Example.

Check Assumptions. We assume a random sample and that the population of CFI complacency scores is normally distributed.

Calculate the Test Statistic. Given $\mu = 20$, $\sigma = 5$, $n = 16$, and a treated sample mean of $M = 18$:

$$z* = \frac{M - \mu}{\dfrac{\sigma}{\sqrt{n}}} = \frac{18 - 20}{\dfrac{5}{\sqrt{16}}} = \frac{-2}{\dfrac{5}{4}} = \frac{-2}{1.25} = -1.6$$

As shown in Figure 4.10(a), the treated sample mean is 1.6 standard errors below the hypothesized population mean $\mu = 20$ and falls in the noncritical region.

Determine p. As shown in Figure 4.10(b), the area under the curve relative to the test statistic is .4452, and the corresponding *p* value is $p = .0548$.

Step 4. Decide whether to reject or fail to reject the null hypothesis, and make a concluding statement relative to the RQ.

Decision. Because z^* falls in the noncritical region, we *fail to reject the null hypothesis.* Although $p = .0548$, which is very close to the preset alpha of $\alpha = .05$, some students might say that the result "approaches" significance. This is incorrect. An observed outcome either is significant or it is not significant relative to the preset alpha level.

Concluding Remark. The treated sample's mean complacency score from attending the 15-hour workshop was 2 points lower than that of the untreated population, but the difference was not sufficiently large enough to be statistically significant. In the context of inferential statistics, if we were to administer this workshop to the entire untreated population of CFIs, we would indirectly infer from the treated sample mean that the treated population mean is $\mu = 18$. However, this inferred treated population mean would not be significantly different from the untreated population mean of $\mu = 20$.

(a)

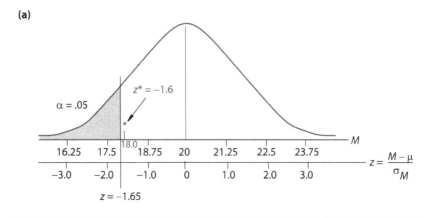

(b)

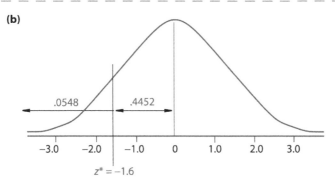

Figure 4.10 Contrasting views of the alpha-level approach (a) vs. the *p*-value approach (b) to significance based on CFIs' complacency scores from the file Ch_4 Complacency Scores for Hypothesis Testing Example.

Decision Errors Associated with Hypothesis Testing

As demonstrated in the BIM example, the decision to reject or fail to reject H_0 is based on the treated sample data coupled with a preset alpha level that marks the boundary between significance and nonsignificance. Although the decision we make is grounded in sound statistics, it does not necessarily mean we made the correct decision. Sometimes our decision will be wrong. When conducting a hypothesis test, there are four possible outcomes related to the decision we make with respect to H_0. Two are correct decisions and two are incorrect decisions. These four decisions are summarized in Table 4.1 and a brief discussion follows.

Type I Error

A Type I decision error occurs when we *incorrectly reject H_0* and infer that the treatment effect observed in the sample also exists in the population. Thus, we are falsely claiming something is true in the population. A Type I error is analogous to a *false positive* in the medical profession. If you submit to a drug test and the results show drugs in your system when in fact there are none, then the result is a false positive: The result is not true in the population (you do not have drugs in your system), yet the evidence (the sample data, which is your drug test) indicates it is true.

To address a Type I error, we assign a probability factor that specifies the chance of committing the error, and this probability is the preset alpha we use to establish the level of significance. If the preset level of significance is $\alpha = .05$, and the decision is to reject H_0, then there is a 5% chance that this was the wrong decision. The alpha level we use should be commensurate with the context of the research setting and the consequences of making a false claim. In situations where it is extremely critical that the results do not yield a false positive, we might use a more stringent alpha level such as $\alpha = .01$ or $\alpha = .001$, whereas in other situations where a false claim might not be so critical, we could set alpha to something less stringent such as $\alpha = .10$.

Type II Error

A Type II error occurs when we *incorrectly fail to reject the null hypothesis* and infer that a treatment effect does not exist in the population because we did not observe this effect in the sample. This leads to the conclusion that the treatment effect was not statistically

Table 4.1 Possible Outcomes and Decisions Relative to Testing the Null Hypothesis

Case Number	Was the sample mean in the critical region?	What was the decision with respect to hypothesis testing?	Does the effect truly exist in the population?	What is the result of your decision?
1	Yes	Reject H_0	Yes	**Power** (Strong Correct Decision)
2	Yes	Reject H_0	No	**Type I Error** (False Positive)
3	No	Fail to reject H_0	Yes	**Type II Error** (False Negative)
4	No	Fail to reject H_0	No	**Wasted Time** (Weak Correct Decision)

significant. A Type II error is analogous to a *false negative* result in the medical profession. Revisiting the drug testing example, if you did indeed have drugs in your system but the drug test did not detect them, then the result is true in the population (you have drugs in your system), but the evidence (the drug test) indicates it is false. As was the case for a Type I error, we assign a probability factor that specifies the chance of committing a Type II error, and this probability factor is called beta (β). For example, if β = .20, then there is a 20% chance of committing a Type II error. We will discuss the concept of β shortly.

Type I Error vs. Type II Error

Depending on the context, either a Type I or Type II error can be more serious than the other. For example, in the context of our legal system, a Type I error could result in sending an innocent person to prison for committing a crime he or she did not commit (false positive), whereas a Type II error could result in not sending a guilty person to prison who did indeed commit the crime (false negative). Thus, a Type I error is more serious. In the context of aviation, though, a Type II error could be more serious. For example, consider a study that is examining the effect of alcohol on pilot performance. A Type II error could result in incorrectly claiming that consuming alcohol prior to a flight has no effect on pilot performance even though we know that it does. Thus, from an aviation safety perspective, a Type II error could have a devastating effect.

Power

It also is conceivable that the decision we make about the null hypothesis is indeed correct. For example, we can *correctly reject the null hypothesis*, which means that the effect we observed in the sample also truly exists in the population. This is called power, which is the probability of making a *strong correct decision*. Power is the inverse of beta (β). For example, if β = .20, then power = 1 − .20 = .80. In other words, a 20% chance of not finding an effect in the sample when the effect truly exists in the population (Type II error) is equivalent to having an 80% chance of finding a true sample effect.

We also can *correctly* fail to reject the null hypothesis, which means that the effect we observed in the sample was due to chance and that this also is the case in the population. Although this is a correct decision, we qualify it as a *weak correct decision* because it suggests that we wasted our time and resources trying to find an effect that really does not exist.

The β–α Ratio

When we examine beta and alpha using their traditional settings of β = .20 and α = .05, note that the corresponding *beta–alpha* ratio is 4 to 1. This means we are four times more likely to commit a Type II error (false negative) than a Type I error (false positive). This ratio is grounded in the notion that a Type I error is more serious than a Type II error: *'Tis better to not find a treatment effect in the sample even though we believe it exists in the population, than to find a treatment effect in the sample that does not exist in the population.* Remember, though, the seriousness of an error is context driven, and it is up to us as researchers to determine what settings should be applied to alpha and beta.

The Size of an Effect: Cohen's d

As an indirect approach to statistical inference, the results of a hypothesis test tell us whether or not we can infer the observed outcome from the sample to the parent population. A hypothesis test, however, does not provide any real information about the *absolute size* of the treatment effect. It only makes a *relative* comparison where the size of the treatment effect is evaluated relative to the standard error. If the standard error is small, then the treatment effect also can be very small yet large enough to be significant. Thus, a significant effect does not necessarily mean a large treatment effect. A key to how large of a treatment effect is observed is sample size. If the sample size is large enough, then any treatment effect, no matter how small, can be sufficient to rejecting the null hypothesis. This is one reason why we distinguish between statistical and practical significance.

One way to reconcile this matter is to report the effect size (*ES*), which is a measurement of the absolute magnitude of a treatment effect independent of the sample size. There are several ways in which to describe the *ES*. One of the simplest for a single-sample *z* test of the mean is Cohen's *d*, which measures the size of the treatment relative to standard deviation.

$$\text{Cohen's } d = \frac{M - \mu}{\sigma}$$

For example, if *d* = 0.50, then the treatment increased the untreated population mean by one-half of a standard deviation, and if *d* = 1.00, then the size of the treatment effect is one full standard deviation. The following labels are commonly associated with Cohen's *d*:

- A *small effect* is when $d \leq 0.20$.
- A *medium effect* is when *d* is approximately around 0.5 (i.e., $0.20 < d < 0.80$).
- A *large effect* is when $d \geq 0.80$.

Applying this concept to the hypothesis test we conducted involving CFIs' complacency scores where $M = 18$, $\mu = 20$, and $\sigma = 5$, the corresponding *ES* is $d = (18–20)/5 = (–2)/5 = –0.40$. Thus, the magnitude of the treatment effect is $d = 0.40$ standard deviation, which is a medium effect, and means that the 15-hour workshop decreased the untreated population mean by four-tenths of a standard deviation. The negative symbol indicates the direction was to the left. This *ES* is illustrated in Figure 4.11, which shows the before-and-after treatment distributions.

Note the role of standard deviation when computing Cohen's *d*. If $\sigma = 2$ instead of 5 in the complacency score example, then $d = –1.0$, which is a large effect. So, although in both cases there is a 2-point difference between the untreated and treated population means, there is a difference in the magnitude of the respective effect sizes. Thus, reporting the effect size of a treatment provides more meaningful information about the treatment.

Also observe that *the relationship between effect size and sample size is inverse*. To detect a large treatment effect does not require a large sample, but to detect a small treatment effect, a large sample size is needed. This relationship is best understood using Brewer's (1996) analogy of searching for a needle vs. a basketball in a haystack: To detect a needle (small *ES*), we must divide the haystack into lots of segments, which implies a large *n*. To detect a basketball (large *ES*), we do not have to divide the haystack into as many segments, which implies a small *n*. A researcher must know or estimate the degree of

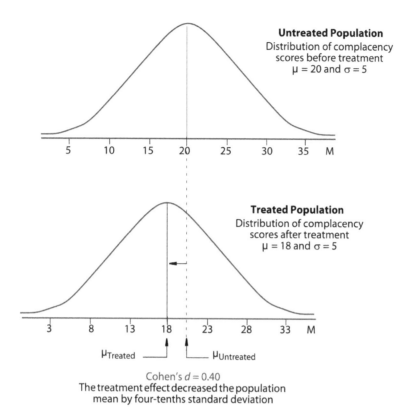

Untreated Population
Distribution of complacency
scores before treatment
$\mu = 20$ and $\sigma = 5$

Treated Population
Distribution of complacency
scores after treatment
$\mu = 18$ and $\sigma = 5$

$\mu_{Treated}$ ⎯⎯ $\mu_{Untreated}$

Cohen's $d = 0.40$
The treatment effect decreased the population
mean by four-tenths standard deviation

Figure 4.11 The treatment effect of the 15-hour workshop designed to reduce CFIs' level of complacency "shifted" the untreated population's distribution four-tenths of a standard deviation to the left. This is considered a medium effect.

effect (needle or basketball) desired before setting n, which is a function of sample size planning.

Power Analysis and Sample Size Planning

We initially introduced the concept of sample size planning in Section 4.2 in the context of confidence intervals where the objective was to determine the minimum sample size needed relative to the desired width of a CI: Narrow CIs have higher accuracy in parameter estimation, whereas broad CIs have lower AIPE. Thus, the focus was on the precision of a CI.

When the concept of sample size planning is applied to hypothesis testing, the objective is to determine the minimum sample size needed that will lead to high power. Recall that power is the probability of correctly rejecting the null hypothesis, which means that the effect observed in the sample truly exists in the population. In this context, sample size planning is referred to as power analysis. Thus, power analysis is a formal procedure, completed a priori, that helps determine the appropriate sample size of a study involving hypothesis testing.

To perform a power analysis, we need to know the: *preset alpha level*, which is traditionally set at $\alpha = .05$, but can vary depending on context; *minimum power* $(1 - \beta)$, which is commonly set at .80 and means we will have an 80% chance of correctly rejecting the null hypothesis; and *anticipated effect size (ES)*. We then consult a "power table" such as those provided in Cohen (1988); the power analysis feature of a statistics program; an independent power program such as *G*Power* (Faul et al., 2007, 2009); or an online sample size calculator such as those from Statistics Kingdom (2022).

As an example, let's use Statistics Kingdom's (2022) calculator to determine what size sample is needed to have an 80% chance of correctly rejecting the null hypothesis for our hypothesis test example:

- Go to the web site: https://www.statskingdom.com/sample_size_all.html
- In the left column under Calculator, click on "#2 T sample size. Z Sample size."
- In the pull-down boxes, select Tails = Left, Distribution = Normal, $\alpha = .05$, Effect = Medium, Effect Size = 0.5, Digits = 4, Sample = One, Power = .8, and Effect type = Cohen's *d*.
- Click on Calculate.

The result is $n^* = 25$ for power = .803765. Thus, to have an 80.4% probability of correctly rejecting the null hypothesis, we would need a minimum sample of 25 participants. In our example, we only had $n = 16$ participants. Therefore, one plausible explanation as to why we did not have statistical significance is that we did not have a sufficiently large sample size.

4.5 Confidence Intervals vs. Hypothesis Tests

Although CIs and hypothesis tests use data from the treated sample to make inferences about the parent population, CIs are preferable. This is because a hypothesis test answers one yes–no question: "Did the treatment have a statistically significant effect in the population?" A CI, however, is much more informative. It (a) provides an interval that contains probable values of the corresponding population parameter, including a point estimate that can be used as an approximation of the population parameter; (b) has an associated degree of confidence, and (c) is able to answer the question posed by a hypothesis test.

To illustrate this last point, let's revisit the CFI complacency example, which we presented to demonstrate the four steps of hypothesis testing to determine the effectiveness of a 15-hour workshop designed to reduce CFIs' level of complacency where H_0: $\mu_{Workshop} = 20$ and H_1: $\mu_{Workshop} < 20$. Based on the given sample data (Ch_4 Complacency Scores for Hypothesis Testing Example), we failed to reject H_0. Let's now calculate the corresponding 95% CI:

- From Step 3 of the hypothesis test example:

$$n = 16, M = 18, \mu = 20, \sigma = 5, \text{ and standard error is } \sigma_M = 1.25.$$

- The lower and upper bounds of the 95% CI are:

$$\text{Lower Limit: } M - (\sigma_M)(1.96) = 18 - (1.25)(1.96) = 18 - 2.45 = \mathbf{15.55}$$
$$\text{Upper Limit: } M + (\sigma_M)(1.96) = 18 + (1.25)(1.96) = 18 + 2.45 = \mathbf{20.45}$$

Therefore, the 95% CI for μ = [15.55, 20.45].

Note that a score of 20 is within this interval. This means that among the probable mean complacency scores, it is possible to acquire a sample that has a mean score of 20, which is the basis for H_0. As a result, we fail to reject the null hypothesis because it is possible for the treated population mean to be μ = 20. If a score of 20 were absent from this interval, then we would reject the null hypothesis.

Although we prefer CIs over hypothesis tests as an approach to statistical inference, keep in mind that they each have a different objective. The purpose of a CI is to use sample statistics to estimate an unknown population parameter, whereas the purpose of a hypothesis test is to determine if an effect observed in a sample also exists in the parent population. Both approaches are appropriate for aviation research. As an aviation researcher, you also should recognize there is a considerable difference between reporting a significant effect that has a broad CI (poor AIPE) vs. reporting a significant effect that has a very narrow CI (high AIPE). There also is a considerable difference between a hypothesis test that did not yield a significant effect but has a very narrow CI. As a result, we will include CIs as a complement to the results of hypothesis testing in subsequent chapters.

Chapter Summary

1. Parameter estimation is a direct approach to statistical inference that uses sample statistics to approximate the corresponding population parameter. This is done using point and interval estimates. The former consists of the sample statistic, and the latter consists of an interval of probable values within which the parameter lies.
2. A confidence interval is an interval estimate with an assigned probability that reflects the degree of confidence the population parameter lies within the interval. The precision of a CI is the interval's width. CIs with narrow widths have higher AIPE. A CI also can be used to determine the results of a hypothesis test.
3. Sample size planning is a formal process conducted a priori to determine the minimum sample size needed for a study. If the study's goal is parameter estimation, then the focus is the precision of a CI. If the goal is hypothesis testing, then the focus is power analysis.
4. A research hypothesis answers the research question by expressing the researcher's prediction about a variable. A statistical hypothesis is used for hypothesis testing and consists of the null hypothesis, H_0, which presumes that the current state of affairs in the population is true, and H_1, which is an alternative to H_0.
5. A hypothesis test is an indirect approach to statistical inference that is based on the logic that an effect observed in a sample also will be present in the population. This approach involves formulating H_0 and H_1, establishing test criteria, collecting/analyzing data, and making a decision and concluding comment relative to H_0 and the RQ.
6. The significance, or alpha level, is the probability assigned a priori that specifies the metric used to define an "unlikely" outcome, given a true H_0. The z score that corresponds to alpha partitions the normal distribution into critical and noncritical regions, where the critical region contains extreme values that represent "unlikely" outcomes relative to a true H_0.

7. Statistical significance is a term used to indicate that an observed outcome from a sample is unlikely to have occurred, relative to preset criteria, given a true H_0. The alpha-level approach to significance involves setting alpha a priori, determining the z boundary for the critical region, obtaining z^* from sample data, and comparing z^* to z. The p-value approach to significance uses a post hoc probability that represents the likelihood of obtaining a certain result given a true H_0. A small p infers the observed outcome is unlikely and means the outcome is inconsistent with the statistical model relative to H_0. A large p infers the observed outcome is consistent with the statistical model relative to H_0 and is what would be expected given a true H_0.

8. A Type I error is a false positive and occurs when a researcher concludes that an effect observed in the sample also is present in the population, when in fact it is not present. The chance of committing a Type I error is governed by α. A Type II error is a false negative and occurs when a researcher concludes that no effect was observed in the sample and therefore is not present in the population, when in fact it is present. The chance of committing a Type II error is governed by β.

9. Power = $(1 - \beta)$ and is the probability of correctly rejecting H_0, which means an effect observed in the sample truly is present in the population. Power analysis is a sample size planning procedure for hypothesis tests to determine the minimum size sample needed to correctly reject H_0.

10. Cohen's d standardizes a treatment effect by expressing it relative to standard deviation units.

Vocabulary Check

Accuracy in parameter estimation (AIPE)	Power
Alpha level	Power analysis
Alternative hypothesis (H_1)	Practical significance
Beta–alpha ratio	Precision
Cohen's d	Research hypothesis
Confidence interval	Sample size planning
Critical region	Significance level
Effect size (ES)	Statistical hypotheses
Hypothesis testing	Statistical significance
Inferential statistics	Test statistic
Interval estimate	Treated population
Noncritical region	Treated sample
Null hypothesis (H_0)	Two-tailed test
One-tailed test	Type I Error
p value	Type II Error
Parameter estimation	Untreated population
Point estimate	z test

Review Exercises

A. Check Your Understanding

In 1–10, choose the best answer among the choices provided.

1. Which of the following statements describes the role of the null hypothesis in research?

a.   It enables researchers to determine if there is a real relationship between variables.
b.   It enables researchers to prove there is no real relationship between variables.
c.   It enables researchers to prove that an effect exists in the population.
d.   It enables researchers to determine the probability of an event occurring through chance alone when there is a real relationship between variables.

2. If the results of a research study are statistically significant, then this means that the results
   a.   are extremely important relative to the context of the research setting.
   b.   most likely also will be true in the parent population.
   c.   are of practical use.
   d.   are of no practical use.

3. Any shift to a CI with a higher degree of confidence produces a
   a.   wider, less precise CI.
   b.   wider, more precise CI.
   c.   narrower, less precise CI.
   d.   narrower, more precise CI.

4. As sample size increases, a CI approaches
   a.   standard error
   b.   population range
   c.   point estimate
   d.   infinity

5. If H_0 is true, then the hypothesized and true sampling distributions will
   a.   differ
   b.   have a mean of zero
   c.   overlap
   d.   be the same

6. Let's assume we conduct a hypothesis test where we eliminate the rejection region—that is, instead of $\alpha = .05$ or $\alpha = .01$, we set $\alpha = 0$ or let α approximately equal 0. What do you think will happen?
   a.   A true null hypothesis never would be rejected.
   b.   A true null hypothesis never would be retained.
   c.   A false null hypothesis never would be rejected.
   d.   A false null hypothesis never would be retained.

7. The analysis of a random sample of 500 pilots who fly into and out of JFK indicates that a 98% CI for their mean annual income is [$41,300, $58,630]. Could this informa-tion be used to conduct a hypothesis test given H_0: $\mu = $40,000$ and H_1: $\mu \neq $40,000$ for $\alpha = .02$?
   a.   No. It is not known whether the data are normally distributed.
   b.   No. The entire data set is needed to do this test.
   c.   Yes. Because $40,000 is not within the 98% CI, the decision would be to reject H_0 and conclude the mean annual income is significantly different from $40,000 for $\alpha = .02$.
   d.   Yes. Because $40,000 is not within the 98% CI, the decision would be to fail to reject H_0 and that the mean annual income is not significantly different from $40,000 for $\alpha = .02$.

8. A researcher collected data from a random sample of 50 flight students to create a CI to estimate the mean time to PPL. Let's assume that the sample mean time to PPL

does not change. Which of the following best describes the anticipated effect on the width of the CI relative to AIPE if the researcher were to use data from a random sample of 200 rather than 50 flight students?

   a. The width of the new interval would be narrower than the width of the original interval.

   b. The width of the new interval would be wider than the width of the original interval.

   c. The width of the new interval would be about the same as the width of the original interval.

   d. The width of the new interval cannot be determined based on the given information.

9. Let's assume Gleim claims that students who use its software to prepare for the FAA IRA exam will score above the national average. To test this claim, a researcher randomly selects 100 flight students and has them prepare for the exam using Gleim's software. The results show no statistically significant difference in exam scores compared to the national average based on $\alpha = .05$. Gleim complained that the study did not have sufficient statistical power. If we were to assume that Gleim's software is indeed effective, which of the following would be an appropriate method for increasing power?

   a. Use a two-tailed test, not a one-tailed test.

   b. Use a one-tailed test, not a two-tailed test.

   c. Use $\alpha = .01$ instead of $\alpha = .05$.

   d. Increase the sample size to 200 students.

10. Goldberg and Williams' (1988) General Health Questionnaire (GHQ) is a 12-item instrument that measures psychological distress such as depression and anxiety. Scores could range from 0 to 36, with lower scores reflecting lower levels of distress. Let's assume the national mean GHQ score for aircraft maintenance technicians (AMTs) is $\mu = 28$, and you propose that if AMTs were to listen to music during their shift, their level of distress would be lower than the national average. For this study, the treatment is _____ and the parameter of interest is _____.

   a. listening to music, AMTs

   b. AMTs, GHQ scores

   c. AMTs, mean

   d. listening to music, mean

B. Apply Your Knowledge

Use the following research description to answer Exercises 1–10.

Let's assume a published op-ed article by a critic of low-cost carriers (LCCs) claims that the aircraft LCCs fly are at least 12 years old and therefore are too old and unsafe. After discussing this article with AMTs who work for various MROs, an aviation researcher believes that the critic is misinformed because LCCs fly newer aircraft that are more fuel efficient and require less maintenance, which helps save costs. To investigate her belief, the researcher randomly selects 20 LCCs from Skytrax (2021), and then consults Airfleets.net (n.d.) to acquire fleet age data for these airlines. The data are provided in the file Ch_4 Review Exercises Part B Data. Assume the population mean is $\mu = 12$ years, the standard deviation is $\sigma = 10$ years, and the parent population is normally distributed.

1. Write the corresponding RQ and research hypothesis and include any needed operational definitions.
2. Formulate H_0 and H_1 in words and symbols relative to the researcher's belief.
3. Identify the test criteria that correspond to the hypothesis test.
4. Compute the test statistic.
5. Determine if you will reject or fail to reject H_0 and explain why. As part of your response, determine and interpret the p value.
6. Report and interpret the *ES*.
7. Report and interpret the 95% CI.
8. Explain what it would mean if the researcher committed a Type I or Type II error.
9. Write a concluding comment about the findings relative to the critic's claim as well as with respect to what you hypothesized.
10. What is one plausible explanation for the results?

References

Airfleets.net (n.d.). *Search a plane, an airline.* https://www.airfleets.net/home/

Brewer, J. K. (1996). *Everything you always wanted to know about statistics, but didn't know how to ask* (2nd ed.). Kendall/Hunt.

Cohen, J. (1988). *Statistical power analysis for the behavioral sciences* (2nd ed.). Lawrence Erlbaum Associates.

Dunbar, V. L. (2015). *Enhancing vigilance in flight instruction: Identifying factors that contribute to flight instructor complacency* (Publication No. 3664585) [Doctoral dissertation, Florida Institute of Technology]. ProQuest Dissertations and Theses Global.

Faul, F., Erdfelder, E., Buchner, A., & Lang, A.-G. (2009). Statistical power analyses using G*Power 3.1: Tests for correlation and regression analyses. *Behavior Research Methods, 41*, 1149–1160. [Download software at http://www.gpower.hhu.de]. https://www.psychologie.hhu.de/fileadmin/redaktion/Fakultaeten/Mathematisch-Naturwissenschaftliche_Fakultaet/Psychologie/AAP/gpower/GPower31-BRM-Paper.pdf

Faul, F., Erdfelder, E., Lang, A.-G., & Buchner, A. (2007). G*Power 3: A flexible statistical power analysis program for the social, behavioral, and biomedical sciences. *Behavior Research Methods, 39*, 175–191. [Download software at http://www.gpower.hhu.de]. https://www.psychologie.hhu.de/fileadmin/redaktion/Fakultaeten/Mathematisch-Naturwissenschaftliche_Fakultaet/Psychologie/AAP/gpower/GPower3-BRM-Paper.pdf

Goldberg, D., & Williams, P. (1988). *A user's guide to the general health questionnaire.* NFER-Nelson.

Greenland, S., Senn, S. J., Rothman, K. J., Carlin, J. B., Poole, C., Goodman, S. N., & Altman, D. G. (2016). Statistical tests, p values, confidence intervals, and power: A guide to misinterpretations. *European Journal of Epidemiology, 31*, 337–350. https://link.springer.com/article/10.1007/s10654-016-0149-3

Kachigan, S. K. (1991). *Multivariate statistical analysis: A conceptual introduction* (2nd ed.). Radius Press.

Kettley, N., Whitehead, J. M., & Raffan, J. (2008) Worried women, complacent men? Gendered responses to differential student funding in higher education, *Oxford Review of Education, 34*(1), 111–129. https://www.tandfonline.com/doi/abs/10.1080/03054980701565360

Skytrax (2021). *World's best low-cost airlines 2021.* https://www.worldairlineawards.com/worlds-best-low-cost-airlines-2021/

Statistics Kingdom (2022). *Sample size calculators.* https://www.statskingdom.com/sample_size_all.html

Wasserstein, R. L., & Lazar, N. A. (2016). The ASA statement on p-values: Context, process, and purpose, *The American Statistician, 70*(2), 129–133. https://doi.org/10.1080/00031305.2016.1154108

Zippia (n.d.). *Certified flight instructor demographics and statistics in the U.S.* https://www.zippia.com/certified-flight-instructor-jobs/demographics/

Part C

Analyzing Research Data Involving a Single Sample

5 Single-Sample *t* Test

Student Learning Outcomes

After studying this chapter, you will be able to do the following with respect to a single-sample *t* test:

1. Determine the critical *t* values from a *t* distribution table or an online critical value calculator.
2. Determine and interpret the 95% confidence interval for the mean.
3. Perform and interpret the results of a hypothesis test of the mean.
4. Engage in pre-data analysis, data analysis, and post-data analysis activities as described in Section 5.5.

5.1 Chapter Overview

The research studies we presented in Chapter 4 assumed we knew the population standard deviation, σ. Although σ is unknown in most practical research settings, by providing an assumption of what σ is equal to, we were able to develop the underlying conceptual foundation of inferential statistics. In this chapter—which is the first of three chapters that consist of statistical procedures for analyzing research data involving a single sample or group—we introduce the *single-sample t test*, which is similar to the *z* test statistic from Chapter 4. The difference, though, is that the *t* test makes no assumption of σ, but instead uses the sample standard deviation, *SD*, as an estimate of σ. As part of our discussion, we introduce the *t distribution*, which is the basis for the *t* test, and the concept of *degrees of freedom*. We also examine statistical inference with respect to the single-sample *t* test, and we conclude the chapter with a guided example that demonstrates the application of the single-sample *t* test in a practical research setting.

5.2 The *t* Test Statistic and *t* Distribution

The *t* Test Statistic

In Chapter 3, we converted raw scores into standardized scores using the formula $z = \dfrac{X - \mu}{\sigma}$, which requires knowing the population mean and standard deviation and requires a

DOI: 10.4324/9781003308300-8

normally distributed population. When all the scores of a sample are converted to z scores, we get a standard normal distribution with a mean of $\mu = 0$ and standard deviation of $\sigma = 1$. We also introduced in Chapter 3 the concept of a sampling distribution and examined one specific sampling distribution, the sampling distribution of sample means, which involves (a) selecting all possible samples of the same size from the same population, (b) calculating the means of these samples, and (c) distributing these means. We also introduced the central limit theorem and discovered that the mean of the sample means—the grand mean—is equal to the population mean, and the standard error is equal to the standard deviation of the population divided by the square root of the sample size: $\sigma_M = \dfrac{\sigma}{\sqrt{n}}$.

This latter discussion effectively changed the context from comparing a single raw score of a sample relative to all the other scores in the sample to comparing a sample mean relative to all the other sample means. To reflect this change in context, we modified the z equation in Chapter 4 so that the denominator was with respect to the standard error of the sampling distribution, and then applied this equation to calculate the z test statistic for hypothesis testing:

$$z = \frac{M - \mu}{\sigma_M} = \frac{M - \mu}{\sigma / \sqrt{n}}$$

Suppose, though, we do not know σ, which is the case in most practical research settings. To make inferences about μ, our only alternative is to estimate σ using the sample standard deviation, *SD*. When we replace σ with *SD*, the standard error is now referred to as the *estimated standard error,* denoted SE_M. Thus, the estimated standard error is defined as $SE_M = \dfrac{SD}{\sqrt{n}}$. If we replace σ in the z equation with *SD*, we get a new variable called Student's *t*, which is defined as

$$t = \frac{M - \mu}{SE_M} = \frac{M - \mu}{SD / \sqrt{n}}$$

Although it is perfectly reasonable to make this substitution, it does change things a bit. For example, in the z formula, the numerator varies, but the denominator is constant because σ_M is based on σ, which does not change among the samples that comprise a sampling distribution of the mean. In *t*, though, both the numerator and denominator vary because SE_M is based on standard deviation, which is not constant. This makes *t* more variable than *z*. Furthermore, because the estimated standard error, SE_M, is based on the sample standard deviation, it is now considered a sample statistic and not a population parameter. Note also that when using sample data to make inferences about the population mean, μ, the z statistic relies on one value from the sample, namely, the sample mean, *M*. On the other hand, if the population standard deviation is unknown, then we must use the sample *SD* as an estimate of σ, which means the *t* statistic requires two values from the sample, namely, *M* and *SD*.

The t Distribution

Because of the variability in SE_M, we cannot claim that t is normally distributed with mean $\mu = 0$ and standard deviation $\sigma = 1$ as is the case with z. As a result, this introduces a new distribution called the *Student's t distribution*, or more simply, the t distribution, which is the basis for the t statistic. The t distribution is similar in shape to the normal distribution except that it tends to be flatter with the tails "pushed out" farther. Unlike the standard normal distribution, though, there is no single t distribution. Instead, there is a family of t distributions, which are indexed by a characteristic called *degrees of freedom*, denoted *df*.

To understand the concept of degrees of freedom, consider the simple data set of 4, 6, 8, 12, 15, which we presented in Chapter 2 when we demonstrated how to calculate standard deviation.

Score (X)	Deviation from Mean
4	$4 - 9 = -5$
6	$6 - 9 = -3$
8	$8 - 9 = -1$
12	$12 - 9 = 3$
15	$15 - 9 = 6$
Sums 45	0

Focusing on the second column where each score is subtracted by the mean, $M = 9$, note that when we add the first four deviation scores, we get a sum of -6:

$$(-5)+(-3)+(-1)+3=-6$$

Because the sum of the deviations from the mean for the entire data is equal to zero, we now know that the deviation from the mean for the last score must be 6 because $-6 + 6 = 0$. Furthermore, given that the mean of this data set is 9, we also know that the last score must be 15 because there is only one way the deviation from the mean can be 6 given a mean of 9, namely, $15 - 9 = 6$.

This demonstrates that if we know the size of a sample (n) and the corresponding sample mean, then all but one of the n scores (i.e., $n - 1$) are free to vary—that is, they can assume any numerical value without restriction. The nth score, however, must be fixed because it must equal whatever numerical value forces the sum of the deviations from the mean to equal zero. The term "degrees of freedom" is very descriptive because it refers to the number of scores in a sample that are *free* to vary after we compute the mean. So, for the current demonstration where $n = 5$ and $M = 9$, $df = 5 - 1 = 4$.

As shown in Figure 5.1, when compared to the normal distribution, the shape of the t distribution is flatter in form for small *df*. As the degrees of freedom increase, though, the t distribution approaches the shape of a normal distribution, and it will closely approximate the normal distribution for $df = 30$. To find the proportion of area under the curve for the t statistic, we use Table 2 in Appendix A. This table is known as the t distribution table and is similar to the unit normal table we used for z.

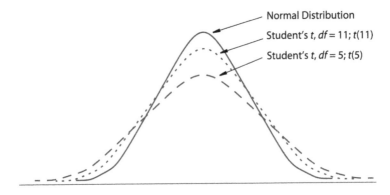

Figure 5.1 A comparison of different *t* distributions relative to the normal distribution.

Example 5.1: *Using the t Distribution Table for a One-Tailed Test*

In the hypothesis test example presented in Chapter 4, we examined the effectiveness of a workshop to reduce CFIs' level of complacency to flight instruction using a hypothetical set of CFI complacency scores for significance where H_0: $\mu = 20$, H_1: $\mu < 20$, $n = 16$, and $\alpha = .05$. Determine the corresponding critical *t* value using Table 2/Appendix A.

Solution

To use Table 2/Appendix A, we need to know *df* and α, which represents the area under the curve that reflects *a one-tailed test to the right*. Given $n = 16$, $df = 16 - 1 = 15$, and because H_1: $\mu < 20$, we have a one-tailed test to the left. Based on the *t* table entry, the critical *t* value for $df = 15$ for $\alpha = .05$ is 1.753. However, because we have a one-tailed test to the left, the corresponding *t* value is $t(15) = -1.753$. This is illustrated in Figure 5.2. Therefore, if we were to replicate the hypothesis test example from Chapter 4 but used *SD* as an approximation for σ, then the boundary value that separates the critical and noncritical regions for $\alpha = .05$ would be $t(15) = -1.753$. Compare this to the corresponding *z* boundary, which was $z = -1.65$ (Figure 4.7).

Example 5.2: *Using the t Distribution Table for a Two-Tailed Test*

If in Example 5.1, H_1: $\mu \neq 20$ and $n = 45$, but all the other information holds, determine the corresponding *t* critical values from Table 2/Appendix A.

Solution

- Given $n = 45$, $df = 44$, but the *t* table does not contain an entry for $df = 44$. Therefore, we use the next *smaller* table value, which is $df = 40$.
- Because H_1: $\mu \neq 20$, we have a two-tailed test.

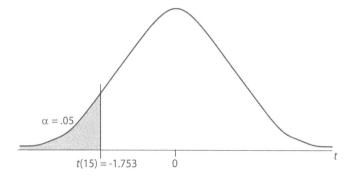

Figure 5.2 Illustration of a one-tailed *t* test to the left for Example 5.1.

- An $\alpha = .05$ for a two-tailed test means the noncritical region will be in the middle of the curve and the critical regions will be in the tails. Thus, the noncritical region will occupy 95% of the curve, and each tail will occupy 2.5% of the curve. Therefore, $t(40)$ for a two-tailed test with $\alpha = .05$ is ± 2.021. This is illustrated in Figure 5.3.
- As an alternative to Table 2/Appendix A, we could use Omni Calculator (n.d.) to acquire critical *t* values for degrees of freedom that are not listed in the *t* table. For $df = 44$, the critical *t* value for $\alpha = .05$ is $t(44) = 2.0154$.

5.3 Confidence Intervals and the *t* Distribution

The process of constructing and interpreting confidence intervals for a population mean, μ, where σ is unknown is done exactly as we did in Chapter 4 where σ was given. The only differences are: we use the sample standard deviation, *SD*, as an approximation for σ when calculating standard error; and we use the critical *t* value instead of *z*. This is summarized in Table 5.1.

Example 5.3: *Confidence Intervals (σ Unknown)*

Let's revisit Example 4.1 from Chapter 4, which involved constructing the 95% CI for the mean CFI complacency score based on the data given in the file Ch_4 CFI Complacency

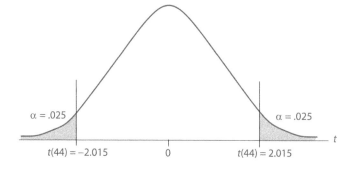

Figure 5.3 Illustration of a two-tailed *t* test for Example 5.2.

Table 5.1 Comparison of Confidence Intervals Relative to the Status of σ

	σ Given	σ Not Given
Standard Error of the Mean	$\sigma_M = \dfrac{\sigma}{\sqrt{n}}$	$SE_M = \dfrac{SD}{\sqrt{n}}$
Critical Variable Associated with Confidence Level	z (based on standard normal distribution)	$t_{critical}$ (based on df)
% CI for the Population Mean • 90% • 95% • 99%	• $M \pm (\sigma_M)\,(1.65)$ • $M \pm (\sigma_M)\,(1.96)$ • $M \pm (\sigma_M)\,(2.58)$	$M \pm (SE_M)(t_{critical})$

Scores. Recall from Example 4.1 that we assumed σ = 5, and that the population compla-cency scores were normally distributed in the population. Replicate Example 4.1, but now assume that σ is unknown.

Solution

- From Example 4.1, the target population is all CFIs in the United States and $n = 255$.
- With respect to the given data set, the descriptive statistics from our statistics pro-gram are $M = 15.7$, $SD = 3.3$, $SE_M = 0.21$, and the 95% CI for the mean is [15.3, 16.1].
- Because σ is unknown, we use SD as an approximation for standard error of the mean:

$$SE_M = \frac{SD}{\sqrt{n}} = \frac{3.3}{\sqrt{255}} = \frac{3.3}{16} = 0.206$$

- Given $n = 255$, $df = 254$. Because Table 2/Appendix A does not have an entry for $df = 254$, we either use the next smaller entry ($df = 250$) or consult Omni Calculator (n.d.). We choose the latter. As a result, $t(254) = 1.969$.
- We now construct the lower and upper bounds of the 95% CI.

$$\text{Lower Limit}: M - \left(SE_M\right)\left(t_{critical}\right) = 15.7 - (0.206)(1.969) = 15.7 - 0.406 = \mathbf{15.3}$$
$$\text{Upper Limit}: M + \left(SE_M\right)\left(t_{critical}\right) = 15.7 + (0.206)(1.969) = 15.7 + 0.406 = \mathbf{16.1}$$

Therefore, the 95% CI for μ = [15.3, 16.1]. Based on the sample data, we believe with 95% certainty that the mean complacency score of all CFIs in the United States will fall within the interval [15.3, 16.1].

The result in Example 5.3 is the same as what was provided by our statistics program. This is because the statistics program is using SD to calculate standard error. Recall from Example 4.1 that we got a slightly different result because we used the assumed population standard deviation, σ, when calculating standard error. Further, the sample size of $n = 255$ yielded a crit-ical t value of $t(254) = 1.969$, which was nearly the same as the corresponding z (1.96) used in Example 4.1. This is because we have a very large sample size, and as the number of degrees of freedom increases, the t distribution approaches the standard normal distribution.

5.4 Hypothesis Testing and the *t* Distribution

A hypothesis test of the mean where the population standard deviation, σ, is unknown is performed using the same four-step procedure presented in Chapter 4. The only difference is with respect to the second step where the test statistic is now *t*, which means we use the sample standard deviation, *SD*, to estimate the standard error of the mean.

Example 5.4: *Hypothesis Test of the Mean (σ Unknown)*

Let's revisit Example 4.3(a) from Chapter 4 and the corresponding hypothesis test, which involved examining the effectiveness of a 15-hour flight instruction safety workshop designed to reduce complacency in flight instruction among CFIs. The hypothetical data set was given in the file Ch_4 Complacency Scores for Hypothesis Testing Example, and we assumed the mean complacency score in the population was $\mu = 20$, which we believed was too high, and we assumed $\sigma = 5$. We also noted that complacency was measured using Dunbar's (2015) instrument where scores could range from 7 to 35, with higher scores reflecting a higher level of complacency in flight instruction. Replicate this example, but now assume that σ is unknown.

Solution

Step 1. Identify the targeted parameter, and formulate the null and alternative hypotheses. The targeted parameter is the mean, and the null and alternative hypotheses are:

$$H_0 : \mu_{Workshop} = 20$$
$$H_1 : \mu_{Workshop} < 20$$

Step 2. Determine the test criteria. The test criteria involve three components:

Level of Significance (α). We will set alpha to $\alpha = .05$.

Test Statistic. The test statistic is *t* because σ is unknown:

$$t^* = \frac{M - \mu}{SD / \sqrt{n}}$$

The Critical Boundary. The critical *t* value is acquired from Table 2/Appendix A. Because we have a one-tailed test to the left and $n = 16$, the corresponding *t* boundary for $df = 16 - 1$ is $t(15) = -1.753$.

Step 3. Collect data, check assumptions, calculate the test statistic, and determine the corresponding *p* value.

Collect Data. The data are provided in the file Ch_4 Complacency Scores for Hypothesis Testing Example.

Check Assumptions. There are two primary assumptions for the *t* statistic.
- Independence. We assume that the data were randomly selected.
- Normality. We assume that the population of CFI complacency scores is normally distributed. (Note that we cannot apply the CLT because we have a relatively small sample size of $n = 16$.)

Calculate the test statistic. Based on the given data set, the treated sample mean is $M = 18$ and $SD = 3.67$:

$$t^* = \frac{M-\mu}{SD/\sqrt{n}} = \frac{18-20}{\dfrac{3.67}{\sqrt{16}}} = \frac{-2}{\dfrac{3.67}{4}} = \frac{-2}{0.9175} = -2.18$$

As shown in Figure 5.4, the treated sample mean of $M = 18$ is being inferred to the population to reflect that the treated population mean would be $\mu = 18$. This is 2.18 standard errors below the hypothesized untreated population mean of $\mu = 20$ and falls inside the critical region.

Determine p. We will not calculate p as we did in Chapter 4, but instead acquire it from our statistics program. Based on the output provided by our statistics program, $p = .0228$.

Step 4. Decide whether to reject or fail to reject the null hypothesis, make a concluding statement relative to the RQ, and interpret the p value.

Decision. t^* falls in the critical region. Therefore, reject H_0: $\mu_{Workshop} = 20$.

Concluding statement. Relative to the preset significance level of $\alpha = .05$, the sample data provide sufficient evidence that the treated sample's mean complacency score of $M = 18$, which is 2 points lower than that of the untreated population mean and the result of the 15-hour workshop, is statistically significant. Therefore, based on the sample data, the treatment effect of the 15-hour workshop was significant, which means that if we were to administer this workshop to the entire population of CFIs, then the treated population mean as inferred from the sample would be $\mu = 18$.

Interpret p. $p = .0228$ is a relatively small probability (close to 0) and indicates that the sample data are not consistent with the hypothesized model, which is based on the presumption that the true mean complacency score in the population really is $\mu = 20$. This p value provides further support for the decision to reject H_0.

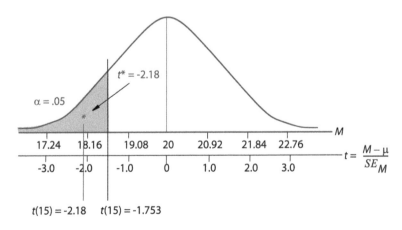

Figure 5.4 Illustration of t test for Example 5.4.

Note that a two-tailed hypothesis test of the mean with σ unknown is performed in the same manner as we did in Example 5.4. The only difference is that the alpha level would be divided by 2 and the corresponding critical *t* value from Table 2/Appendix A would be assigned both positive and negative signs to reflect the left and right tails, respectively, of the distribution. For example, if in Example 5.4 the alternative hypothesis were $H_1: \mu \neq 20$, then assuming $\alpha = .05$, the critical *t* values for $df = 15$ would be ± 2.131. As shown in Figure 5.5, this also would have led to the decision to reject H_0.

5.5 Using the Single-Sample *t* Test in Research: A Guided Example

We now provide a guided example of a research study that involves the use of the single-sample *t* test. The example replicates a former student's end-of-course project that examined the effect of using a test preparation software package to prepare for the 60-item FAA IRA exam. Flight students identified the resources they used to prepare for the exam and the percentage grade they received, which was then converted to a raw score. For example, a reported grade of 90% was converted to a raw score of $0.90 \times 60 = 54$. The student then focused on one specific package to see if there was a significant difference between the mean score of this group and the 2020 national average of 85.83% as reported by the FAA (2020). This national average corresponds to a raw score mean of $\mu = 51.5$. For this guided example, we will refer to the software package as BIM, which we introduced as a fictitious company earlier, and we will use the scores provided in the file Ch_5 Guided Example Data. We also will structure this guided example into three distinct parts: (a) pre-data analysis, (b) data analysis, and (c) post-data analysis.

Pre-Data Analysis

Pre-data analysis activities occur before data collection and involve: (a) stating the research question, operational definitions, and research hypothesis; (b) determining the research methodology to answer the RQ; and (c) engaging in sample size planning, which focuses on the precision of a confidence interval if the objective is parameter estimation, or power analysis if the objective is hypothesis testing. Because most research studies

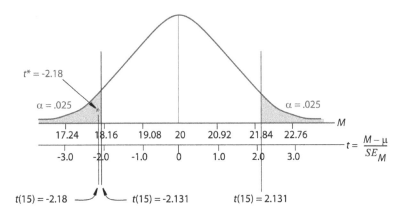

Figure 5.5 Illustration of a two-tailed test for Example 5.4.

involve hypothesis testing, we will focus on power analysis. However, as part of post-data analysis, we also will examine the precision of the 95% CI.

Research Question and Operational Definitions

What is the effect of using BIM's test preparation software with respect to the national average score of the FAA IRA exam? Key terms or phrases that need to be defined are test preparation software and IRA exam, both of which were defined in earlier discussions.

Research Hypothesis

Recall that when we first introduced the fictitious company BIM, we indicated that it was new to the marketplace. As a result, we do not know the impact the software will have on the national average. Therefore, we predict that preparing for the IRA exam using BIM's software will result in a mean score that is *different* than the national average.

Research Methodology

Based on the description of how the student conducted the study, the research methodology is causal comparative (ex post facto) because the treatment has already occurred.

Sample Size Planning (Power Analysis)

To determine the minimum sample size needed, we consult Statistics Kingdom (2022a) and enter the following inputs:

- Tails: Two (H1: $\mu = \mu_0$)
- Distribution: Student's t
- Significance level (α): .05
- Effect: Medium
- Effect size: 0.5

- Digits: 4
- Sample: One sample
- Power: .8
- Effect type: Cohen's d

The result is $n^* = 34$ for power = .8078. Thus, to have an 80.8% probability of finding a medium effect and correctly rejecting the null hypothesis—that is, the effect observed in the sample also exists in the population—we need a minimum of 34 participants. Because the given data set has $n = 100$ participants, which is nearly three times the minimum needed, this suggests there is a greater than 80.78% of correctly rejecting H_0.

Data Analysis

We now analyze the given data set relative to a hypothesis test of the mean.

Step 1. Identify the targeted parameter, and formulate null/alternative hypotheses. The targeted parameter is the mean, and the null/alternative hypotheses are:

H_0: $\mu = 51.5$. Using BIM's software to prepare for the FAA IRA exam will not result in a treated population mean that is significantly different than the 2020 national average.

H_1: $\mu \neq 51.5$. Using BIM's software to prepare for the FAA IRA exam will result in a treated population mean that is significantly different from the 2020 national average.

Step 2. Determine the test criteria.

Level of Significance. The preset alpha level is $\alpha = .05$.

Test Statistic. The test statistic is the *t* test.

Boundary of the Critical Region. Given $n = 100$, $df = 99$, $\alpha = .05$, and a two-tailed test, we discover there is no entry for $df = 99$ in Table 2/Appendix A. Therefore, we can either use the next *smaller* table entry, which yields a critical *t* value of $t(60)$ = ±2.00, or consult Omni Calculator (n.d.), which yields critical *t* values of $t(99)$ = ±1.9842. We choose the latter, as illustrated in Figure 5.6.

Step 3. Collect data, check assumptions, run the analysis, and report the results.

Collect the Data. The data are in the file Ch_5 Guided Example Data.

Check Assumptions. We assume (a) the data were collected randomly; (b) the parent population is normally distributed, although this is not a concern because the sample size of $n = 100$ is sufficiently large to apply the CLT; and (c) σ is unknown.

Run the Analysis and Report the Results. The results from performing a single-sample *t* test for the given data set are given below, and a graphical illustration is provided in Figure 5.7.
- Sample mean: $M = 43.66$
- Sample standard deviation: $SD = 9.40$
- Estimated standard error of the mean: $SE_M = 0.94$
- Calculated *t* test statistic: $t^*(99) = -8.34$ for $\alpha = .05$ and a two-tailed test (the reader is encouraged to confirm this result by calculating t^* directly).
- *p* value: $p < .0001$

Step 4. Decide whether to reject or fail to reject the null hypothesis, and make a concluding statement relative to the RQ.

Decision. The calculated $t^* = -8.34$ lies in the critical region. Therefore, the decision is to reject H_0.

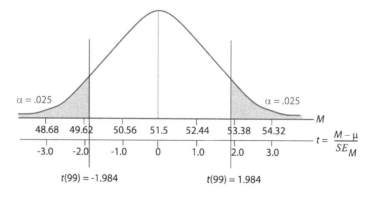

Figure 5.6 Critical regions for Chapter 5's Guided Example.

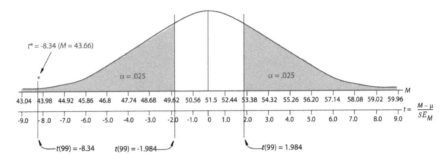

Figure 5.7 Illustration of results of *t* test for Chapter 5's Guided Example.

Concluding Remark. Based on the sample data of *n* =100, it appears that using BIM's test preparation software has a significant *negative* effect relative to the national average. Students who used this software to prepare for the IRA exam scored on average 8.34 points *lower* than the national average, and this 8.34-point difference is statistically significant.

Post-Data Analysis

Post-data analysis activities include determining and reporting the corresponding effect size, 95% confidence interval, and power. We also discuss plausible explanations for the results.

Effect Size

There are two ways to measure the effect size involving a *t* test: Cohen's estimated *d*, which is with respect to standard deviation, and r^2, which is with respect to explained variance.

Cohen's estimated d. In Chapter 4, we learned that Cohen's *d* reflects the comparison between a sample mean and the population mean, and measures the size of a treatment relative to standard deviation:

$$\text{Cohen's } d = \frac{M - \mu}{\sigma}$$

In a single-sample *t* test, because we do not know σ, we use the sample standard deviation, *SD*, as an estimate of σ and refer to the modified formula as Cohen's estimated *d*. Applying this formula to the current example:

$$\text{Estimated } d = \frac{M - \mu}{SD} = \frac{43.66 - 51.5}{9.4} = \frac{-7.84}{9.4} = -0.83$$

According to Cohen's effect size labels, an effect size of *ES* = −0.83 is a *large effect*. More specifically, the treatment (BIM's test preparation software) reduced the national mean FAA IRA score by nearly a full standard deviation.

Explained variance (r^2*).* An alternative to measuring effect size with respect to standard deviation, we also can determine the effect size of a treatment by examining how much of the variability in the dependent variable is being explained by the independent variable. This approach to effect size is called explained variance, denoted r^2, and is determined by the formula:

$$r^2 = \frac{t^2}{t^2 + df}$$

To interpret the magnitude of r^2, Cohen (1988) provided the following guidance: $r^2 = .01$ is a small effect, $r^2 = .09$ is a medium effect, and $r^2 = .25$ is a large effect. Applying this formula to the current example:

$$r^2 = \frac{t^2}{t^2 + df} = \frac{-8.34^2}{-8.34^2 + 99} = \frac{69.5556}{69.5556 + 99} = \frac{69.5556}{168.5556} = 0.41$$

Thus, 41% of the variability in the IRA exam scores is being explained by the group membership variable, which represents the treated group that prepared for the IRA exam using BIM's test preparation software. Based on Cohen's guidance, this is a large effect.

95% Confidence Interval

The 95% CI reported by our statistics program is (the reader is encouraged to confirm this result by calculating the confidence interval directly):

$$95\% \text{ CI: } \mu = \left[41.8, 45.5\right]$$

Therefore, 95% of the time we can expect the mean of the treated population—that is, flight students who use BIM's test preparation software for the IRA exam—to lie within the interval [41.8, 45.5]. Alternatively, if we were to collect 100 random samples of size $n = 100$ from the sample population, then 95 of these samples would have a mean IRA exam score between 41.8 and 45.5. Because the hypothesized mean of $\mu = 51.5$ is *not* within the interval, this confirms our decision to reject the null hypothesis. Also note that this interval is relatively narrow. Its width is 3.7 units, which means we can estimate the true treated population mean within (3.7/2) = 1.85 points. This infers a very precise interval that has high accuracy in parameter estimation (AIPE).

Power (Post Hoc)

Recall that power refers to the probability of making a strong correct decision such that the effect found in the sample truly exists in the parent population. To determine the actual power of this study we conduct a post hoc power analysis using Statistics Kingdom's (2022b) power calculator with the following inputs:

- Tails: Two (H₁: $\mu = \mu_0$)
- Distribution: T
- Significance level (α): .05
- Effect type: Standardized effect size
- n_1: 100

- Digits: 4
- Sample: One sample
- Effect: Large
- Effect size: 0.83

The result is a power of 1.0. Because we do not believe in a power probability of 1.0, we report that the power of our hypothesis test was greater than .99. What this means is there is a greater than 99% probability (or chance) that we made the correct decision to reject H_0. This also infers there is a greater than 99% chance that the 0.83 standard deviation effect found in the sample—which equates to a mean score that was nearly 8 points lower than the national average—also exists in the population. Remember: There still is a 5% chance of a false effect (Type I error).

Plausible Explanations for the Results

In addition to reporting the results of data analysis, researchers also should report plausible explanations for a study's results. One possible reason that flight students who used BIM's test preparation software to prepare for the IRA exam had a significantly lower average compared to the national average is maybe the students did not use the software properly as instructed, or possibly did not use it at all. Without any verification of "treatment fidelity," this is reasonable.

A second plausible explanation is perhaps the students were stressed or fatigued when they took the exam. Without capturing any information about their stress or fatigue levels, this too seems reasonable.

It also is possible that outliers might have influenced the results. Recall that we did not conduct an outlier analysis prior to hypothesis testing. As a result, it would be prudent to first determine if the data set contains outliers. If so, we would then examine their impact by calculating the t statistic in Step 3 of hypothesis testing twice: once in the presence and once in the absence of outliers. If the results are different, then we would have to decide whether to keep the outliers (assuming they are rare cases) or remove them.

Chapter Summary

1. When the population standard deviation, σ, is unknown, we use the sample standard deviation, SD, to estimate standard error, denoted SE_M. In doing so, this introduces the Student's t statistic, which is based on the z statistic but uses SD in place of σ:

$$t = \frac{M - \mu}{SD_M} = \frac{M - \mu}{SD / \sqrt{n}}$$

2. The Student's t distribution is symmetrical in shape and based on degrees of freedom (df), which is a function of sample size, n. For $n = 30$, a t distribution closely approximates the normal distribution.

3. When computing confidence intervals of the mean involving a t distribution, the critical t value is used instead of z. This value is obtained from Table 2/Appendix A or from an online critical value calculator.

4. When conducting a hypothesis test of the mean (σ unknown), the steps are the same as the z test (σ known) except the test statistic is t^* and not z^*. Furthermore, the assumptions are (a) the data are collected randomly; (b) the parent population is normally distributed, and if not, then should be sufficiently large to apply the CLT; and (c) σ is unknown.

5. When using Cohen's *d* to determine effect size, σ is replaced by *SD*, and the resulting effect is referred to as the estimated *d*.
6. The effect size of a study involving *t* can be reported relative to explained variance instead of standard deviation units via Cohen's estimated *d*. Explained variance for *t* is computed by

$$r^2 = \frac{t^2}{t^2 + df}$$

Vocabulary Check

Degrees of freedom (*df*)	Single-sample *t* test
Estimated *d*	Student's *t* distribution
Estimated standard error	Student's *t* test
Explained variance, r^2	*t* distribution
Post hoc power analysis	*t* test

Review Exercises

A. Check Your Understanding

In 1–10, choose the best answer among the choices provided.

1. When performing a test of significance about a population mean, a *t* distribution instead of a normal (*z*) distribution often is used. Which of the following is the most appropriate explanation for this?
   a. The population standard deviation is unknown.
   b. The sample does not follow a normal distribution.
   c. There is an increase in the variability of the test statistic due to the estimation of the population standard deviation.
   d. The sample standard deviation is unknown.
2. The major difference between a *z* ratio and a *t* ratio is that the *t* ratio
   a. contains an extra component.
   b. is entirely different.
   c. really is not a ratio.
   d. contains two sources of chance sampling variability.
3. If, for a one-tailed test, critical *t* values of 2.048, 2.763, and 3.674 correspond to α levels of .05, .01, and .001, respectively, the *p* value for the test statistic *t** = 2.25 could be expressed as
   a. *p* > .05
   b. *p* < .05
   c. *p* < .01
   d. *p* < .001
4. Given a critical *t* of −2.718 and an observed *t* of −2.69, you should
   a. reject the null hypothesis
   b. conduct another test using a larger sample size
   c. postpone any decision
   d. fail to reject the null hypothesis

5. Given $n = 30$, $\alpha = .01$, σ unknown, and a two-tailed test, which of the following is correct?

  a. $t = \pm 2.457$
  b. $t = \pm 2.756$
  c. $t = \pm 2.750$
  d. $t = \pm 2.462$

6. Which of the following choice is true about the *t* distribution?

  I.   It is symmetric regardless of *df*.
  II. It has more variability than the *z* distribution regardless of *df*.
  III. The variance associated with the *t* distribution is independent of *df*.

  a. I only
  b. II only
  c. III only
  d. I and II only

7. BIM's test preparation software for the FAA IRA exam claims it will enable flight students to score above the national average, which is $\mu = 51.5$ (σ unknown). A researcher tests this claim by randomly selecting 100 flight students, having them prepare for the exam using BIM's software, and then taking the exam after preparing for it for 30 days. Based on the results, the researcher failed to reject H_0 for $\alpha = .05$. The reason for this decision is because

  a. *t* critical > *t* statistic
  b. *t* critical < *t* statistic
  c. The sample mean was considered a highly improbable outcome.
  d. The researcher used a one-tailed test to the right.

8. In the airline industry, a measure of passenger carrying capacity is available seat kilometers (ASK), which is equal to "seats available × distance flown." For example, an aircraft with 100 seats that flies 250 km generates 25,000 ASKs for that flight. Let's assume that a revenue analyst for a low-cost carrier needs to estimate the mean ASK for one of the carrier's aircraft flown on a particular day and randomly selects 11 flights for that day involving this aircraft. Let's further assume that the ASK distribution in the population is unknown but is believed to be approximately normal in shape, but there is no historical data about its variability. Which of the following is the primary reason the revenue analyst should use a *t* confidence interval rather than a *z* confidence interval when estimating the targeted day's mean ASK?

  a. She is estimating μ using *M*.
  b. She is using *SD* as an estimate of σ.
  c. She is using data from only one specific day.
  d. The *z* confidence interval should never be used with a small sample

9. A flight test engineer endeavors to estimate the mean Mach Number of a particular model aircraft. (*Note:* Mach Number = Aircraft Speed divided by the speed of sound, where the speed of sound is a function of altitude.) The engineer randomly selects and tests 11 of the targeted aircraft and records the Mach Number for each one, with $M = 0.83$ Mach and $SD = 0.25$. If we were to assume that the sample's distribution is near-normal in form, which of the following statements is correct? Assume all aircraft were tested at an altitude of 9,144 m (30,000 ft.).

a. 95% CI for μ is $0.83 \pm \dfrac{0.25}{\sqrt{11}} \times 2.228$

b. 95% CI for μ is $0.83 \pm \dfrac{0.25}{\sqrt{11}} \times 2.201$

c. 95% CI for μ is $0.83 \pm \dfrac{0.25}{\sqrt{10}} \times 2.228$

d. 95% CI for μ is $0.83 \pm \dfrac{0.25}{\sqrt{10}} \times 2.201$

10. Research studies of melatonin, an over-the-counter sleep aid supplement, have shown that it can reduce the amount of time it takes to fall asleep. Let's assume that a causal-comparative study involving $n = 100$ flight attendants who fly internationally took 5 mg of melatonin daily 2 hours before they went to sleep for 30 days. Let's further assume that the flight attendants self-reported how much faster (or slower) it took them to fall asleep when using melatonin compared to when they did not use it and $M = 15$ min. and $SD = 10$. If the average time it takes international-bound flight attendants to fall asleep is $\mu = 15$ min., then a hypothesis test of the mean (σ unknown) where H_0: $\mu = 15$, H_1: $\mu < 15$, and $\alpha = .05$ would render which p value?
    a. $p = 0$
    b. $p = 1$
    c. $p < .05$
    d. $p = .05$

B. Apply Your Knowledge

Use the following research description and the corresponding data set to conduct the activities given in Parts A–C.

Although all airlines have luggage size restrictions for carry-on bags, not all airlines have weight restrictions for carry-ons. For example, according to luggagepros.com (n.d.), with the exception of Frontier Airlines, which has a weight limit of 35 lb. (15.88 kg), the major U.S. domestic carriers, including Alaska Airlines, American, Delta, JetBlue, Southwest, United, and Spirit do not have weight limits for carry-ons. However, many international airlines such as Aer Lingus, Air Canada, Air France, Air New Zealand, Alitalia, Japan Airlines, KLM, Lufthansa, and Singapore Airlines have weight restrictions ranging from 15 lb. (6.8 kg) to 26 lb. (11.8 kg). Let's assume that the average weight of carry-ons is $\mu = 17$ lb. (7.7 kg). Given that many airlines have instituted checked bag fees but still offer one free carry-on bag, a concern is that passengers who regularly check their bags will replace their checked bag with a carry-on, and this in turn would increase the mean weight of carry-ons. The file "Ch_5 Exercises Part B Data" contains fictitious data of carry-on weights from a random sample of airlines that have carry-on weight restrictions. You are to import this data set into your statistics program and conduct a single-sample t test.

A. Pre-Data Analysis

1. Specify the research question and corresponding operational definitions.
2. Specify the research hypothesis.
3. Determine the appropriate research methodology/design and explain why it is appropriate.
4. Conduct an a priori power analysis to determine the minimum sample size needed. Compare this result to the size of the given data set and explain what impact the size of the given sample will have on the results relative to the minimum size needed.

B. Data Analysis

1. Conduct a hypothesis test of the mean by applying the four steps presented in Section 5.5.

C. Post-Data Analysis

1. Determine and interpret the estimated effect size using Cohen's *d*.
2. Determine and interpret the effect size relative to explained variance (r^2).
3. Determine and interpret the power of the study from a post hoc perspective.
4. Determine and interpret the 95% confidence interval, including its precision and AIPE.
5. Present at least two plausible explanations for the results.

References

Cohen, J. (1988). *Statistical power analysis for the behavioral sciences* (2nd ed.). Lawrence Erlbaum Associates.

Dunbar, V. L. (2015). *Enhancing vigilance in flight instruction: Identifying factors that contribute to flight instructor complacency* (Publication No. 3664585) [Doctoral dissertation, Florida Institute of Technology]. ProQuest Dissertations and Theses Global.

Federal Aviation Administration (2020). *2020 airmen knowledge tests.* https://www.faa.gov/data_research/aviation_data_statistics/test_statistics/media/2020/annual/2020_Annual_Statistics.pdf

Luggagepros.com (n.d.). *Airline luggage restrictions: Carry-on items.* https://www.luggagepros.com/pages/carry-on

Omni Calculator (n.d.). *The critical value calculator.* https://www.omnicalculator.com/statistics/critical-value

Statistics Kingdom (2022a). *Z-test and T-test sample size calculator.* https://www.statskingdom.com/sample_size_t_z.html

Statistics Kingdom (2022b). *Normal, T—Statistical power calculators.* https://www.statskingdom.com/32test_power_t_z.html

6 Examining Bivariate Relationships: Correlation

Student Learning Outcomes

After studying this chapter, you will be able to do the following with respect to bivariate correlation:

1. Determine and interpret from a scatter plot and line of best fit key characteristics of the corresponding correlation.
2. Determine and interpret the Pearson r, the 95% confidence interval, and the coefficient of determination.
3. Conduct and interpret the results of a hypothesis test of the correlation coefficient.
4. Engage in pre-data analysis, data analysis, and post-data analysis activities as described in Section 6.6.

6.1 Chapter Overview

In this second of three chapters that present statistical procedures for analyzing research data involving a single sample or group, we introduce the concept of bivariate correlation, which examines the relationship between two variables. We begin by visualizing relationships graphically via scatter plots and then present the Pearson r, which is a single numerical metric called a correlation coefficient that denotes the strength and direction of a relationship. We follow this with a discussion on various statistical aspects of correlation, including confidence intervals and hypothesis testing involving the correlation coefficient. We conclude this chapter with a guided example that demonstrates the application of correlation to a correlational research study.

6.2 Correlation Fundamentals

Correlation in Research

In Chapter 1, we introduced *correlational research*, which is a research methodology that involves examining data on two or more research factors from a single group to determine if the variables share something in common. If they do, then we say that the variables are *correlated* (or *co-related*). For example, the director of a flight school might be interested in determining if there is a relationship between flight students' age and time to PPL. Other examples we might be interested in studying include examining the relationship between the number of flight hours as PIC and aviation accident rate per 10,000 flight hours, home

DOI: 10.4324/9781003308300-9

prices and distance from an airport, salt intake and blood pressure of pilots, and stress and number of maintenance errors committed by aviation maintenance technicians (AMTs).

For a more concrete example, let's assume that we want to conduct a study that examines the relationship between airport managers' salary and their career satisfaction. In doing so, we would state the study's purpose, operationally define key terms and variables, pose the research question, and then state the research hypothesis, which would be our prediction of the relationship between the two variables: This is summarized below.

> **Purpose.** To determine the relationship between airport managers' annual salary and their career satisfaction.
>
> **Operational Definitions.** The key terms/variables that need to be defined are airport managers and career satisfaction. We will use Byers' (2004) definitions:
>
> - *Airport managers* are defined as "individuals currently employed as a Director, Manager, Assistant Director, Director of Operations or similar senior-level management position at a commercial service or large general aviation airport…with at least 5 years' experience in airport management" (p. 8).
> - *Career satisfaction* is defined as scores on Byers' 17-item *Career Satisfaction in Airport Management* questionnaire (pp. 188–189, items 16–32), which uses a traditional 5-point Likert response scale. Thus, scores could range from 17 to 85, with higher scores reflecting higher levels of career satisfaction.
>
> **Research Question.** What is the relationship between airport managers' annual salary and their career satisfaction?
>
> **Research Hypothesis.** There will be a positive relationship between airport managers' annual salaries and their level of career satisfaction: As salaries increase, level of career satisfaction also will increase.

We would then identify our target and accessible populations, select a sample from the accessible population, administer Byers' career satisfaction questionnaire, ask participants to report their annual salaries, and then analyze the data.

Note that this study has two variables (salaries and satisfaction scores), and the data will be collected from one group (airport managers). Although we could report the mean and standard deviation of each variable, the two means and standard deviations will not answer our research question because they will not describe the extent to which the two variables are related. To answer our RQ, we use a statistical procedure called *correlation analysis*, or simply, *correlation*, which produces a *correlation coefficient*. Just as the mean and standard deviation are statistical measures that describe the center and spread of a distribution, a correlation coefficient is a statistical measure that describes the relationship between sets of observations.

A correlation analysis that examines the relationship between exactly two variables is called *bivariate correlation*, and the distribution of scores on the two variables is called a *bivariate distribution*. A bivariate distribution implies that participants have scores on each of two variables. Thus, the data are called *bivariate data* and are treated as *pairs of scores* instead of individual scores. This is illustrated in Table 6.1, which contains responses from a random group of 30 large-hub airport managers located in the southern United States who were asked to report their annual salary and complete Byers' (2004) career satisfaction questionnaire. This data set is provided in the file Ch_6 Salary-Career Satisfaction

Data. Note that each participant has a paired score. For example, Participant 1's paired observation is ($61,000, 24), and Participant 20's paired observation is ($69,000, 64).

Visualizing Relationships: Scatter Plots

Although distribution tables such as the one given in Table 6.1 are helpful for organizing bivariate data, they are not helpful for recognizing possible relationships, patterns, or trends between variables. For example, Figure 6.1 illustrates the general form of four different relationships involving paired data: linear, quadratic, exponential, and logarithmic. Given the data presented in Table 6.1, can you detect if the relationship between X and Y reflects any of these forms, or if there appears to be a pattern or trend between the variables? Your answer most likely is "no."

To visualize the relationship between paired data from a correlation, we construct and examine a scatter plot. Because bivariate correlation requires one score from each of two variables, we identify the variables as X and Y and list the corresponding pairs of scores in a table as we did in Table 6.1. If there is a clear understanding that one variable is an independent variable and the other is a dependent variable, then we identify X as the IV and Y as the DV. For our running example, we labeled annual salary as the IV and career satisfaction as the DV because it is more meaningful to examine this relationship from the perspective of salary influencing career satisfaction. On the other hand, if there is no such understanding—for example, the heights and weights of Air Force pilots—then it does not matter which variable is identified as X or Y.

Although most statistics programs provide a scatter plot as part of their results, it is beneficial to understand how scatter plots are constructed. To construct a scatter plot involving bivariate data, we place X scores along the horizontal axis and Y scores along the vertical axis. Each pair of scores is then plotted as a single point (x, y). For example, in Figure 6.2, the point that corresponds to Case 1 from Table 6.1 represents the paired observation (61000, 24), which reflects an annual salary of $x = \$61,000$ and a career satisfaction score of $y = 24$. Similarly, Case 2 is represented by the point (57000, 24).

Table 6.1 Annual Salary and Career Satisfaction Scores from 30 Large-Hub Airport Managers in the Southern United States

Case #	Annual Salary (X)	Career Satisfaction (Y)	Case #	Annual Salary (X)	Career Satisfaction (Y)
1	$ 61000	24	16	$ 68500	72
2	57000	24	17	59500	40
3	55500	24	18	68000	80
4	62000	48	19	68000	72
5	62000	40	20	69000	64
6	55000	16	21	63000	64
7	61000	48	22	62000	64
8	58000	32	23	61000	40
9	57000	24	24	62000	40
10	59000	24	25	62000	48
11	66000	72	26	61000	49
12	60000	32	27	57000	24
13	65000	64	28	66000	56
14	68000	72	29	64000	56
15	71000	80	30	68000	64

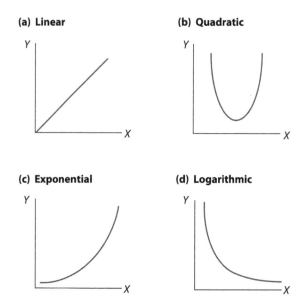

Figure 6.1 General forms of various bivariate relationships.

After we plot the paired observations, we draw an imaginary line, called the line of best fit, through the middle of the cluster of points to help identify the general trend of the data. This is illustrated in Figure 6.3. This line, which we will discuss in more detail in Chapter 7, represents the center of the relationship similar to the way a mean represents the center of a distribution.

If the line of best fit "rises" from left to right, then the values of X and Y tend to increase and decrease together. This appears to be the general trend for the salary–career satisfaction data: high annual salaries tend to be associated with high career satisfaction scores, and low annual salaries tend to be associated with low career satisfaction scores. On the other hand, if the line of best fit "falls" from left to right, the general trend is for high values of X to be associated with low values of Y, and vice versa. Note that the line of best fit is not necessarily a "line" per se. In situations where the paired data form a nonlinear relationship such as those shown in Figures 6.1(b, c, d), then the line of best fit will follow the trend of the data. An example of a nonlinear relationship is given in Figure 6.4.

Characteristics of Relationships

When examining the relationship between variables, we endeavor to describe three primary characteristics: the *direction* of the relationship, the *form* of the relationship, and the *strength* of the relationship.

The Direction of a Relationship

The direction of a relationship refers to whether the relationship is positive or negative. In a *positive correlation* the two variables tend to move in the same direction: they

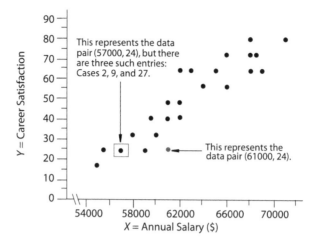

Figure 6.2 Bivariate scatter plot of the raw data given in Table 6.1. (*Note*: The scatter plot does not show duplicate entries.)

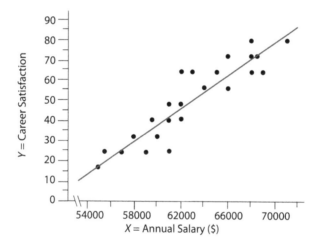

Figure 6.3 Bivariate scatter plot of the raw data given in Table 6.1 with the "line of best fit" inserted.

increase or decrease together. A positive relationship is characterized graphically in a scatter plot by a line of best fit that "rises" from left to right. A positive correlation also is referred to as a *direct relationship*. In Figure 6.2, the correlation between salary and career satisfaction is positive because scores on both variables increase together.

In a *negative correlation* the two variables tend to move in opposite directions: as one increases, the other decreases. A negative relationship is characterized graphically in a scatter plot by a line of best fit that "falls" from left to right. A negative correlation also is referred to as an *inverse relationship*. An example of an inverse relationship would be between altitude and fuel consumption of an airliner shown in Figure 6.5: as altitude increases, fuel consumption decreases.

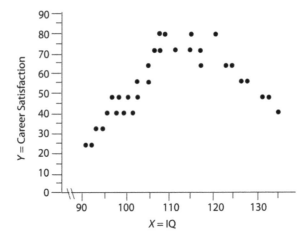

Figure 6.4 Example of a nonlinear relationship (parabolic). As IQ scores increase, career satisfaction scores also increase but up to a certain point where they level off and then
begin to decrease.

Figure 6.5 Scatter plot of altitude versus fuel consumption relative to jet aircraft. The line of
best fit "falls," which signifies a negative, or inverse, relationship.

The Form of a Relationship

The form of a relationship refers to whether the relationship is linear or nonlinear. Although we restrict our discussion to linear relationships, it also is possible to have
nonlinear relations as illustrated previously in Figure 6.4, which is an extension of the salary–career satisfaction example involving airport managers, but now includes their IQ
scores as measured by the IQ Test Academy (n.d.). These data are provided in the file Ch_6
Nonlinear Example Data. The relationship initially is positive with low IQ scores associated
with low career satisfaction scores. As IQ scores continue to increase, though, they reach a
plateau around 110 with career satisfaction scores between 80 and 90, and then the relationship becomes negative with higher IQ scores associated with lower career satisfaction
scores. Thus, the relationship Y has with X changes from positive to negative, resulting in

low career satisfaction being associated with low *and* high IQ scores. In general, when the rate at which *Y* increases or decreases changes as *X* changes, the scatter plot results in a "curved" pattern as opposed to a "straight" pattern, and therefore, nonlinear relationships are commonly called *curvilinear* relationships.

The Strength of a Relationship

In addition to providing a visual representation of a relationship, a scatter plot also gives an indication of the strength of the relationship, which is a function of how consistent the association is between variables. There are three general levels of consistency: perfect, zero, and intermediate, which is somewhere between zero and perfect. These are illustrated in Figure 6.6 and summarized below.

- A *perfect correlation* means that for every unit increase in *X*, there is a consistent and predictable increase in *Y*. When viewed from a scatter plot, all the data points of a perfect correlation will fall on the line of best fit as shown in Figures 6.6(a) and 6.6(b). A perfect correlation may be positive or negative.
- A *zero correlation* means there is no consistency between variables. When viewed from a scatter plot, the data points are scattered randomly throughout the graph and appear as a cloud of points without any discernible pattern or trend such as a line or curve. An example of a zero (or near zero) correlation is shown in Figure 6.6(f), which is between pilot height and the number of FAA ratings. The line of best fit for a zero relationship is a horizontal line.
- An *intermediate correlation* is neither perfect nor zero and indicates that as values of *X* increase, values of *Y* increase but they also might decrease. Thus, changes in *Y* with respect to changes in *X* are neither consistent nor predictable. When viewed from a scatter plot all the data points will not fall on the line of best fit. They instead will vary in how "close" they are to the line of best fit. In some cases, the points might be very close to the line as shown in Figure 6.6(c). In other cases, the points might be more spread out as shown in Figures 6.6(d) and 6.6(e).

6.3 Quantifying Relationships: The Pearson *r*

To quantify the strength of a relationship, we examine the corresponding *correlation coefficient*, which is a numerical index derived from the raw data. One such index is the Pearson product moment coefficient of correlation, *r*, or more simply, the *Pearson r*, which is used to measure linear relationships. The Pearson *r* can vary between −1.0 and +1.0, and it is non-dimensional, which means it has no units such as millimeters, inches, or kilograms:

- A perfect correlation has a coefficient of $r = 1$ or $r = -1$.
- A zero correlation has a coefficient of $r = 0$.
- An intermediate correlation has a coefficient between −1 and +1, that is: $-1 < r < 1$.

In addition to quantifying the strength of a relationship, the Pearson *r* also provides information about the direction of a relationship. If $0 < r \leq 1$, then the direction of the relationship is positive or direct, and if $-1 \leq r < 0$, then the direction of the relationship is negative

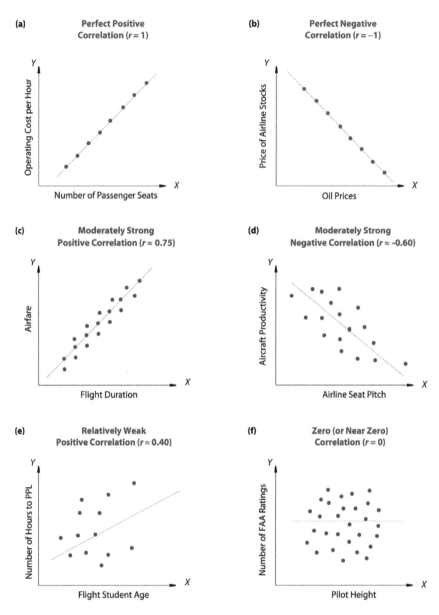

Figure 6.6 Scatter plots of six different data sets showing different degrees of linear relationship.

or inverse. Because *r* is with respect to linear relationships, the assumption is that the form of the relationship is linear, but we need a scatter plot to confirm this assumption.

In addition to the Pearson *r*, there are several other approaches to measuring relationships. These include (a) the *Spearman correlation*, which is used for data measured on an ordinal scale (i.e., ranked data); (b) the *point-biserial coefficient*, which is used when one variable consists of numerical data and the second variable is dichotomous; and (c) the *phi-coefficient*, which is used when both variables are dichotomous. We will not discuss these approaches.

We can estimate the correlation coefficient between two variables from their scatter plot by observing how close or far the cluster of points lie with respect to the line of best fit. For example, in Figures 6.6(a) and 6.6(b), because all the points that represent the pairs of scores on the two variables lie exactly on the line of best fit, this indicates a perfect positive correlation, $r = +1$ and a perfect negative correlation, $r = -1$, respectively. As the points become more scattered and are spaced farther away from the line of best fit, the correlation becomes less than perfect and therefore weaker. For example, in Figures 6.6(c, d, e, f), given the proximity of the points to the line of best fit, we would estimate the strength of the respective relationships to be approximately $r = .7$, $r = -.6$, $r = .4$, and $r = 0$. Note how the line of best fit becomes more horizontal as the relationship approaches zero. Using these examples as a guide, we would estimate the correlation coefficient for the salary–career satisfaction data depicted in Figure 6.3 to be $r = .9$.

In practice, we do not calculate r manually but instead rely on our statistics program to provide us with r as part of the output of a correlation analysis. Nevertheless, it is instructional to demonstrate the manual calculation of r, and we do so here using the raw scores formula:

$$r = \frac{N(\text{Sum } XY) - (\text{Sum } X)(\text{Sum } Y)}{\sqrt{N(\text{Sum } X^2) - (\text{Sum } X)^2} \times \sqrt{N(\text{Sum } Y^2) - (\text{Sum } Y)^2}}$$

where

- N = the number of paired scores
- Sum X = the sum of scores in the X distribution
- Sum Y = the sum of scores in the Y distribution
- Sum XY = the sum of the products of paired X and Y scores
- Sum X^2 = the sum of the squared scores in the X distribution
- Sum Y^2 = the sum of the squared scores in the Y distribution

The best way to apply this formula is to build a table, column-by-column, for each of the terms given in the formula:

1. Determine the total number of paired scores. This is N.
2. Add the X values. This is Sum (X).
3. Add the Y values. This is Sum (Y).
4. Multiply each XY data pair and then add these products. This is Sum (XY).
5. Square each X value and then add these squares. This is Sum (X^2).
6. Square each Y value and then add these squares. This is Sum (Y^2).
7. Substitute the values for N and the respective sums into the Pearson r formula and perform the indicated operations. Round results to two decimal places.

To demonstrate the application of the raw scores formula, we will use the small data set given in Table 6.2. This data set consists of hypothetical data where X = the number of FAA type ratings an airline pilot has, and Y = the number of pilot deviations committed by these pilots that an air traffic controller observed and recorded. A scatter plot of these data is given in Figure 6.7. Based on this plot, we estimate $r = -.85$.

Before we apply the raw scores formula, we first put this example into a research context by presenting the corresponding purpose statement, operational definitions, research question, and research hypothesis.

Purpose. To determine the relationship between the number of FAA type ratings and the number of pilot deviations among airline pilots.

Operational Definitions.

- *Airline pilot* is defined as a Part 121 ATP.
- *Type rating* is a designation added to a pilot license/certificate and issued by an authorizing agency such as the FAA or EASA that provides the holder with additional privileges relative to operating certain types of aircraft. Examples include the AirBus A320 and Boeing 737 type ratings.
- *Pilot deviation*, as defined by the FAA (2014), refers to "the actions of a pilot that result in the violation of a Federal Aviation Regulation or a North American Aerospace Defense Command Air Defense Identification Zone (ADIZ) tolerance."

Table 6.2 **Hypothetical Data Used to Demonstrate Raw Scores Formula for Calculating Pearson *r***

Case	Number of FAA Type Ratings (X)	Number of Pilot Deviations (Y)
1	4	7
2	5	8
3	8	3
4	6	4
5	7	2
6	9	1
7	6	5
8	7	4

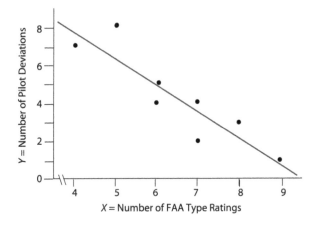

Figure 6.7 Bivariate scatter plot for the data given in Table 6.2.

Research Question. What is the relationship between the number of FAA type ratings and the number of pilot deviations among airline pilots?
Research Hypothesis. There will be an indirect (negative) relationship: As FAA type
ratings increase, the number of pilot deviations will decrease.

Let's now calculate r for the hypothetical data given in Table 6.2 using the seven steps
listed above. Table 6.3 contains the results of the first six steps. Substituting the data from
Table 6.3 into the formula, we now do the calculations in stages:

- Determine the numerator:

$$N(\text{Sum} XY) - (\text{Sum} X)(\text{Sum } Y)$$
$$= 8(197) - 52(34)$$
$$= 1576 - 1768$$
$$= -192$$

- Determine the first square root term in the denominator:

$$\sqrt{N(\text{Sum } X^2) - (\text{Sum } X)^2}$$
$$= \sqrt{8(356) - (52)^2}$$
$$= \sqrt{2848 - 2704}$$
$$= \sqrt{144}$$
$$= 12$$

- Determine the second square root term in the denominator:

$$\sqrt{N(\text{Sum } Y^2) - (\text{Sum } Y)^2}$$
$$= \sqrt{8(184) - (34)^2}$$
$$= \sqrt{1472 - 1156}$$
$$= \sqrt{316}$$
$$\approx 17.8$$

Table 6.3 Summary Data Used for the Raw Scores Formula for Calculating Pearson r

(1)	(2)	(3)	(4)	(5)	(6)
Case	X	Y	XY	X^2	Y^2
1	4	7	28	16	49
2	5	8	40	25	64
3	8	3	24	64	9
4	6	4	24	36	16
5	7	2	14	49	4
6	9	1	9	81	1
7	6	5	30	36	25
8	7	4	28	49	16
Sum	52	34	197	356	184

- Multiply the two square root terms in the denominator: $12 \times 17.8 = 213$

- Divide the numerator by the denominator: $\dfrac{-192}{213} = -.901$

Therefore, the relationship between the two variables is $r = -.901$, which supports the research hypothesis and confirms our estimation from the corresponding scatter plot.

6.4 Statistical Aspects of Correlation

Statistical Significance of r

The concept of statistical significance, which we introduced in Chapter 4, involves de-termining whether the results obtained from analyzing sample data are truly reflective of the parent population. In the context of correlation, a statistically significant corre-lation coefficient implies that the r value calculated from the sample data, denoted r^*, is most likely an accurate representation of the true measure of association between the variables in the parent population, instead of being due to chance. In other words, when we report a statistically significant correlation coefficient, we are implying that it reflects a true positive or negative relationship in the population and not a zero rela-tionship. It also implies that if we were to draw different samples from the same popu-lation and calculate the respective r^* values for each sample, then these r^* values will be similar.

To determine if a sample r is statistically significant, we use the alpha-level approach pre-sented in Chapter 4 for z and then extended to t in Chapter 5. Thus, to determine whether a sample correlation coefficient is statistically significant, we compare it to a critical r value that corresponds to the sample size and a preset alpha level. A table of critical r values is provided in Table 3/Appendix A. As an example in using this table, if $N = 25$, $\alpha = .05$, and we have a two-tailed test, then the critical r value is $r = \pm .396$. Because we have a two-tailed test, this means that the critical region will be to the right of $r = .396$ and to the left of $r = -.396$. If r^* falls in either region, then the correlation is significant and is either positive or negative depending on the direction of r^*. On the other hand, if r^* falls between these boundary values, then r^* is not significant, and there is no significant correlation in the population. This is illustrated in Figure 6.8(a).

To demonstrate this more concretely, let's determine if the correlation coefficient we calculated for the hypothetical data involving the relationship between the number of FAA type ratings of airline pilots and the number of pilot deviations they committed is significant. Given $n = 8$, $df = 6$, $\alpha = .05$, and a two-tailed test, the critical r values from Table 3/Appendix A are $r = \pm .707$. Because $r^* = -.901$, which is less than $-.707$, it falls in the critical region. Thus, there is a statistically significant *negative* relationship between the number of FAA type ratings and the number of pilot deviations among airline pilots: As the num-ber of type ratings increase, the number of pilot deviations decrease. This is illustrated in Figure 6.8(b).

Effect of Outliers

Recall from Chapter 2 that outliers are extreme scores that are not consistent with the other scores within a distribution. The presence of outliers can have a profound effect on the results of any data analysis, particularly those involving small sample sizes. To

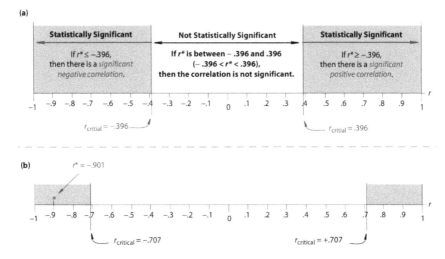

Figure 6.8 Illustration of critical *r* values for a two-tailed test of the correlation coefficient, *r*, based on a sample size of *n* = 25, *df* = 23, and α = .05 (a), and illustration of testing the significance of the correlation coefficient for the hypothetical data given in Table 6.2 (b).

demonstrate this, consider the scatter plots shown in Figure 6.9, which depicts the relationship between the number of credit hours flight students took in Spring 2022 (*X*) and their time to PPL (*Y*) from two different university Part 141 flight school programs. Both sample sizes are *n* = 11.

In Figure 6.9(a), the left scatter plot represents the actual data collected from the first flight program and does not contain any outliers. Note that the correlation coefficient is *r* = .09, which is not statistically significant for α = .05. Thus, there is no significant relationship between the number of credit hours flight students took and their time to PPL. Let's now assume the researcher accidentally recorded the data point (12, 60), which is circled in Figure 6.9(a), as (18, 100) as shown in the scatter plot on the right. This data point is an outlier and the corresponding correlation coefficient is now *r* = .74, which is significant. Thus, the outlier *inflated* the relationship by making it appear stronger than what it was initially.

In Figure 6.9(b), we have just the opposite situation. The left scatter plot represents the actual data collected from the second flight program and does not contain any outliers. Here, the correlation coefficient is *r* = .75, which is statistically significant for α = .05. Thus, there is a significant positive relationship between the number of credit hours students took and their time to PPL. Let's assume that the researcher accidentally recorded the data point (16, 105) as (20, 60). This data point is an outlier as shown in the scatter plot on the right, and the correlation coefficient is now *r* = .15, which is not significant. Thus, the outlier *masked* the relationship between the variables by making it appear weaker than what it was initially.

Because of the impact outliers can have on correlation, one of the assumptions of a correlation analysis is "no outliers," which means we must conduct an outlier analysis. Although there are several ways to do this, we prefer the *Jackknife distances* approach. In practice, if outliers are present, rather than simply delete them to be compliant with the

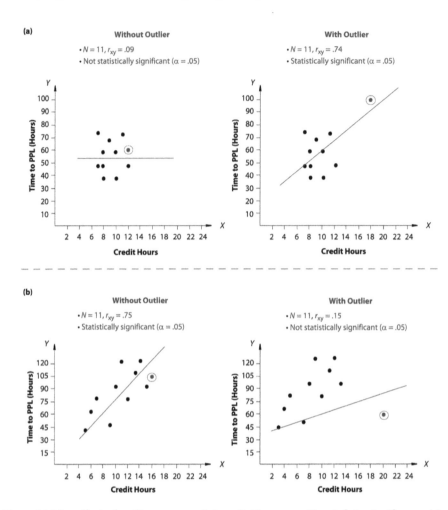

Figure 6.9 The effect of outliers on correlation. Outliers can either inflate significance (a) or mask significance (b).

assumption, we perform two independent analyses: one in the presence and one in the absence of outliers. We also determine if the outliers are rare cases or contaminants. We then decide whether to include or exclude the outliers in the final analysis and inform the reader of our decision and corresponding reason(s).

Effect Size and the Coefficient of Determination

In our discussion of the single-sample *t* test in Chapter 5, we introduced the concept of explained variance, r^2, as an alternative approach to Cohen's *d* for reporting effect size. We now extend this concept of explained variance to the context of correlation. As an alternative to using the Pearson *r* to quantify the relationship between two variables, we can square *r*. This numerical index, r^2, is called the *coefficient of determination* and represents the effect size of a correlation analysis. The coefficient of determination provides a more concrete meaning to the size of a correlation. It gives the percentage of variance in one variable that is associated with the percentage of variance in a second variable.

To illustrate this concept, consider the hypothetical data from Table 6.2 involving the correlation between FAA type ratings and the number of pilot deviations among airline pilots. For $r = -.90$, the corresponding coefficient of determination is $r^2 = .81$ and is interpreted as:

> *81% of the variation in FAA type ratings is being explained by the variation in number of pilot deviations.*

This measure also may be interpreted from a prediction perspective as follows:

> *If we know how many FAA type ratings an airline pilot has, then we have 81% of the information needed to perfectly predict the number of pilot deviations the pilot committed.*

In a broad sense, r^2 is a measure of the amount of information one variable carries about another variable. It also demonstrates that a negative and positive correlation of the same magnitude, such as $r = .6$ and $r = -.6$, are equivalent with respect to prediction because their respective coefficients of determination are equal, namely, $r^2 = .36$.

6.5 Statistical Inferences Involving Pearson *r*

Statistical inference can be approached either directly via confidence intervals or indirectly via hypothesis testing. When applied to the context of bivariate correlation, the focus is with respect to the Pearson *r*, which is a sample statistic, and the corresponding population parameter, which is the Greek letter for "r," namely, *rho*, symbolized as ρ. We now extend our discussion of CIs and hypothesis testing to include correlation.

Confidence Intervals

Recall from Chapter 4 that the general form of a confidence interval relative to a standard normal distribution (*z*) is

$$\text{Population Parameter} = \text{Point Estimate} \pm (SE)(\text{Corresponding } z \text{ score})$$

where the population parameter is the population mean (μ), the point estimate is the sample mean (*M*), *SE* is the standard error of the mean (σ_M), and the corresponding *z* score is the critical *z* value from Table 1/Appendix A. This led to the general structure of the 95% CI for the mean:

$$95\% \text{ CI } \mu = [a, b]$$

where

- *a* is the lower limit and is equal to $M - (\sigma_M)(z_{critical})$
- *b* is the upper limit and is equal to $M + (\sigma_M)(z_{critical})$

When applied to the context of correlation, although the 95% CI for rho (ρ) will be of the same form as the 95% CI for μ, the calculations for the lower and upper limits of the CI are not so straightforward. This is because we are dealing with scores from different distributions. For example, when examining the correlation between airport managers' annual salary and their level of career satisfaction, although we are working with one group (airport managers), we have two disparate distributions with different metrics: annual salary measured in U.S. dollars, and career satisfaction measured on a traditional 5-point Likert response scale.

To address this issue of disparate distributions, a statistical procedure called the Fisher z transformation is performed. A z transformation converts separate distributions into a standardized (z) distribution that allows us to compare scores from disparate distributions. This means that to determine the CI for the correlation coefficient, we must:

(a)   transform the sampling distribution of the Pearson r to a normal distribution using the z transformation,
(b)   compute the CI relative to z as we did in Chapter 4, and
(c)   convert the lower and upper limits of this z-based CI so that it is in terms of r.

Fortunately, our statistics program automatically provides us with the CI for correlation so it is not necessary to perform a z transformation. For those interested, Turner (2021) provides a demonstration of this process using *Excel* and provides a link to download the corresponding *Excel* macro.

For our purposes, we will focus our attention on interpreting the CI for the correlation coefficient. To do this, we will run a correlation analysis for the annual salary–career satisfaction data set given in Table 6.1 and the file Ch_6 Salary-Career Satisfaction Data. The output from this analysis yields $r = .9263$ and 95% CI $\rho = [.8496, .9647]$. You might recall that based on the corresponding scatter plot (Figure 6.3), we estimated the correlation coefficient to be approximately $r = .90$. So, we were not too far off in our estimate. With respect to the 95% CI, based on the lower and upper limits, it appears that in the population of airport managers, the relationship between airport managers' annual salary and their career satisfaction scores is anywhere between .8496 and .9647. More concretely:

> 95% of the time, we can expect the relationship between airport managers' annual salary and their career satisfaction score in the population will be between .8496 and .9647.

An alternative interpretation would be

> If we were to randomly select 100 samples of size n = 30 from the same population of airport managers, then we would expect 95 of these samples to have a correlation coefficient between r = .8496 and r = .9647.

Focusing on the precision of this CI, note that the width is $.9647 - .8496 = .1151$, which is a relatively narrow interval. Because the precision is relatively high, this means that the accuracy in parameter estimation (AIPE) also is high. Therefore, our sample of size $n = 30$ provides us with a relatively good estimate of the true relationship between annual salary and level of career satisfaction among the target population of airport managers.

Hypothesis Testing

To conduct a hypothesis test involving a bivariate linear relationship, we follow the same four-step procedure presented in our previous work, but note the following changes so the procedures reflect the context of correlation:

- The population parameter is rho (ρ).
- The purpose is to determine if the sample correlation coefficient, r, reflects more than a chance deviation from a hypothesized population correlation coefficient, ρ. If it does, then we can infer that the relationship observed in the sample also exists in the population.
- The null hypothesis is H_0: $\rho = 0$, which claims that in the population the variables are *not* significantly related: "There is no significant relationship between X and Y." Recognize that setting ρ to zero does *not* mean there is a zero relationship in the population. It simply means that whatever relationship exists, it is not statistically significant.

In addition to the changes noted above, the test statistic in Step 2, which establishes the critical boundary values, can be either r or t. A brief explanation of the differences follows.

Using r for Critical Values

To use r as the test statistic for critical values, we consult Table 3/Appendix A, which was demonstrated earlier in Figure 6.8. If the sample size is not listed in the table, then we have three choices. We can use the next *smaller* listing, consult an online calculator such as Mathcracker's (n.d.) critical correlation calculator, or manually calculate the critical r value using the formula

$$r_c = \frac{t_c}{\sqrt{t_c^2 + df}}$$

where

- r_c is the critical r
- t_c is the critical t value that corresponds to the given alpha level
- $df = n - 2$

To illustrate these three approaches, consider the example from Figure 6.8(a), which is relative to a two-tailed test for $n = 25$, $df = 23$, and $\alpha = .05$. The critical r value from Table 3/Appendix A is $r_c = .396$. Let's assume there is no table entry for $n = 25$:

- Table 3/Appendix A. The next smaller entry is for $n = 24$, which yields $r_c = .404$.
- Mathcracker's (n.d.) calculator. For $n = 25$, $\alpha = .05$, and a two-tailed test, $r_c = .396$.
- Critical r value formula. We first determine the critical t value from the t table given in Table 2/Appendix A for $df = 25 - 2 = 23$, $\alpha = .05$, and a two-tailed test, which means we focus on the $\alpha = .025$ column. This yields $t_c = 2.069$. We then apply the formula:

$$r_c = \frac{t_c}{\sqrt{t_c^2 + df}} = \frac{2.069}{\sqrt{2.069^2 + 23}} = \frac{2.069}{\sqrt{4.28 + 23}} = \frac{2.069}{\sqrt{27.28}} = \frac{2.069}{5.22} = 0.396$$

Of the three approaches, the online calculator and the manual calculation yielded the correct r_c. Therefore, you are encouraged to use either of these two approaches to determine r_c for sample sizes that are not listed in Table 3/Appendix A.

Using t for Critical Values

As an alternative to r, we could use t to determine the critical value. Note in the critical r value formula given above that the numerator is the critical t value, and therefore, $t_c = 2.069$ for our running example. Thus, if we want to use the t test statistic to determine the critical value for a hypothesis test for bivariate correlation, then we consult Table 2/Appendix A as we did in Chapter 5. However, if we use t to determine the critical value for a correlation-based hypothesis test, then we must also use t to calculate t^*. The problem, though, is that the t formula presented in Chapter 5 is from the perspective of the mean:

$$t = \frac{M - \mu}{SE_M}$$

So, we must derive a new t formula that can be applied to correlation. In general terms, the structure of the t statistic is as follows:

$$t = \frac{\text{Point estimate} - \text{Population parameter}}{\text{Standard error}}$$

When applied to correlation, the point estimate is the Pearson r, the population parameter is ρ, which will equal 0 as noted earlier, and the standard error for correlation is defined as

$$SE_r = \sqrt{\frac{1 - r^2}{n - 2}}$$

Therefore, putting this together, the t statistic formula for correlation is

$$t = \frac{r - \rho}{\sqrt{\dfrac{1 - r^2}{n - 2}}}$$

We will demonstrate the application of this formula to determine the calculated t value (t^*) shortly.

A Hypothesis Test Example

We now conduct a hypothesis test for the correlation coefficient using the salary–career satisfaction data from Table 6.1. For the corresponding research context, recall the purpose statement, operational definitions, research question, and research hypothesis were presented in Section 6.2, and the research hypothesis indicated a positive relationship.

Step 1. Identify the targeted parameter, and formulate the null and alternative hypotheses. The targeted parameter is the correlation coefficient, ρ, and the null and alternative hypotheses are:

H_0: $\rho = 0$. There is no significant relationship between annual salary and level of career satisfaction among large-hub airport managers from the southern United States.

H_1: $\rho > 0$. There is a significant, positive relationship between annual salary and level of career satisfaction among large-hub airport managers from the southern United States.

Step 2. Determine the test criteria. The test criteria involve three components:
Level of Significance (α). We will set alpha to $\alpha = .05$.

Test Statistic. The test statistic may be either r or t. We will use r. (*Note:* A discussion using t as the test statistic is given at the end of this example.)

The Critical Boundary. The critical r value from Table 3/Appendix A for a one-tailed test to the right with $n = 30$, $\alpha = .05$, and $df = 28$ is $r_c = .317$.

Step 3. Collect data, check assumptions, run the analysis, and report the results.
Collect Data. The data are given in Table 6.1 and in the file Ch_6 Salary–Career Satisfaction Data.

Check Assumptions. There are five assumptions for bivariate linear correlation:
- The relationship is assumed to be linear.
- The variables are continuous (i.e., they are measured on an interval or ratio scale).
- The paired data are from a random sample and independent of each other.
- There are no outliers, but if there are, an outlier analysis is to be conducted to determine the effect they are having on the results.
- Both variables should be approximately normally distributed.
We will assume all assumptions have been met.

Run the Analysis and Report the Results. Because we previously demonstrated the application of the formula to manually calculate the critical r value, we will run the analysis using our statistics program and record the results, which is what is done in practice. For the salary–career satisfaction data, $r^* = .9263$ and the corresponding p value is $p < .0001$.

Step 4. Decide whether to reject or fail to reject the null hypothesis, and make a concluding statement relative to the RQ.
Decision. The decision is to *reject the null hypothesis* because $r^* = .9263$ is greater than r critical ($r = .317$) and hence falls in the critical region.

Concluding Statement. Relative to the preset significance level of $\alpha = .05$, the sample data provide sufficient evidence of a statistically significant positive, linear relationship between annual salary and level of career satisfaction among large-hub airport managers from the southern United States: As annual salary increases, level of career satisfaction also increases.

If we used *t* instead of *r* as the test statistic, then we would have the following:

- In Step 2 of the hypothesis test, the critical *t* value from Table 2/Appendix A for *df* = 30 − 2 = 28, α = .05, and a one-tailed test to the right is *t*(28) = 1.701.
- In Step 3 of the hypothesis test, the calculated *t* value would be obtained by applying the derived *t* formula for *r* where *r* = .9263, $r^2 = (.9263)^2 = .858$, ρ = 0, and *n* = 30:

$$t = \frac{r - \rho}{\sqrt{\frac{1 - r^2}{n - 2}}} = \frac{.9263 - 0}{\sqrt{\frac{1 - .858}{30 - 2}}} = \frac{.9263}{\sqrt{\frac{.142}{28}}} = \frac{.9263}{\sqrt{.005}} = \frac{.9263}{.07} = 13$$

- Because *t** = 13 is greater than t_c = 1.701, it falls in the critical region. Therefore, the decision and concluding statement from Step 4 of the hypothesis test hold.

Correlation vs. Cause-and-Effect Relationships

In Chapter 1, we indicated that the results from a properly designed and executed experimental study could be interpreted as a cause-and-effect relationship because such studies incorporate the concepts of manipulation and control. A correlational study, however, is observational and does not involve manipulation and control. Although there are many instances of variables that might seem to be plausibly related to other variables, they fail as causal explanations because there is no manipulation of the independent variable, and there is no control of extraneous variables that could provide alternative explanations for the results. Thus, in the airport managers' annual salary–career satisfaction study, we cannot conclude that salaries *caused* airport managers' level of satisfaction with their career.

If two variables are significantly correlated, it is possible that the observed relationship could be explained by one (or more) alternative explanations. For example, in the airport managers' annual salary–career satisfaction study, it is plausible that the significant relationship is the result of a third variable such as "years of employment." As airport managers' tenure in their profession increases, so too do their salaries, and a long tenure suggests airport managers are satisfied with their career. Thus, it is conceivable that the two variables might not be related at all, but instead are separate results of a third variable: years of employment. This example demonstrates the concept of *common cause*.

Another alternative explanation is the concept of *reverse causality*, which is when the presumed cause (*X*) and the corresponding effect (*Y*) are reversed. For example, if there is a significant negative relationship between flight students' excessive drinking and their GPA, it might not be that excessive drinking causes a lower GPA, but instead it is a lower GPA that causes excessive drinking.

The bottom line is that an observed relationship between variables does *not* imply and must *not* be interpreted as a cause-and-effect relationship regardless of the strength of the relationship. In short: **Correlation does not imply causation.** For a good demonstration of this point, the reader is directed to Vigen (n.d.), which presents several examples of relationships involving correlations of |*r*| > .80.

6.6 Using Bivariate Correlation in Research: A Guided Example

We now provide a guided example of a research study that involves bivariate correlation. The example is based in part on Uhuegho (2017), who examined the safety climate of a

United States-based aviation maintenance, repair, and overhaul (MRO) organization. Our focus is on examining the relationship between AMTs' training perceptions and maintenance errors. The data set is given in the file Ch_6 Guided Example Data. We will structure the guided example into three distinct parts: (a) pre-data analysis, (b) data analysis, and (c) post-data analysis.

Pre-Data Analysis

Pre-data analysis activities include stating the research question, operational definitions, and research hypothesis; determining the research methodology to answer the RQ; and engaging in sample size planning.

Research Question and Operational Definitions

What is the relationship between training perceptions and maintenance errors among AMTs working at a United States-based MRO? *Training perceptions* refer to AMTs' perceptions about the training they received for their job as well as opportunities management provides to undertake further training. This was measured using five items from Fogarty's (2005) Maintenance Environment Survey (MES) and scored on a traditional 5-point Likert response scale, with higher scores reflecting more positive perceptions. *Maintenance errors* refer to the number of maintenance errors AMTs made on the job, including self-detected errors and those flagged by their supervisors. This was the DV and was measured using 13 items from the MES and scored on a traditional 5-point Likert scale, with higher scores reflecting higher levels of maintenance errors.

Research Hypothesis

There will be an inverse relationship between training perceptions and maintenance errors: As training perceptions increase (become more positive), maintenance errors will decrease.

Research Methodology

The research methodology that would best answer this question is *correlational* because we are interested in examining the relationship between the two targeted variables, training perceptions and maintenance errors, relative to a single group (AMTs).

Sample Size Planning (Power Analysis)

To determine the minimum sample size needed, we consult Table 4 in Appendix A. To use this table, we need to know power, which we set at .80, and r, which is what we expect the correlation coefficient to be in the population relative to the alternative hypothesis. In the absence of any guidance from the literature, we will select a *medium effect*. According to Cohen (1988, pp. 79–80): $|r| = 0.10$ is a small effect, $|r| = 0.30$ is a medium effect, and $|r| = .50$ is a large effect. Therefore, the corresponding sample size involving a one-tailed test for $\alpha = .05$ is $n^* = 67$. This is illustrated in Figure 6.10. Given that our sample consists of $n = 80$ cases, we will have at least an 80% chance of finding an effect of $|r| = .30$ for $\alpha = .05$.

Data Analysis

We now analyze the given data set relative to a hypothesis test of the correlation coefficient.

				One-Tailed Test (α = .05)					
					r				
Power	.10	.20	.30	.40	.50	.60	.70	.80	.90
.25	96	26	13	8	6	5	4	4	3
.50	271	68	31	18	12	8	6	5	4
.60	360	90	40	23	14	10	7	6	4
.70	470	117	51	29	18	12	9	6	5
.75	537	133	59	32	20	14	19	7	5
.80	616	153	67	37	23	15	11	8	5
.85	716	177	77	42	26	17	12	8	6
.90	853	211	92	50	31	20	14	10	6
.95	1077	266	115	63	38	25	17	11	8
.99	1569	386	167	91	55	36	24	16	10

Figure 6.10 An illustration of using Table 4/Appendix A for sample size planning for Chapter 6's Guided Example.

Step 1. Identify the targeted parameter, and formulate null/alternative hypotheses. The targeted parameter is the correlation coefficient, and the null/alternative hypotheses are:

H_0: $\rho = 0$. There is no significant relationship between training perceptions and maintenance errors.

H_1: $\rho < 0$. There is a significant inverse relationship between training perceptions and maintenance errors.

Step 2. Determine the test criteria. The test criteria involve three components:

Level of Significance (α). We will set alpha to α = .05.

Test Statistic. The test statistic is the Pearson *r*.

The Critical Boundary. In Table 3/Appendix A, we discover there is no table entry for n = 80 and df = 78. So, we will use Mathcracker's (n.d.) critical correlation calculator, which yields a critical *r* value of $r_c = -.185$.

Step 3. Collect data, check assumptions, run the analysis, and report results.
Collect the Data. The data are in the file Ch_6 Guided Example Data.

Check Assumptions. There are five assumptions.
• Linearity. As shown in Figure 6.11, the scatter plot and line of best fit confirm linearity.
• Continuous variables. Both variables are measured on a traditional 5-point Likert scale and hence are continuous.
• Paired observations. Every data point represents a paired response, and the paired responses are independent of each other.
• No outliers. We use a technique called *Jackknife distances*, which examines both variables simultaneously for outliers instead of using separate boxplots for each variable. Based on the results from our statistics program, the Jackknife technique confirms the data set is free of outliers.

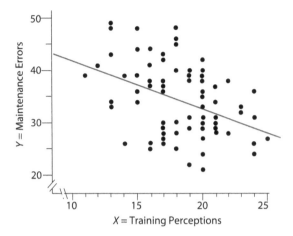

Figure 6.11   Scatterplot and line of best fit confirm linearity assumption for Chapter 6's Guided Example.

- Normality. Rather than assume normality as we have done previously, we use a formal normality test called the *Shapiro-Wilk Goodness of Fit* test, which is a hypothesis test where

  H_0: The variable is normally distributed
  H_1: The variable is not normally distributed

  If $p > \alpha$, then we *fail to reject* H_0, which confirms normality. When applied to the Guided Example, $p = .0551$ for $Y =$ Maintenance Errors and $p = .1086$ for $X =$ Training Perceptions. Therefore, the normality assumption is met.

  Run the Analysis and Report the Results. The calculated r value is $r* = -.4230$ and $p < .0001$.

Step 4. Decide whether to reject or fail to reject the null hypothesis, and make a concluding statement relative to the RQ.

Decision. Reject H_0 because $r* < r_c$ and falls in the critical region.

Concluding Remark. There is a significant negative relationship between AMTs' perceptions of training and maintenance errors: As perceptions of training increase (become more positive), the number of maintenance errors significantly decreases.

Post-Data Analysis

Post-data analysis activities include determining and reporting the corresponding effect size, the 95% confidence interval, and power of the study. We also discuss plausible explanations for the results.

Effect Size

To measure the effect size, we use the coefficient of determination, which is r^2. If $r = -.4230$, then $r^2 = .179$. This may be interpreted from either an explained variance or prediction perspective. With respect to explained variance, $r^2 = .179$ means that approximately 18% of the variance in maintenance error scores is being explained by the variance in training perception scores (and vice versa). With respect to prediction, $r^2 = .179$ means that if we know an AMT's training perception score, then we have approximately 18% of the information needed to perfectly predict this person's maintenance error score.

The 95% CI

The 95% CI reported by our statistics program is

$$95\% \; \rho \; \text{is} \; [-.5880, -.2241]$$

Therefore, 95% of the time we can expect the true relationship between AMTs' training perceptions and their maintenance errors to range between −.5880 and −.2241. Alternatively, if we were to randomly select 100 samples of size $n = 80$ from the same population, then 95 of these samples would have a calculated r value between −.5880 and −.2241. Furthermore, given H_0: $\rho = 0$, because 0 is not within this interval, we reject H_0. We also note that the width of the interval is somewhat large, which infers that the corresponding accuracy in parameter estimation (AIPE) is not that great. Therefore, the sample data are not very good for accurately estimating the true relationship between the targeted variables.

Power (Post Hoc)

To determine the approximate power of the study, we consult Table 4/Appendix A, which is demonstrated in Figure 6.12. As a result, the power of our study is between .95 and .99. Thus, the probability we made the correct decision to reject H_0 is between 95% and 99%, and the probability that the effect found in the sample ($r^2 = .179$) truly exists in the population is between 95% and 99%. Keep in mind, though, there is still a 5% chance this effect is false (Type I error).

Plausible Explanations for the Results

One plausible explanation is it is reasonable to assume that if AMTs believe they are adequately trained to do their job and receive encouragement to pursue further training,

				One-Tailed Test ($\alpha = .05$)					
					r				
Power	.10	.20	.30	.40	.50	.60	.70	.80	.90
.25	96	26	13	8	6	5	4	4	3
.50	271	68	31	18	12	8	6	5	4
.60	360	90	40	23	14	10	7	6	4
.70	470	117	51	29	18	12	9	6	5
.75	537	133	59	32	20	14	19	7	5
.80	616	153	67	37	23	15	11	8	5
.85	716	177	77	42	26	17	12	8	6
.90	853	211	92	50	31	20	14	10	6
.95	1077	266	115	63	38	25	17	11	8
.99	1569	386	167	91	55	36	24	16	10

Figure 6.12 Illustration of using Table 4/Appendix A to determine the power of Chapter 6's Guided Example.

then they would become more proficient at their job, which would result in fewer errors. A second plausible explanation is sample size. We had a larger sample size than required via a priori power analysis, which increases the likelihood of finding a significant effect.

Chapter Summary

1. A correlational study examines the relationship between variables and involves collecting data on two or more variables from a single group.
2. A correlation analysis is a statistical procedure for identifying, describing, and measuring the relationship between variables. If the analysis involves exactly two variables, it is called a bivariate correlation, the distribution of scores on the two variables is called a bivariate distribution, and the corresponding data are called bivariate data.
3. A scatter plot is a graphical illustration of a bivariate relationship and consists of a set of ordered pairs (x, y), which represent the paired data for each observation. A line of best fit is used in tandem with a scatter plot to help identify the general trend of the relationship.
4. The characteristics of a relationship include its direction (positive or negative), form (linear or nonlinear), and strength, which is quantified by the Pearson r, called the correlation coefficient. The Pearson r can vary between -1 and $+1$: $r = 1$ or $r = -1$ is a perfect relationship, $r = 0$ is a zero relationship, and $-1 < r < 1$ is an intermediate relationship. The corresponding population parameter is rho (ρ).
5. Outliers can mask a significant relationship or inflate a nonsignificant relationship. An outlier analysis strategy called Jackknife distances is used to determine the presence of outliers.
6. The coefficient of determination, r^2, is an effect size metric that can be interpreted from the perspective of explained variance or prediction.
7. A significant correlation coefficient does *not* imply causation because it is possible that the relationship could be explained by common cause, reverse causality, or other factors that were not considered.

Vocabulary Check

Bivariate correlation	Form of a relationship
Bivariate data	Intermediate correlation
Bivariate distribution	Inverse relationship
Cause-and-effect relationship	Jackknife distances
Coefficient of determination	Line of best fit
Common cause	Negative correlation
Correlation	Pearson r
Correlation coefficient	Perfect correlation
Correlation vs. Causation	Positive correlation
Correlation vs. Correlational research	Reverse causality
Critical r	Rho
Curvilinear relationship	Scatter plot
Direct relationship	Strength of a relationship
Direction of a relationship	Zero correlation
Fisher z transformation	

Review Exercises

A. Check Your Understanding

In 1–10, choose the best answer among the choices provided.

1. Which of the following describes a bivariate distribution?
   a. Annual incomes of air traffic controllers.
   b. Aviation Safety Locus of Control scores of 50 GA pilots.
   c. Medical examination results for male pilots.
   d. Number of hours flown and years of experience of Delta Airline pilots.
2. An outlier in a correlation analysis is a score that
   a. makes $r = 1$.
   b. makes $r = 0$.
   c. increases or decreases r.
   d. is usually of no concern.
3. When compared to the correlation coefficient, r, the coefficient of determination, r^2
   a. is always larger than r.
   b. can be larger or smaller than r.
   c. is always smaller than r.
   d. is negative if r is negative.
4. Which one of the following correlation coefficients has the most predictive value?
   a. $r = .88$
   b. $r = .50$
   c. $r = -.23$
   d. $r = -.92$
5. One advantage a scatter plot has over the Pearson r is that a scatter plot can confirm
   a. that the form of the relationship is linear.
   b. whether the relationship between the variables is positive, negative, or zero.
   c. which is the independent variable, and which is the dependent variable.
   d. that all the variables are being considered.
6. Let's assume that the correlation between two variables, X and Y, is -1.00, where $X = $ scores on the Test of English for Aviation (TEA) and $Y = $ scores on the TOEFL iBT. Let's further assume that your score on the TEA is 1.3 standard deviations below the mean. What do you predict your score will be on the TOEFL iBT?
   a. A raw score of -1.3.
   b. A raw score of 1.3.
   c. A z score of 1.3.
   d. A z score of -1.3.
7. An investigator found strong evidence that elderly GA pilots and young GA pilots scored low on a general self-efficacy scale while middle-aged GA pilots scored high on this scale. This is an example of:
   a. high positive Pearson r correlation.
   b. high negative Pearson r correlation.
   c. curvilinear relationship.
   d. lack of relationship between variables.

8. An air delivery service company places packages into large containers before fly-ing them across the country. These filled containers vary greatly in their weight. Suppose the company's airplanes always transport two such containers on each flight. The two containers are chosen so their combined weight is close to but does not exceed a specified weight limit. A random sample of flights with these containers is taken, and the weight of each of the two containers on each se-lected flight is recorded. The weight of the two containers on the same flight will have a
   a. correlation of 0.
   b. negative correlation.
   c. correlation of 1.
   d. positive correlation less than 1.

9. Based on a sample size of $n = 9$ air traffic controller trainees, a researcher found that the relationship between the number of times ATCs took a break during their training session and the number of times they identified a potential hazard was $r = -.25$. Given this result, what is an appropriate inference?
   a. Most ATC trainees who identify potential hazards also took breaks.
   b. Taking breaks during a training session causes a reduction in the number of po-tential hazards that are identified.
   c. ATC trainees who identify potential hazards often need to take fewer breaks dur-ing their training session.
   d. The correlation in the population of all ATC trainees most likely is $r = 0$.

10. If the relationship between reaction time and attention span among a sample ($n = 150$) of GA pilots is $r = -.42$, then we may conclude:
   a. There is no significant relationship between pilots' reaction time and attention span.
   b. Having a long attention span causes a decline in reaction time.
   c. If I know a pilot's reaction time, I will have about 16% of the information needed to perfectly predict the pilot's attention span.
   d. As pilots' reaction time increases, the more likely their attention span will increase.

B. Apply Your Knowledge

Use the following research description and corresponding data set to conduct the activities given in Parts A–C.

Bell et al. (1995) examined the relationship between pilot experience and aviation safety. They defined pilot experience as pilots' total flight hours in all aircraft, and they defined aviation safety as scores on the Aviation Safety Index (ASI), which con-sisted of five safety-related statements that correspond to behaviors associated with preflight preparation. The five statements are given below.

   1. I do a thorough walk-around inspection of an aircraft before I fly it.
   2. I check the weather thoroughly before I fly (even on local flight).
   3. I compute fuel requirements with an eye toward a 30- or 45-minute fuel reserve.
   4. I use a checklist for interior and exterior inspection of an aircraft.
   5. I compute takeoff and landing distances as well as runway lengths at airports of intended use for each flight.

Respondents were asked to self-assess the degree to which they performed each behavior using a 7-item bipolar adjective scale with "Never" at one end of the scale, which was scored 1, and "Always" at the other end of the scale, which was scored 7. Thus, aggregate scores could range from 5 to 35, with higher scores indicating more safety conscious pilots. The file Ch_6 Exercises Part B Data contains a set of hypothetical data that represents a pseudo replication of Bell et al. You are to import this data set into your statistics program and do the following (see also Section 6.6):

A. Pre-Data Analysis

1. Specify the research question and corresponding operational definitions.
2. Specify the research hypothesis.
3. Determine the appropriate research methodology/design and explain why it is appropriate.
4. Conduct an a priori power analysis to determine the minimum sample size needed. Compare this result to the size of the given data set and explain what impact the size of the given sample will have on the results relative to the minimum size needed.

B. Data Analysis

1. Conduct a hypothesis test of the correlation coefficient by applying all four steps as presented in Section 6.6.

C. Post-Data Analysis

1. Determine and interpret the effect size, r^2, from both explained variance and prediction perspectives.
2. Determine and interpret the 95% confidence interval, including its precision and AIPE.
3. Determine and interpret the power of the study from a post hoc perspective.
4. Present at least two plausible explanations for the results.

References

Bell, B. D., Robertson, C. L., & Wagner, G. S. (1995). Aviation safety as a function of pilot experience: Rationale or rationalization. *The Journal of Aviation/Aerospace Education & Research, 5*(3), 27–38.

Byers, D. L. (2004). *The making of the modern airport executive: Causal connections among key attributes in career development, compromise, and satisfaction in airport management.* (Publication No. 3664585) [Doctoral dissertation, Florida Institute of Technology]. ProQuest Dissertations and Theses Global.

Cohen, J. (1988). *Statistical power analysis for the behavioral sciences* (2nd ed.). Lawrence Erlbaum Associates.

Federal Aviation Administration (2014). *FAASTeam notice topic of the month: Pilot deviations.* [Notice No. NOTC5597]. https://www.faasafety.gov/spans/noticeView.aspx?nid=5597

Fogarty, G. J. (2005). Psychological strain mediates the impact of safety climate on maintenance errors. *International Journal of Applied Aviation Studies, 5*(1), 53–63. https://eprints.usq.edu.au/434/1/IJAAS_paper_Mar_05_revised.pdf

IQ Test Academy (n.d.). *Welcome to the IQ test academy™.* https://www.iqtestacademy.org/

Mathcracker (n.d.). *Critical correlation calculator.* https://mathcracker.com/critical-correlation-calculator

Turner, A. (2021, January 7). *Calculating confidence intervals for correlations* [Video]. https://www.youtube.com/watch?v=W7oCCw9gGvw

Uhuegho, K. O. (2017). *Examining the safety climate of U. S. based aviation maintenance, repair, and overhaul (MRO) organizations.* [Doctoral dissertation, Florida Institute of Technology]. https://repository.lib.fit.edu/handle/11141/1371

Vigen, T. (n.d.). *Spurious correlations.* https://www.tylervigen.com/spurious-correlations

7 Examining Bivariate Relationships: Regression

Student Learning Outcomes

After studying this chapter, you will be able to do the following with respect to bivariate linear regression:

1. Determine the equation of the corresponding regression line.
2. Determine and interpret the regression coefficient, B, regression constant, B_0, and 95% confidence interval of the regression coefficient.
3. Conduct and interpret the results of a hypothesis test of the regression coefficient.
4. Engage in pre-data analysis, data analysis, and post-data analysis activities as described in Section 7.5.

7.1 Chapter Overview

In this third of three chapters that present statistical procedures for analyzing research data involving a single sample or group, we introduce bivariate linear regression. This statistical strategy describes the linear relationship between variables via a mathematical model called a regression equation, and then uses this equation to predict scores on Y from scores on X. We introduce the regression equation numerically and graphically, and then examine the two main terms of the equation: the regression coefficient and regression constant. We follow this with a discussion on various statistical aspects of regression, including confidence intervals and hypothesis testing involving the regression coefficient, and present information related to the accuracy of prediction. We conclude this chapter with a guided example that demonstrates the application of bivariate linear regression to a research study.

7.2 The Regression Equation

The Equation of a Line

The concept of regression is similar to that of correlation in that both approaches are used to examine linear relationships between variables. The difference, though, is that regression describes the relationship using an equation, which is then used for prediction purposes. For example, in Chapter 6/Section 6.3, we demonstrated in Figure 6.6 how the line of best fit enables us to assess the relative strength of a relationship by examining how far the data points are from the line: The closer the points are to the line, the stronger the relationship, and the farther the points are from the line, the weaker the relationship. In

DOI: 10.4324/9781003308300-10

regression, the line of best fit is critical because the equation of this line is what we use for our mathematical model. Therefore, when placing a line of best fit in a scatter plot, we always try to place it in the center of the cluster of points. No matter how careful we are, though, without an objective criterion for determining the center, the placement of the line is nothing more than an educated guess and is subjective. To eliminate any uncertainty about where the line of best fit should be placed, we need a procedure that always defines a line that is indeed the "best fit" for a given bivariate distribution. Thus, the primary goal of linear regression is to find the equation for the best-fitting straight line, called the *regression line*, and then use this equation, called a *regression equation*, to make predictions.

In linear regression, we use the *slope-intercept form* of a linear equation as the mathematical model for making predictions. As shown in Figure 7.1(a), you might recall from basic algebra that the slope-intercept form of the equation of a line is $y = mx + b$, where m is the slope of the line and b is the y-coordinate of the y intercept, which is where the

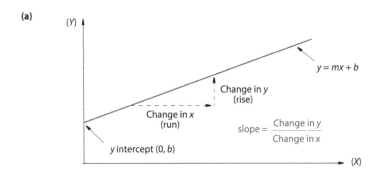

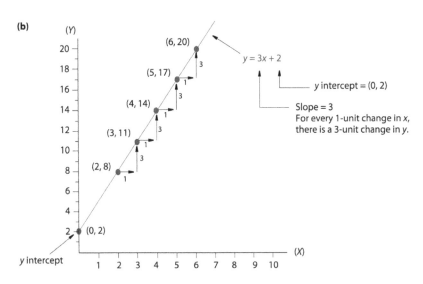

Figure 7.1 Illustration of the general characteristics of a linear equation given in slope-intercept form (a), and a specific application of the slope-intercept form of a linear equation (b).

line crosses the y-axis. Note that slope is defined as the ratio of the change in vertical distance (y), called the "rise," to the change in horizontal distance (x), called the "run."

$$m = slope = \frac{rise}{run} = \frac{Change\ in\ y}{Change\ in\ x}$$

This concept is analogous to what we do when walking up (or down) a set of stairs: For example, when walking upstairs, we first lift our leg up (change in y) and then "slide" it forward (change in x) for each step. An application of the slope-intercept form is illustrated in Figure 7.1(b) for the equation $y = 3x + 2$. Note that the slope is equal to 3, which is $\frac{3}{1}$, and the y intercept is the point (0, 2). Therefore, for every 1-unit change in x, there is a corresponding 3-unit change in y. We can confirm that the slope is $m = 3$ via the coordinates of each point. As illustrated in Figure 7.1(b), each 1-point increase in x corresponds to a 3-point increase in y.

When applied to the context of bivariate linear regression, we alter the notation of the slope-intercept form slightly to represent the general form of the regression equation:

$$\hat{y} = Bx + B_0$$

From this representation, note the following:

- Instead of using y, we use $\hat{y}$ (pronounced y hat) to signify that the equation is a *prediction equation* because we will use it to predict y values from x values.
- The slope, m, is denoted B and is called the *regression coefficient*.
- The y coordinate of the y intercept, b, is denoted B_0 and is called the regression constant.

A common procedure used to determine the regression line is called the *least squares criterion*, which minimizes the distance between the regression line and the points that represent the paired observations. For example, in Figure 7.2, Point 10 is on the line and hence has no deviation from the line, Point 7 has a small amount of deviation, and Point 8 has a considerable amount of deviation. The least squares criterion method produces a regression line that minimizes all these deviations. This leads to a line where the distance each point is from the line is minimal. A regression analysis that uses the least squares criterion method is called ordinary least squares (OLS) regression.

In practice, we do not actually calculate the regression coefficient and regression constant manually but instead rely on our statistics program to provide us with B and B_0 as part of the output of a regression analysis. Nevertheless, it often is instructional to demonstrate the manual calculation of a regression line. A formula for calculating B and B_0 based on raw data is given below. You should note that the factors for calculating B also are used in the regression coefficient formula we presented in Chapter 6 for calculating a correlation coefficient.

$$Regression\ Coefficient: B = \frac{N(Sum\ XY) - (Sum\ X)(Sum\ Y)}{N(Sum\ X^2) - (Sum\ X)^2}$$

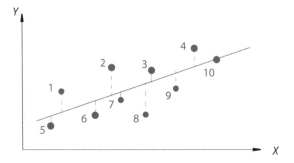

Figure 7.2 Illustration that shows the deviation between observed points and their corresponding points on the regression line.

where

- N = the number of paired scores
- Sum X = the sum of scores in the X distribution
- Sum Y = the sum of scores in the Y distribution
- Sum XY = the sum of the products of paired X and Y scores
- Sum X^2 = the sum of the squared scores in the X distribution

$$\text{Regression Constant}: B_0 = M_Y - (B)(M_X)$$

where

- M_Y = the mean of Y
- M_X = the mean of X

To demonstrate the application of this formula, we will use the hypothetical data from Table 6.2/Chapter 6 when we manually calculated the Pearson r. Recall that this data set related to examining the relationship between X = the number of FAA type ratings of a Part 121 airline pilot, and Y = the number of pilot deviations committed by these pilots. Because we are using the same data set, the corresponding research context is the same except instead of focusing on a "relationship" between variables, our focus is on the "effect" one variable is having on another: For each change in x, what is the corresponding change in y? Following are the purpose statement, operational definitions, research question, and research hypothesis.

> **Purpose.** To determine the effect the number of FAA type ratings (x) has on pilot deviations among airline pilots (y).
> **Operational Definitions.** These are the same from Chapter 6's example.
> **Research Question.** What is the effect of the number of FAA type ratings on the number of pilot deviations among airline pilots?
> **Research Hypothesis.** There will be a negative effect between the number of FAA type ratings and the number of pilot deviations: The slope of the regression line will be negative, which indicates that as the number of FAA type ratings increases, the number of pilot deviations will decrease.

Let's now apply the formulas for B and B_0. Because we are using the same data set, and because the factors in the formula for B were previously calculated in Chapter 6 when we demonstrated the Pearson r formula, please refer to Chapter 6 for these inputs.

$$B = \frac{N(\text{Sum } XY) - (\text{Sum } X)(\text{Sum } Y)}{N(\text{Sum } X^2) - (\text{Sum } X)^2}$$

$$= \frac{8(197) - (52)(34)}{8(356) - 52^2}$$

$$= \frac{1576 - 1768}{2848 - 2704}$$

$$= \frac{-192}{144}$$

$$= -1.33$$

Therefore, $B = -1.33$. We now solve for B_0:

$$B_0 = M_Y - (B)(M_X)$$

$$= 4.25 - (-1.33)(6.5)$$

$$= 4.25 + 8.645$$

$$= 12.895$$

Therefore, $B_0 = 12.895$. When expressed in the form of $y = mx + b$, we get the regression equation, $\hat{y} = -1.33x + 12.895$, which when rounded to one decimal place is $\hat{y} = -1.3x + 12.9$.

Let's now graph this line. Recall that to graph any line, we identify two points that fit the equation and then connect them with a straight line. One point that always will fit a regression equation is the y intercept, which is the y value when $x = 0$—that is, $(0, B_0)$. A second point we always select is defined by the means of the two variables, (M_X, M_Y). Note that both points are readily available from the calculation of the regression constant: $M_X = 6.5$, and $M_Y = 4.25$. Therefore, we will plot the points $(0, 12.9)$ and $(6.5, 4.25)$. It also is prudent to plot a third point that is between $x = 0$ and M_X. We choose $x = 3$. Substituting 3 into the regression equation we get: $\hat{y} = -1.3x + 12.9 = -1.3(3) + 12.895 = -3.9 + 12.9 = 9$. Therefore, the third point we plot is $(3, 9)$. The graph of the regression line is given in Figure 7.3. We plotted the raw scores in black and the three points used to construct the regression line in a second color.

Note that the slope $B = -1.3$ is equivalent to the fraction $\frac{-1.3}{1}$. Therefore, as illustrated in Figure 7.3, for every 1 unit we move to the *right* on x, we move 1.3 units *down* on y. This also is consistent with the way we interpret this slope: "For every 1-unit change in x there will be *negative* 1.3-unit change in y." When comparing the regression line given in Figure 7.3 to the line of best fit given in Figure 6.7/Chapter 6, you might think that the two lines are not the same because they have different y intercepts. For example, the regression line in Figure 7.3 crosses the y-axis at the point $(0, 12.9)$, but the regression line in Figure 6.7 appears to cross the y-axis at the point $(0, 8.2)$. If you examine the horizontal axis in Figure 6.7 more closely, though, you will see that it does not start at 0. Thus, we must remember to check the axes before making any interpretations from a graph.

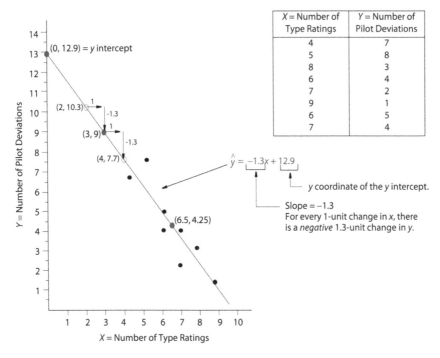

X = Number of Type Ratings	Y = Number of Pilot Deviations
4	7
5	8
8	3
6	4
7	2
9	1
6	5
7	4

$\hat{y} = -1.3x + 12.9$

— y coordinate of the y intercept.

Slope = -1.3
For every 1-unit change in x, there is a *negative* 1.3-unit change in y.

Figure 7.3 Graph of the regression line $\hat{y} = -1.3x + 12.9$. The black points represent the observed scores from the given chart, and the three red points were used to construct the regression line.

Understanding the Regression Equation Relative to the Scatter Plot

When reviewing the regression line that accompanies a scatter plot, you should recognize there are two distinct sets of points. One set consists of the actual cluster of data points from the sample that makes up the scatter plot. These are the *observed scores*. The second set consists of the points that lie on the regression line. Because the equation that corresponds to the regression line is used to predict y scores from x scores, the points on the regression line are the *predicted scores*. For example, in the scatter plot shown in Figure 7.4, each x value corresponds to two points. The first point, (x, y), is an observed score, and the second point, $(x, \hat{y})$, is the corresponding predicted score. It is the predicted y value given x, which is obtained from the regression equation by substituting the numerical value of x into the regression equation and solving for $\hat{y}$.

If the difference between y and $\hat{y}$ is 0 (i.e., $y - \hat{y} = 0$), then the predicted score matches the observed score, which means that the observed score falls on the regression line and there is no error in the prediction. One such point is depicted in Figure 7.4. If the difference between y and $\hat{y}$ is positive (i.e., $y - \hat{y} > 0$), then the observed score lies *above* the regression line, and if $y - \hat{y} < 0$, then the observed score lies *below* the regression line. The nonzero difference between y and $\hat{y}$ is the amount of *error* that exists between an observed score and its corresponding predicted score, and as noted in Figure 7.4, reflects the vertical distance between the observed and predicted scores. This error is called the *residual* and is denoted $y - \hat{y}$.

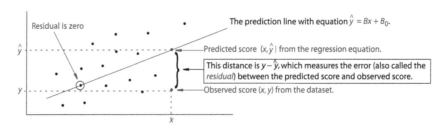

Figure 7.4 Illustration of a regression line that highlights there are always two points that correspond to any given x value: the observed score (x, y), and the predicted score (x, ŷ).

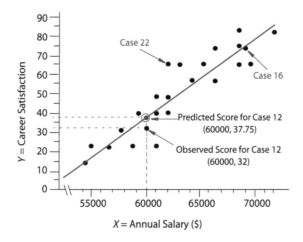

Figure 7.5 Scatter plot of the salary–career satisfaction data from the file Ch_6 Salary–Career Satisfaction Data (see also Table 7.1). The regression equation is $\hat{y} = 0.0041233x - 209.6475$.

To illustrate this concept more concretely, let's examine the scatter plot, regression line, and regression equation for the data set involving airport managers' annual salary and career satisfaction scores, which we presented initially in Chapter 6 (see the file Ch_6 Salary–Career Satisfaction Data). A summary of the output for this regression analysis is shown in Figure 7.5, and an extension of this data set that includes the predicted y values and the residuals is provided in Table 7.1.

Note that the regression equation is $\hat{y} = 0.0041233x - 209.6475$. Let's now focus on Case 12. From Table 7.1, the observed score for this participant is (60000, 32), which means that this airport manager reported an annual salary of $60,000 (x) and scored 32 (y) on the career satisfaction questionnaire. To get the corresponding predicted score, we substitute 60,000 into the regression equation and solve for $\hat{y}$:

$$\hat{y} = 0.0041233x - 209.6475$$
$$= 0.0041233(60000) - 209.6475$$
$$= 247.398 - 209.6475$$
$$= 37.75$$

Table 7.1 Observed Scores, Predicted Scores, and Residuals for the Airport Managers'
Annual Salary–Career Satisfaction Scores Data Set

Case	X	Y	$\hat{Y}$	$(Y - \hat{Y})$
1	61000	24	41.88	−17.90
2	57000	24	25.38	−1.38
3	55500	24	19.20	4.80
4	62000	48	46.00	2.00
5	62000	40	46.00	−6.00
6	55000	16	17.14	−1.14
7	61000	48	41.88	6.12
8	58000	32	29.51	2.49
9	57000	24	25.38	−1.38
10	59000	24	33.63	−9.63
11	66000	72	62.49	9.51
12	60000	32	37.75	−5.75
13	65000	64	58.37	5.63
14	68000	72	70.74	1.26
15	71000	80	83.11	−3.11
16	68500	72	72.80	−0.80
17	59500	40	35.69	4.31
18	68000	80	70.74	9.26
19	68000	72	70.74	1.26
20	69000	64	74.86	−10.90
21	63000	64	50.12	13.88
22	62000	64	46.00	18.00
23	61000	40	41.88	−1.88
24	62000	40	46.00	−6.00
25	62000	48	46.00	2.00
26	61000	40	41.88	−1.88
27	57000	24	25.38	−1.38
28	66000	56	62.49	−6.49
29	64000	56	54.25	1.76
30	68000	64	70.74	−6.74

Therefore, based on the sample data, we predict that an airport manager whose annual salary is $60,000 will have a career satisfaction score of 37.75. Note that the residual for this prediction is $y - \hat{y} = 32 - 37.75 = -5.75$. Because this difference is negative, the observed score lies *below* the regression line. Further note that this nearly 6-point difference represents how far "off" the predicted y value is compared to the observed y value. This is illustrated in Figure 7.5. A word of caution: When we calculated the predicted y value using the regression equation, we did not round the regression coefficient or constant because the rounding error in the final score would have been too large. For example, if we were to round B to 0.004 and B_0 to −209.65, then the predicted score would be $(0.004)(60000) - 209.65 = 240 - 209.65 = 30.35$, which is 7.4 units smaller than the correct predicted score and does not fall on the regression line.

Also note from Figure 7.5 that the closer an observed score is to the regression line, the smaller the error will be (e.g., the residual for Case 16 is −0.8), and the farther away an observed score is from the regression line, the larger the error will be (e.g., the residual for Case 22 is 18). However, if an observed score falls directly on the regression line, then the prediction for that case will be "perfect" because the observed and predicted scores would be equal, and the residual would be zero.

Let's now extend this concept of a perfect prediction for a single case to a perfect prediction for *all* cases in a sample. From a graphical perspective, perfect prediction means that all the observed scores fall exactly on the regression line. It also implies that perfect correlation ($r = 1$ or $r = -1$) leads to perfect prediction. As the amount of error increases, though, the correlation approaches zero ($r \to 0$) and the gap between the predicted and observed scores increases, which leads to less accurate predictions. So, returning to Case 12 of our running example, we predicted that an airport manager whose annual salary is $60,000 would have a career satisfaction score of 37.75, which is "off" by about 6 points when compared to the observed score. The reason the prediction is not accurate is because the correlation between the two variables we reported in Chapter 6 was $r = .9263$, which is not a perfect correlation.

When using a regression equation to make predictions, it also is important to be sensitive to the concept of the interval of predictability, which restricts predictions to the range of *x* values of the observed scores. For example, because the range of salaries in our running example is between $55,000 and $71,000, we should restrict any predictions of career satisfaction scores to this range. Making predictions outside this interval is risky because we do not have any data that represent that part of the relationship. Similarly, as noted in our discussion of correlation, the prediction is only applicable to large-hub airport managers from the southern United States whose annual salaries are within this range.

Interpreting a Regression Equation

In practice, you will enter a set of data into a statistics program that will perform the regression analysis. As part of the output, the program will provide the correlation coefficient, the regression equation, and the scatter plot with the regression line. This means that you will not have to construct a scatter plot, derive the regression equation, or graph the regression line. What you will have to do, though, is interpret the regression equation in the context of the given research setting. The interpretation of a regression equation involves interpreting the regression coefficient, *B*, and the regression constant, B_0. To illustrate how this is done, we will use the regression equation for the annual salary–career satisfaction study, $\hat{y} = 0.0041233x - 209.6475$, which is illustrated in Figure 7.5.

Interpreting the Regression Coefficient (B)

To understand how to interpret a regression coefficient, it is helpful to begin with a generic interpretation and then progress toward an interpretation that is relative to the research setting. It also is helpful to recall that the regression coefficient is the slope of the regression line.

(a)   A generic interpretation of the regression coefficient *B*:

For every <u>1-unit change in x</u> there will be, on average, an estimated change of <u>B units in y</u>.

(b)   We refine this generic interpretation to include the specific regression coefficient, $B = 0.0041233$:

For every <u>1-unit change in x</u> there will be, on average, an estimated change of <u>0.0041233 units in y</u>.

(c)   We refine the previous interpretation even more to reflect the context of the given setting where *x* = annual salary and *y* = career satisfaction score:

For every 1-dollar increase in an airport manager's annual salary, there will be, on average, an estimated increase of 0.0041233 units in a manager's career satisfaction score.

(d) Finally, to interpret it more realistically in the context of the given research setting where annual salary is expressed in *thousands of dollars*, we multiply "1-dollar" by a factor of 1,000, which means we also must multiply B by a factor of 1,000:

For every $1,000-dollar increase in an airport manager's annual salary, there will be, on average, an estimated increase of 4.1233 points in a manager's career satisfaction score.

More succinctly, for every $1,000 increase in the annual salary of a large-hub airport manager from the southern United States, we predict that on average this manager's career satisfaction score will increase by approximately 4 points. Note that this is nothing more than interpreting the slope of a line, except now we are doing so with respect to a specific application.

Interpreting the Regression Constant (B_0)

The regression constant is equal to the *y* value when $x = 0$. In the regression equation for the annual salary–career satisfaction study, $\hat{y} = 0.0041233x - 209.6475$, the regression constant is $B_0 = -209.6475$ and is interpreted as follows:

When annual salary (X) is $0, we predict that, on average, the career satisfaction score will be – 209.6475.

Note that in the context of the current study, this constant does not make sense for two reasons: an airport manager's salary is not going to be $0, and the possible range of career satisfaction scores from Byers' (2004) questionnaire is between 17 and 85. As a result, we would simply conclude that the regression constant is meaningless in the context of the given research setting. This is an important observation: Although the regression coefficient is always interpretable, sometimes a regression constant is not interpretable in a particular context. Thus, it is incumbent for us to determine if the constant has meaning for the given research setting.

Example 7.1: *Applying and Interpreting a Regression Equation*

Sheppard (2013) examined the effect aviation fuel costs had on airline passenger itinerary costs with respect to major U.S. airlines and hypothesized there would be a positive effect. He operationally defined *aviation fuel costs* as the quarterly average cost in U.S. dollars of Jet A fuel per gallon between 1996 and 2012 for major U.S. carrier airlines, *airline passenger itinerary price* as the U.S. domestic quarterly average fare in dollars of tickets purchased during the same time period for major U.S. carrier airlines, and *major U.S. airlines* as an airline that profited over $20 million a year in revenue in 2013. A copy of his data is provided in the file Ch_7 Example 7.1 Data.

(a) Write the corresponding research question and research hypothesis.
(b) Use your statistics program to generate the regression equation, scatter plot, regression line, and correlation coefficient.

(c) Interpret the regression coefficient (B), regression constant (B_0), and coefficient of determination (r^2) in the context of the given research setting.

(d) Use the regression equation to predict the price of an airline ticket if the average per quarter cost of Jet A fuel is $4.77 per gallon, which was the average cost for Q2 2021, and interpret the result.

Solution

(a) **Research Question.** What is the effect of aviation fuel costs on airline passenger itinerary costs with respect to major U.S. airlines?

Research Hypothesis. Aviation fuel costs will have a positive effect on airline passenger itinerary costs (the slope of the regression line will be positive).

(b) From our statistics program:

- The scatter plot and regression line are provided in Figure 7.6.
- The regression equation is $\hat{y} = 46.552471x + 275.25807$
- The correlation coefficient is $r = .64$.

(c) Interpretations:

- $B = 46.552471$: For every $1 increase in quarterly average fuel price (x), there is an approximate $46.55 increase, on average, in quarterly average airfares.
- $B_0 = 275.25807$: If the quarterly average fuel price is $0 (i.e., the airlines receive it for free), then the quarterly average airfare will be approximately $275.26. Although it is highly unlikely that an airline will receive free fuel, the regression constant in the current context is interpretable.
- $r^2 = .64^2 = .41$. This means that 41% of the variance in quarterly average airfare is being explained by the variance in quarterly average fuel price. It also means that if we know the quarterly average fuel price, then we have 41% of the information needed to perfectly predict the quarterly average airfare.

(d) If $x = \$4.77$, then

$$\hat{y} = 46.552471x + 275.25807$$
$$= 46.552471(4.77) + 275.25807$$
$$= 222.0552867 + 275.25807$$
$$= 497.3133567$$

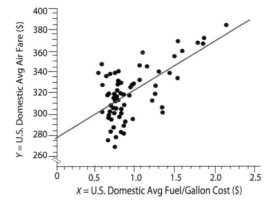

Figure 7.6 Scatter plot of the fuel cost–air fare data for Example 7.1.

Therefore, we predict that if the quarterly average aviation fuel cost is $4.77 per gallon, then the quarterly average ticket cost for all U.S. major airlines will be, on average, $497.31. The reader should note, though, that this prediction is relative to the second quarter of 2021, which is outside the targeted period of 1996–2012, and therefore might be spurious.

7.3 Statistical Aspects of Regression

Statistical Significance of Regression

The concept of statistical significance relative to regression involves deciding if the sample regression coefficient, B, holds in the parent population. Because B is the slope of a regression equation, if there is no relationship between x and y, then we would expect the slope to be zero, which means that the regression line is horizontal. On the other hand, if the relationship between x and y is positive or negative, then the slope will be positive or negative, respectively, and the regression line will rise from left to right ($B > 0$) or fall from left to right ($B < 0$). To test the significance of the regression coefficient, we use the t statistic with $df = n - 2$. This means we consult Table 2/Appendix A to determine the critical t value. This is demonstrated in Section 7.4 when we discuss hypothesis testing involving the regression coefficient.

Accuracy of a Prediction

Although a regression equation enables us to make predictions, it does not tell us anything about the accuracy of the predictions. Because a regression equation is derived from sample data, the slope, B, and regression constant, B_0, are considered sample statistics and subject to *sampling error* just like any other sample statistic such as the mean. As a result, B and B_0 will be estimates of the corresponding population parameters, which are expressed in Greek letters as β (*beta*) and β_0 (*beta sub-zero*), respectively. This means that if we use a sample regression equation to predict y from x in the population, there will be errors in the prediction. This raises the question: How accurate will the predictions be?

To measure the accuracy of a prediction, we examine the *standard error of estimate* ($SE_{Residuals}$), which is provided as part of the output of a regression analysis. The standard error of estimate provides a measure of the distance between a regression line and the actual data points in the population, and it uses the *standard deviation of the residuals* from the sample. This is why we denote the subscript of the standard error of estimate as "residuals" ($y - \hat{y}$). It is the estimated population standard deviation of the residuals from estimating y from x, and therefore is an indicator of the variability of errors, which gives us an idea of the accuracy of a prediction. For example, in Example 7.1, the standard error of estimate reported by our statistics program was 19.32. This means that when we use the regression equation, $\hat{y} = 46.552471x + 275.25807$ to predict the quarterly average ticket price for U.S. major airlines based on the per gallon quarterly average aviation fuel cost as we did in Example 7.1(d), the predicted quarterly average ticket price will be "off" on average by about $19.32.

It is important to observe that although this metric is called "standard error of estimate," it is not the same concept as "standard error of the mean," which we introduced in Chapter 3. A more appropriate term for standard error of estimate would be *standard deviation of regression* because it represents the standard deviation of y for each corresponding x. For example, in Chapter 2, we calculated standard deviation by: (a) determining how far each score deviated from the mean; (b) squaring each deviation score; (c) adding the deviation scores and calling the total the "sum of the squared deviation scores," which we denoted SS; (d) dividing SS by df, which we called variance; and (e) taking the square root of variance, which is standard deviation. To calculate the standard error of estimate for regression, we apply this same procedure except now the deviation scores are the residuals, which represent how far each predicted $\hat{y}$ value is from the observed y value; that is, $(y - \hat{y})$. This can be seen more clearly by examining how standard error of estimate is calculated:

$$SE_{Residuals} = \sqrt{\frac{Sum(y_i - \hat{y})^2}{n-2}}$$

Note that in the numerator we: (a) calculate the residual for each score, $(y_i - \hat{y})$; (b) square each residual, $(y_i - \hat{y})^2$; then (c) add the squared residuals, Sum $(y_i - \hat{y})^2$. Thus, the numerator is referred to as the *sum of the residuals squared*, which is commonly referred to as the *sum of the squared errors (SSE)* or more simply, *sum of squares (SS_Y)*. We then divide SSE by $df = n - 2$ and take the square root of the quotient. As a result, many statistical programs report the standard error of estimate as *root mean square error (RMSE)* where "root" stands for square root, "mean" indicates we are taking the average squared deviations when we divide by df, and "error" to indicate this represents the "average error" we get when we use the regression equation to predict y from x.

To demonstrate the application of this formula, we will use the hypothetical data involving number of FAA type ratings and number of pilot deviations as illustrated in Figure 7.3. Shown in Table 7.2 for each of the $n = 8$ cases are the observed x and y values, the predicted y values ($\hat{y}$), the residuals $(y - \hat{y})$, and the squared residuals $(y - \hat{y})^2$. You will note that the sum of the squared residuals is 7.4999, which we rounded to 7.5. Substituting this into the formula, we get:

$$SE_{Residuals} = \sqrt{\frac{Sum(y_i - \hat{y})^2}{n-2}} = \sqrt{\frac{7.5}{8-2}} = \sqrt{\frac{7.5}{6}} = \sqrt{1.25} = 1.118$$

Therefore, if we were to use the regression equation to predict the number of pilot deviations based on the number of type ratings of airline pilots, our prediction would be "off" on average by about 1 pilot deviation, which suggests that we have a good prediction model. For example, if an airline pilot has five type ratings ($x = 5$), then we would predict that this pilot would have

$$\hat{y} = -1.3x + 12.9$$
$$= -1.3(5) + 12.9$$
$$= -6.5 + 12.9$$
$$= 6.4$$

or approximately 7 pilot deviations, and our prediction would be "off," on average, by about 1 deviation.

This result should not be surprising given that the corresponding correlation coefficient was $r = .90$ and the coefficient of determination was $r^2 = .81$, which means that if we know how many type ratings an airline pilot has, then we have 81% of the information needed to perfectly predict the number of pilot deviations. This relationship can be confirmed

Table 7.2 Inputs for Calculating the Standard Error of Estimate ($SE_{Residuals}$) for the Hypothetical Data Involving Number of Type Ratings and Number of Pilot Deviations

Case	X	Y	$\hat{Y}$	$(Y - \hat{Y})$	$(Y - \hat{Y})^2$
1	4	7	7.5833	−0.583	0.3403
2	5	8	6.25	1.75	3.0625
3	8	3	2.25	0.75	0.5625
4	6	4	4.9167	−0.917	0.8403
5	7	2	3.5833	−1.583	2.5069
6	9	1	0.9167	0.0833	0.0069
7	6	5	4.9167	0.0833	0.0069
8	7	4	3.5833	0.4167	0.1736
				Sum	7.4999

Note. X = Number of FAA type ratings; Y = Number of pilot deviations; $\hat{Y}$ = the predicted y values from the regression equation, $\hat{y} = -1.3x + 12.9$; $(Y_i - \hat{Y})$ = the residuals, which are the difference between each observed y value and respective predicted value; and $(Y_i - \hat{Y})^2$ = residuals squared.

visually by observing from Figure 7.3 how close each score is to the regression line. Remember: If all the observed scores fall directly on the regression line, then we would have perfect correlation ($r = 1$ or $r = -1$), and the corresponding coefficient of determination would be $r^2 = 1$, which means that information about x would provide 100% of the information needed to perfectly predict y.

7.4 Statistical Inferences Involving the Regression Coefficient

Recall from our earlier discussions that statistical inference can be approached either directly via confidence intervals or indirectly via hypothesis testing. When applied to bivariate linear regression, the focus is with respect to the regression coefficient, B, and the corresponding population parameter, β. We now extend our discussion of CIs and hypothesis testing to include regression.

Confidence Intervals

As discussed in previous chapters, the general form of a CI is:

$$Population\ Parameter = Point\ Estimate \pm (SE)\ (Critical\ t\ Score)$$

When applied to regression:

- The point estimate is the regression coefficient, B.
- SE is the standard error of the regression coefficient, denoted SE_B, and is equal to

$$SE_B = \frac{SE_{Residuals}}{\sqrt{Sum\ (x_i - M_x)^2}}$$

where
- $SE_{Residuals}$ = the standard deviation of the residuals, which is an index of the accuracy of a prediction presented earlier.

- M_X = the mean of X.
- $(x_i - M_X)$ = the difference between each x value and the mean.
- The critical t score is acquired from Table 2/Appendix A for $df = n - 2$.

As a result:

$$95\% \text{ CI for } \beta = [a, b]$$

where

- a is the lower bound and is equal to $B - (SE_B)(t_{critical})$
- b is the upper limit and is equal to $B + (SE_B)(t_{critical})$

To illustrate the use of this formula, let's determine the 95% CI for the regression coefficient for the hypothetical data from Table 7.2, which involved number of type ratings and
number of pilot deviations among airline pilots. Recall from our earlier discussion that B
= −1.33, $SE_{Residuals}$ = 1.118, and n = 8, which means that the critical t value for α = .05 (two-
tailed test) from Table 2/Appendix A for $df = n - 2$ is t(6) = 2.447. We now need to determine M_X and the square of the sum of the differences between x and the mean. These are
summarized in Table 7.3. As a result:

$$SE_B = \frac{SE_{Residuals}}{\sqrt{\text{Sum } (x_i - M_X)^2}}$$

$$= \frac{1.118}{\sqrt{18}}$$

$$= \frac{1.118}{4.2426}$$

$$= 0.264$$

We now use these values to calculate the 95% CI:

- Lower bound: −1.33 − (0.264)(2.447) = −1.33 − 0.646 = −1.97
- Upper bound: −1.33 + (0.264)(2.447) = −1.33 + 0.646 = −0.684

Thus, the 95% CI for β = [−1.97, −0.68]. The output for the 95% CI from our statistics program
is [−1.97, −0.69], so our manual calculation is accurate. Interpreting this CI, we would report
that with 95% certainty, in the population of airline pilots, the slope of the regression line

Table 7.3 Inputs for Calculating the Standard Error of the Regression
Coefficient (SE_B) for the Hypothetical Data Involving Number of
Type Ratings and Number of Pilot Deviations

Case	X	$X - M_X$	$(X - M_X)^2$
1	4	−2.5	6.25
2	5	−1.5	2.25
3	8	1.5	2.25
4	6	−0.5	0.25
5	7	0.5	0.25
6	9	2.5	6.25
7	6	−0.5	0.25
8	7	0.5	0.25
Sum	52	0.0	18.00

Note. The mean of X is 52/8 = 6.5.

that reflects the effect the number of type ratings (X) has on number of pilot deviations (Y) is between −1.97 and −0.68. More concretely:

> In the population of airline pilots, 95% of the time we can expect that for every 1-unit increase in the number of type ratings an airline pilot earns, the number of pilot deviations attributed to this pilot will be <u>reduced</u> on average anywhere between approximately 1 and 2.

An alternative interpretation would be

> If we were to randomly select 100 samples of size $n = 8$ from the same population of airline pilots, then we would expect 95 of these samples to have a regression line with a regression coefficient (i.e., slope) between $B = -1.97$ and $B = -0.68$.

Focusing on the precision of this CI, note that the width is $-0.68 - (-1.97) = -0.68 + 1.97 = 1.29$, which is a relatively narrow interval. Because the precision is relatively high, this means that the accuracy in parameter estimation (AIPE) also is high. Therefore, our sample of size $n = 8$ provides us with a relatively good estimate of the effect the number of type ratings has on the number of pilot deviations. Furthermore, this CI does not include 0, which would reflect a zero slope (i.e., horizontal regression line). Therefore, if we were to set up a hypothesis test that investigated the null hypothesis, $H_0: \beta = 0$, then based on the 95% CI, we would reject H_0 because 0 is not in the interval.

Hypothesis Testing

To conduct a hypothesis test involving a bivariate linear regression, we follow the same four-step procedure we presented in our previous work except now the context is with respect to the regression coefficient, B. As a result, the following changes are noted:

- The population parameter will be beta (β).
- The purpose of the test is to determine if the sample regression coefficient, B, reflects more than a chance deviation from a hypothesized population regression coefficient, β. If it does, then we can infer that the slope of the regression line observed in the sample also exists in the population.
- Because we are examining the effect X has on Y via the slope of the regression line, the null hypothesis will claim that X does *not* have a significant effect on Y in the population, which means that the regression line is horizontal, and the slope is zero. In other words, $H_0: \beta = 0$, which states, "X has no significant effect on Y."
- The test statistic in Step 2, which establishes the critical boundary value, is t.

We now conduct a hypothesis test for the regression coefficient using the salary–career satisfaction data from Table 6.1 in Chapter 6. Recall that we presented in Section 6.2 the purpose statement, operational definitions, research question, and research hypothesis, which expressed a positive relationship between the variables. Also recall from Figure 7.5 that the corresponding regression equation is $\hat{y} = 0.0041233x + 209.6475$.

Step 1. Identify the targeted parameter, and formulate null/alternative hypotheses. The targeted parameter is the regression coefficient, β, and the null and alternative hypotheses are:

$H_0: \beta = 0$. Annual salary does not have a significant effect on the level of career satisfaction among large-hub airport managers from the southern United States (i.e., the slope of the regression line is zero).

$H_1: \beta > 0$. Annual salary has a significant, positive effect on the level of career satisfaction among large-hub airport managers from the southern United States (i.e., the slope of the regression line is positive).

Step 2. Determine the test criteria. The test criteria involve three components:

Level of Significance (α). We will set alpha to α = .05.

Test Statistic. The test statistic is *t*.

The Critical Boundary. The critical *t* value is acquired from Table 2/Appendix A. Because the salary–career satisfaction data set consists of *n* = 30 cases, the critical *t* value for a one-tailed test to the right for α = .05 and *df* = 28 is *t*(28) = 1.701. This is shown in Figure 7.7.

Step 3. Collect data, check assumptions, run the analysis, and report the results.

Collect Data. The data are in the file Ch_6 Salary–Career Satisfaction Data.

Check the Assumptions. There are four assumptions for bivariate linear regression:
- Linearity. The relationship between *X* and *Y* must be linear.
- Constant variance of the residuals. For any value of *X*, the variance of the residuals around the regression line in the population is assumed to be constant. This idea of "equal or constant variance of the residuals" is known as *homoscedasticity* (*homo* = same; *scedastic* = scattered).
- Independence of the residuals. The residuals of the observations must be independent of one another. In other words, there must be no relationship among the residuals for any subset of cases in the analysis. This assumption will be met for any random sample.
- Normality of the residuals. For any value of the IV, the residuals around the regression line are assumed to have a normal distribution.

The single best way to confirm these assumptions is by examining a *residual plot* involving the standardized (*z*) scores of the dependent variable. We will demonstrate this in Section 7.5 when we present a guided example on how to conduct a bivariate regression analysis for a research study. For the current discussion, we will assume all assumptions have been met.

Run the Regression Analysis and Report the Results. In practice, we will run the analysis using our statistics program and record the results. However, for instructional purposes, we will derive the formula for the *t* statistic for regression and manually calculate the *t* value.

Recall that the general structure of the *t* statistic is as follows:

$$t = \frac{\text{Point estimate} - \text{Population parameter}}{\text{Standard error}}$$

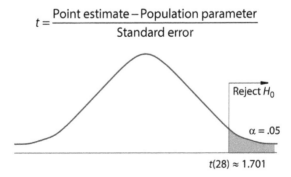

Reject H_0

α = .05

t(28) ≈ 1.701

Figure 7.7 Critical region for one-tailed *t* test to the right for α = .05 with *df* = 28 (*n* = 30).

When applied to regression, the point estimate is the regression coefficient B, the population parameter is β, which will equal 0, and the standard error for regression was defined earlier as $SE_B = \dfrac{SE_{Residuals}}{\sqrt{Sum\ (x_i - M_X)^2}}$. Therefore,

$$t* = \frac{B - \beta}{SE_B}\ for\ df = n - 2$$

Applying this formula to the salary–career satisfaction data, recall that $B = 0.0041233$. We now calculate the standard error, SE_B. The entries for this formula are given in Table 7.4.

Table 7.4 Inputs for Calculating the Standard Error of Estimate ($SE_{Residuals}$) and the Standard Error of Regression (SE_B) for the Annual Salary–Career Satisfaction Data

Case	X	Y	$\hat{Y}$	$(Y - \hat{Y})$	$(Y - \hat{Y})^2$	$(X - M_X)$	$(Y - M_X)^2$
1	61000	24	41.88	−17.90	319.41	−1550	2402500
2	57000	24	25.38	−1.38	1.90	−5550	30802500
3	55500	24	19.2	4.80	23.04	−7050	49702500
4	62000	48	46.00	2.00	4.00	−550	302500
5	62000	40	46.00	−6.00	36.00	−550	302500
6	55000	16	17.14	−1.14	1.30	−7550	57002500
7	61000	48	41.88	6.12	37.45	−1550	2402500
8	58000	32	29.51	2.49	6.20	−4550	20702500
9	57000	24	25.38	−1.38	1.90	−5550	30802500
10	59000	24	33.63	−9.63	92.74	−3550	12602500
11	66000	72	62.49	9.51	90.44	3450	11902500
12	60000	32	37.75	−5.75	33.06	−2550	6502500
13	65000	64	58.37	5.63	31.70	2450	6002500
14	68000	72	70.74	1.26	1.59	5450	29702500
15	71000	80	83.11	−3.11	9.67	8450	71402500
16	68500	72	72.80	−0.80	0.64	5950	35402500
17	59500	40	35.69	4.31	18.58	−3050	9302500
18	68000	80	70.74	9.26	85.75	5450	29702500
19	68000	72	70.74	1.26	1.59	5450	29702500
20	69000	64	74.86	−10.90	118.81	6450	41602500
21	63000	64	50.12	13.88	192.65	450	202500
22	62000	64	46.00	18.00	324.00	−550	302500
23	61000	40	41.88	−1.88	3.53	−1550	2402500
24	62000	40	46.00	−6.00	36.00	−550	302500
25	62000	48	46.00	2.00	4.00	−550	302500
26	61000	40	41.88	−1.88	3.53	−1550	2402500
27	57000	24	25.38	−1.38	1.90	−5550	30802500
28	66000	56	62.49	−6.49	42.12	3450	11902500
29	64000	56	54.25	1.76	3.10	1450	2102500
30	68000	64	70.74	−6.74	45.43	5450	29702500
Sum	1876500			0	1572.03	0	558675000
	$M_X = 62550$						

$$SE_{Residuals} = \sqrt{\frac{Sum(y_i - \hat{y})^2}{n-2}} = \sqrt{\frac{1572.03}{30-2}} = \sqrt{\frac{1572.03}{28}} = \sqrt{56.14392857} = 7.493$$

Note. X = Annual salary; Y = Career satisfaction score; $\hat{Y}$= the predicted y values acquired from the regression equation $\hat{y} = 0.0041233x - 209.65$; $(Y_i - \hat{Y})$ = the residuals, which are the difference between each observed y value and its respective predicted value; and $(Y_i - \hat{Y})^2$ = the residuals squared.

$$SE_B = \frac{SE_{Residuals}}{\sqrt{Sum\ (x_i - M_x)^2}} = \frac{7.493}{\sqrt{558675000}} = \frac{7.493}{23636.30682} = 0.000317$$

$$\text{Therefore, } t^* = \frac{0.0041233}{0.000317} = 13.00$$

Step 4. Decide whether to reject or fail to reject the null hypothesis, and make a concluding statement relative to the RQ.

Decision. The calculated t^* is $t(28) = 13.00$, which is greater than $t_{critical} = 1.701$ and lies in the critical region. Therefore, the decision is to *reject* H_0.

Concluding Statement. Relative to the preset significance level of $\alpha = .05$, the sample data provide sufficient evidence that the slope of the regression line in the population differs significantly from zero. Based on the results of data analysis, annual salary has a significant and positive effect on level of career satisfaction among large-hub airport managers from the southern United States. More specifically, given a slope of $B = 0.0041233$, for every $1,000 increase in annual salary, career satisfaction scores increase, on average, by approximately four points. This effect of salary on career satisfaction was significant, $t(28) = 13.00$ for $\alpha = .05$.

7.5 Using Bivariate Linear Regression in Research: A Guided Example

We now provide a guided example of a research study that involves a bivariate linear regression analysis. The study involves pilots at a low-cost carrier (LLC), and the focus is on examining the effect pilots' biological sex (female or male) has on their annual salary. The ultimate goal is to generate a prediction equation that could be used to predict pilots' salary based on their sex. A random selection of $n = 62$ full-time airline transport pilots (ATPs) from the LLC self-reported their annual salary and biological sex. A copy of the data set is given in the file, Ch_7 Guided Example Data. We will structure this guided example into three distinct parts: (a) pre-data analysis, (b) data analysis, and (c) post-data analysis.

Pre-Data Analysis

Pre-data analysis activities include stating the research question, operational definitions, and research hypothesis; determining the research methodology to answer the RQ; and engaging in sample size planning.

Research Question and Operational Definitions

"What is the effect of biological sex on the annual salaries of ATPs?" In the context of the current study, *biological sex* refers to female or male, *annual salary* is defined in U.S. dollars, and an ATP is defined as a full-time pilot working for the targeted LLC and holds an FAA-issued ATP certificate.

Research Hypothesis

Biological sex will have some effect on salaries, but the direction of this effect is unknown, which means that the slope of the regression line will *not* be zero.

Research Methodology

Given that the study's goal is to generate a prediction equation that can be used to predict salaries based on biological sex, this is a prediction study, and therefore, the appropriate research methodology/design is *prediction correlational research*. This is because we endeavor to estimate the effect on a dependent measure (annual salary) relative to a change in a predictor variable (biological sex). As a result, the focus is on the regression coefficient (B) because it reflects causal effects. A correlational research methodology also is appropriate because the study involves a single group (ATPs) and is examining the relationship between multiple variables (salary and biological sex) of this single group.

Sample Size Planning (Power Analysis)

To determine the minimum sample size needed, we consult Statistics Kingdom (2022) and enter the following:

- Type: Regression
- Power: 0.8
- Effect: Medium
- Effect size: 0.15

- Significance level (α): .05
- Predictors (p): 1
- Effect type: f^2
- Digits: 4

This returns a minimum sample size of $n = 58$ based on Green's rule of thumb for a medium effect that tests the entire model. With respect to this power analysis, observe the following:

- In regression, the "effect type" is also referred to as the "effect size index" and is equal to $f^2 = \dfrac{R^2}{1 - R^2}$.
- Cohen (1988) provided metric-free effect sizes of "small" ($f^2 = 0.02$), "medium" ($f^2 = 0.15$), and "large" ($f^2 = 0.35$).
- Given that the data set for the current study has $n = 62$ cases, we have a sufficiently large sample size to have at least an 80% chance of correctly rejecting the null hypothesis.

Data Analysis

We now analyze the given data set via a hypothesis test of the regression coefficient.

Step 1. Identify the targeted parameter, and formulate null/alternative hypotheses. The targeted parameter is the regression coefficient, and the null/alternative hypotheses are:

H_0: $\beta = 0$. Biological sex (X) has no significant effect on annual salary (Y) in the population of full-time ATPs. (The slope of the regression line in the population is zero—horizontal line.)

H_1: $\beta \neq 0$. Biological sex (X) has a significant effect on annual salary (Y) in the population of full-time ATPs. (The slope of the regression line in the population differs significantly from zero—oblique line.)

Step 2. Determine the test criteria. The test criteria involve three components:

> *Level of Significance* (α). We will set alpha to α = .05.

> *Test Statistic.* The test statistic is *t*.

> *The Critical Boundary.* Using Table 2/Appendix A for *n* = 62, *df* = 60, and a two-tailed test, the critical *t* value is *t* = ± 2.0. This is depicted in Figure 7.8(a).

Step 3. Collect data, check assumptions, run the analysis, and report the results.

> *Collect Data.* The data are in the file Ch_7 Guided Example Data.

> *Check Assumptions.* As noted earlier, there are four assumptions for bivariate linear regression, and we examine these assumptions via *residual plots*.
> - Linearity. To confirm the linearity assumption, we examine a scatter plot with the *y* scores on the *y*-axis and the residuals of *y* on the *x*-axis as shown in Figure 7.9. Alternatively, we place the residuals on the *y*-axis and the predicted *ŷ* values on the *x*-axis. Because the scatter plot does not suggest a nonlinear pattern, this assumption is met.
> - Homoscedasticity of residuals. To confirm this assumption, we examine the same scatter plot we used for the linearity assumption to determine if there is any discernible pattern. Examples of such patterns include the data "fanning out" to resemble the shape of a "V" on its side or a curvilinear form. From Figure 7.9, there appears to be no detected systematic trend (other than the linear relationship), and therefore this assumption is met.
> - Independence of residuals. The best strategy for detecting violations of this assumption is to examine a plot of the residuals vs. the case numbers. If there is no detectable/discernible pattern, then there is a good indication that the residuals are independent. To do so, we plot the residuals on the *y*-axis and the

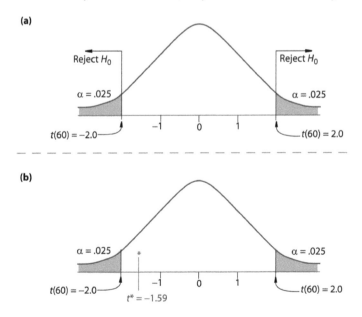

Figure 7.8 Illustration of the critical *t* values for Chapter 7's Guided Example (a), and the location of the calculated *t* statistic (b) with respect to the critical values.

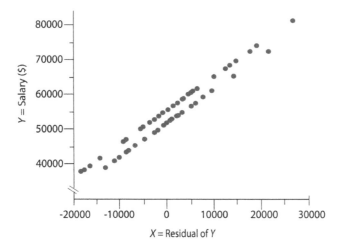

Figure 7.9 Residual plot that examines (and confirms) the linearity assumption for Chapter 7's Guided Example.

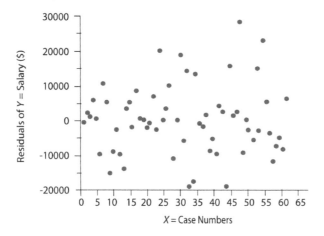

Figure 7.10 Residual plot that examines the independence of the residuals assumption for Chapter 7's Guided Example. The residual scores of Y are placed on the vertical axis and the case numbers (1–62) on the horizontal axis. The result is a scatter plot that depicts no discernible pattern.

case numbers on the x-axis. As shown in Figure 7.10, there appears to be no discernible pattern, and therefore, this assumption is met.

- Normality of residuals. To test for this assumption, we report the corresponding Shapiro–Wilk Goodness of Fit test. Recall from Chapter 6 that Shapiro-Wilk tests for normality has a null hypothesis of "The variable is normally distributed." Therefore, if $p > \alpha$, then we *fail to reject* H_0, which confirms normality. When applied to the Guided Example, $p = .2523$, and therefore, the normality assumption is met.

Run the Analysis and Report the Results. It is important to recognize that the biological sex variable is categorical and not continuous. Therefore, before we run the regression analysis, we must code this variable to make it continuous. To do

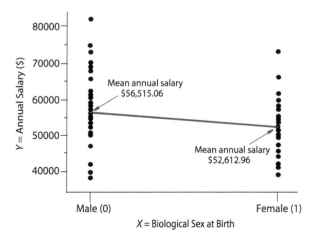

Figure 7.11 Scatterplot from the regression analysis for Chapter 7's Guided Example. The end-points of the regression line represent the mean annual salary for each group, respectively.

this we assign 1 to Female and 0 to Male. We now run the analysis and report the results from our statistics program:

- The scatter plot is given in Figure 7.11. We will make additional observations of the scatter plot and results in the Post-Data Analysis section after completion of the hypothesis test.
- The calculated $t*$ is $t(60) = -1.59, p = .1173$.
- The regression equation is $\hat{y} = -3902.094x + 56515.057$.
- The correlation coefficient is $r = -.20$.

Step 4. Decide whether to reject or fail to reject the null hypothesis, and make a concluding statement relative to the RQ.

Decision. Fail to reject H_0 because $t* = -1.59$ is not in the critical region as shown in Figure 7.8(b).

Concluding Remark. In the population of ATPs, and based on the sample data, biological sex does *not* have a significant effect on pilots' salaries.

Post-Data Analysis

We now examine various aspects of the findings. For regression analysis, this includes discussing and interpreting the regression equation, effect size, coefficient of determination, 95% confidence interval, the standard error of estimate, and post hoc power. We also discuss plausible explanations for the results.

The Regression Equation

The regression equation is $\hat{y} = -3902.094x + 56515.057$. The slope is $B = -3902.094$, and the y value of the y intercept is $B_0 = 56515.057$. From a generic perspective, B is interpreted as for every 1-unit change in x we can expect on average a B-unit change in y. However, because x represents a dichotomy with $0 =$ Males and $1 =$ Females, the only change in the slope is between 0 and 1. This can be seen from the scatter plot given in Figure 7.11. Notice how the slope of the line is negative: It decreases from Males to Females, which indicates that the average annual salary of males is higher than that of females. This can be confirmed by substituting the coded values into the regression equation:

- When we substitute 0 (Males) for x into the equation, we get, $\hat{y} = -3902.094(0) + 56515.057 = 0 + 56515.057 = 56515.057$, which is the regression constant. Thus, the predicted average salary for Males is $56,515.06.
- When we substitute 1 (Females) for x into the equation, we get $\hat{y} = -3902.094(1) + 56515.06 = -3902.094 + 56515.057 = 52612.963$. Thus, the predicted average salary for Females is $52,612.96.
- Note that the difference in mean salaries between Females and Males is $52,612.96 - 56,515.06 = -\underline{3,902.10}$, which is equal to the slope, B, of the regression equation, and indicates that female ATPs at the targeted LLC are earning, on average, $3,902.10 less than male ATPs at the targeted LLC. Thus, when performing a regression analysis involving a dichotomy where one group is coded 1 and the second group is coded 0, the regression coefficient, B, represents the difference in group means, and the regression constant, B_0, represents the mean of the group that was coded 0. For the current example, although male ATPs' mean salary is $3,902.10 more than female ATPs' mean salary, this mean difference is not statistically significant.

Effect Size

Recall from our sample size planning that in regression, the effect size is $f^2 = \dfrac{R^2}{1 - R^2}$. Based on the results of the analysis, $r^2 = .04$. Therefore, the effect size is $f^2 = .04/.96 = 0.042$, which according to Cohen (1998) is considered a small effect size.

Coefficient of Determination

Recall that the coefficient of determination, r^2, can be interpreted from either an explained variance or prediction perspective.

Explained Variance. $r^2 = .04$ means that 4% of the variance in salaries (y) is being explained by the variance in biological sex (x). Because the variance in biological sex is the difference between males and females, we would conclude that 4% of the variance in salaries is being explained by whether a pilot is male or female. This implies that 96% of the variance in salaries is being explained by some other factor(s).

Prediction. $r^2 = .04$ means that if we know the biological sex of a pilot (male or female), then we have 4% of the information needed to perfectly predict his or her annual salary.

The 95% CI

As reported from our statistics program, the 95% CI for β is [−8814.04, 1009.85], which tells us that 95% of the time we can expect the true effect biological sex has on salaries to lie within the interval [−8814.04, 1009.85]. In other words, 95% of the time we can expect female ATPs from the targeted LLC to earn anywhere from $8,814.04 *less* to $1,009.85 *more* than their male counterparts. Another way of looking at this is to say that if we were to randomly select 100 samples of size $n = 62$ from the same population, then in 95 of the samples, the regression coefficient would be between −8,814.04 and 1009.85. Furthermore, given H_0: $\beta = 0$, and because 0 is within the 95% CI, we fail to reject H_0. Finally, we conclude that based on the width of the 95% CI, the corresponding AIPE is not that good. Thus, the given data set does not provide an accurate estimate of the true effect biological sex has on salaries in the population.

The Standard Error of Estimate (Accuracy of Prediction)

The standard error of estimate is a metric that provides an idea of how far "off" we will be on average when the regression equation is used to predict y from x, and that one such index is the RMSE. Based on the results from our statistics program, $RMSE = 9586.92$, which indicates that if we use the regression equation to predict pilots' annual salary based on their biological sex, we would be "off" on average by approximately $9,600.

Power (Post Hoc)

To determine the approximate power of the study, we consult Soper (2022) with the following inputs:

- Number of predictors: 3
- Observed R^2: .04
- Probability level: .05
- Sample size: 62

This returns a power of .35883642, which indicates there is a 35.9% probability that the small effect found in the sample truly exists in the population.

Plausible Explanations for the Results

One plausible explanation is that the targeted LLC might have made a sincere effort to address the gender gap among its ATPs prior to the study being conducted. A second plausible explanation is sample size. Our a priori analysis was based on a medium effect size of $f^2 = 0.15$ but the sample data yielded an effect size of $f^2 = 0.04$, which means the sample size was insufficient. Consulting Statistics Kingdom's (2022) online calculator, a minimum sample size of $n = 199$ would be needed to detect an effect of $f^2 = 0.04$.

Chapter Summary

1. Regression is a statistical procedure for describing the nature of the relationship between two variables via a mathematical model called the regression equation. The graph of a regression equation is called the regression line, which has the general form $\hat{y} = Bx + B_0$: $\hat{y}$ is the predicted score; B is the regression coefficient, which is the

slope and reflects the amount of change in y relative to changes in x; and B_0 is the regression constant, which is equal to the predicted score when $x = 0$.

2. The data points in a scatter plot are called observed scores, and the points that fall on the regression line are called predicted scores. When a regression line is drawn within a scatter plot, there will be two y scores for every x score. One is the observed score (x, y), and the second is the corresponding predicted score $(x, \hat{y})$. The difference between an observed score and corresponding predicted score (i.e., $y - \hat{y}$) is the residual (error), which is the vertical distance between the two points.

3. To use a regression equation for prediction, we substitute an x score into the equation and perform the arithmetic. The result is the predicted score, $\hat{y}$, and its corresponding point $(x, \hat{y})$ fall on the regression line. The standard error of estimate ($SE_{Residuals}$), which is generally reported as the RMSE, provides an index of the accuracy of a prediction. When using a regression equation for prediction, we should not use scores of x that are outside the interval of predictability.

4. The effect size of a regression analysis is given by $f^2 = \dfrac{R^2}{1 - R^2}$. Effect sizes of $f^2 = 0.02$, 0.15, and 0.35 are considered small, medium, and large, respectively.

Vocabulary Check

Beta (β)	Regression equation
Effect size (f^2)	Regression line
Homoscedasticity of the residuals	Residual plots
Independence of the residuals	Residuals ($y - \hat{y}$)
Interval of predictability	Root mean square error (*RMSE*)
Least squares criterion	Slope
Normality of the residuals	Slope-intercept form
Observed scores	Standard deviation of regression
Ordinary least squares (OLS) regression	Standard deviation of the residuals
Perfect prediction	Standard error of estimate ($SE_{Residuals}$)
Predicted scores	Standard error of the regression coefficient (SE_B)
Prediction equation	y intercept
Regression coefficient (B)	y hat ($\hat{y}$)
Regression constant (B_0)	

Review Exercises

A. Check Your Understanding

In 1–10, choose the best answer among the choices provided.

1. The slope of the regression line
   a. is denoted B_0.
   b. is always positive.
   c. means that if the value of x increases by 1 unit, then y increases by the value of the slope units.
   d. is the value of the line where the line crosses the y-axis.

2. The y value of the y intercept, B_0, of a regression equation represents the
   a. average amount of change in y per unit change in x.
   b. average distance between observed and predicted scores.
   c. predicted value of y when $x = 0$.
   d. predicted value of y.

3. If the slope of a regression equation is $B = 0$, then the line will be displayed in a scatter plot as a
   a. vertical line.
   b. horizontal line.
   c. line that rises from left to right.
   d. line that falls from left to right.

4. The regression line involving a dichotomously coded nominal variable always
   a. has a slope that is denoted B_0.
   b. has a positive slope.
   c. has endpoints that represent the mean scores of the coded groups.
   d. is represented by an equation with a constant of 0.

5. From a conceptual perspective, the *RMSE* may be considered an estimator of
   a. σ
   b. R^2
   c. σ^2
   d. Y

6. Which correlation coefficient corresponds to a regression line with a negative slope?
   a. $r = 0$
   b. $r = .5$
   c. $r = -.45$
   d. $r^2 = 1$

7. Which statement about linear regression is NOT correct?
   a. The regression line maximizes the distance between observed and predicted scores.
   b. The regression line fits the data in the best place possible.
   c. A regression line can be drawn for a curvilinear relationship.
   d. The regression line visually shows how much y will change as a result of a change in x.

8. In a bivariate linear regression study, the researcher reported 95% CI β: $[-0.75, 3.25]$. What would a test for H_0: $\beta = 0$ vs. H_1: $\beta \neq 0$ conclude?
   a. Reject the null hypothesis at $\alpha = .05$ and all smaller α.
   b. Fail to reject the null hypothesis at $\alpha = .05$ and all smaller α.
   c. Reject the null hypothesis at $\alpha = .05$ and all larger α.
   d. Fail to reject the null hypothesis at $\alpha = .05$ and all larger α.

9. A hot air balloonist is interested in the relationship between the number of times he engages the blast valve per minute (y) and the temperature inside the envelope of the balloon. Based on the collected data, the least squares regression line is $\hat{y} = 3.41x + 10.53$, where x is the number of degrees by which the temperature exceeds 45° C (113° F). Which of the following best describes the meaning of the slope of the least squares regression line?
   a. For each increase in temperature of 1° C, the estimated number of times the balloonist engages the blast valve per minute increases by 10.53.

b. For each increase in temperature of 1° C, the estimated number of times the bal-loonist engages the blast valve per minute increases by 3.41.

c. For each increase in the number of times the balloonist engages the blast valve per minute, there is an estimated increase in temperature of 10.53° C.

d. For each increase in the number of times the balloonist engages the blast valve per minute, there is an estimated increase in temperature of 3.41° C.

10. A research study examined the extent to which education level (measured in total years of formal education) predicted career satisfaction among a group of commercial airline pilots. The results of data analysis are as follows: $n = 204$, $r^2 = .175$, $RMSE = .618$, $t(202) = -0.708$, $p = .48$, and $\hat{y} = -0.027x + 2.77$. Which of the following statements is true?

a. Education level provides nearly 20% of the information needed to perfectly predict a commercial airline pilots' career satisfaction score.

b. Education level is a major predictor of commercial airline pilots' career satisfaction.

c. The prediction equation suggests that pilots with more formal education are also more satisfied in their career.

d. The regression constant is interpretable in the context of the given research setting.

B. Apply Your Knowledge

Use the following research description and corresponding data set to conduct the activities given in Parts A–C.

The U.S. Air Force (USAF) periodically strengthens and refines its physical fitness testing standards. The most recent updates, which took effect January 2022, include a 1.5-mile (2.4 km) timed run, the number of push-ups completed in 1 minute, and the number of sit-ups completed in 1 minute (Airforce Magazine, 2021). The data file Ch_7 Exercises Part B Data contains a set of data acquired from the records of members of a local state-based Air National Guard, which makes up one arm of the reserve component to the USAF. The file consists of the group's timed run (in seconds) and number of push-ups and sit-ups. You are to develop a regression model to predict the targeted group's timed run by the number of push-ups to see if in the next evolution of the USAF's changes to its physical fitness standards can eliminate the timed run component. Your analysis is to be done as follows (see also Section 7.5):

A. Pre-Data Analysis

1. Specify the research question and corresponding operational definitions.
2. Specify the research hypothesis.
3. Determine the appropriate research methodology/design and explain why it is appropriate.
4. Conduct an a priori power analysis to determine the minimum sample size needed. Compare this result to the size of the given data set and explain what impact the size of the given sample will have on the results relative to the minimum size needed.

Data Analysis

1. Conduct a hypothesis test of the regression coefficient by applying all four steps as presented in Sections 7.4 and 7.5.
2. Conduct an outlier analysis (there should be four outliers) and redo the analysis in the absence of the outliers.
3. Compare the results between outliers present and outliers absent. Determine if you will keep or delete the outliers and explain why.

C. Post-Data Analysis

1. Report and interpret the regression equation, including B and B_0.
2. Determine and interpret the effect size (f^2).
3. Determine and report the coefficient of determination (r^2) from both explained variance and prediction perspectives.
4. Determine and interpret the 95% confidence interval, including its precision and AIPE.
5. Determine and interpret the standard error of estimate.
6. Determine and interpret the power of the study from a post hoc perspective.
7. Use the regression equation to predict a person's timed run score if the number of push-ups is 50.
8. Present at least two plausible explanations for the results.
9. Interpret the findings from a practical perspective that could be used to help inform the USAF relative to keeping or eliminating the timed run component.

References

Byers, D. L. (2004). *The making of the modern airport executive: Causal connections among key attributes in career development, compromise, and satisfaction in airport management.* (Publication No. 3664585) [Doctoral dissertation, Florida Institute of Technology]. ProQuest Dissertations and Theses Global.

Cohen, J. (1988). *Statistical power analysis for the behavioral sciences* (2nd ed.). Lawrence Erlbaum Associates.

Hadley, G. (2021, November 12). Here are the scoring charts for the Air Force's new PT test exercises—minus the walk. *Airforce Magazine.* https://www.airforcemag.com/air-force-pt-test-scoring-alternate-exercises/

Sheppard, R. (2013). *Examining the relationship between aviation fuel costs and airline ticket prices.* [Unpublished manuscript for AVT 4002]. College of Aeronautics, Florida Institute of Technology.

Soper, D. (2022). *Post-hoc statistical power calculator for multiple regression.* https://www.danielsoper.com/statcalc/calculator.aspx?id=9

Statistics Kingdom (2022). *Regression and ANOVA—Sample size calculator.* https://www.statskingdom.com/sample_size_regression.html

Part D

Analyzing Research Data Involving Two Independent Samples

8 Independent-Samples *t* Test

Student Learning Outcomes

After studying this chapter, you will be able to do the following with respect to analyzing data from two independent samples:

1. Identify the group membership variable both generally and from the perspective of a single factor with n levels.
2. Determine if the homogeneity of variances assumption is met and which independent-samples *t* test to use—pooled or separate variances—relative to this result.
3. Determine and interpret the 95% CI for the difference in two group means.
4. Conduct and interpret the results of a hypothesis test for the difference in two group means.
5. Engage in pre-data analysis, data analysis, and post-data analysis activities for a study involving two independent samples as described in Section 8.4.

8.1 Chapter Overview

In the previous three chapters, we focused on statistical strategies for analyzing data from research studies that consisted of a single sample. Although such studies are commonly used in research, they are limited because their focus is on a single group and hence cannot be used, for example, to examine the difference in time to PPL between flight students enrolled in a Part 141 vs. a Part 61 program. In this chapter—the first of three that deals with two or more groups—we introduce the independent-samples *t* test, which is an extension of the single-sample *t* test from Chapter 5, and apply previously developed statistical concepts such as confidence intervals and hypothesis testing to this statistical strategy. We conclude the chapter with a guided example that demonstrates the use of the independent-samples *t* test for a research study that examines the difference in mean scores between two independent groups.

8.2 The Concept of Independent Samples

The Group Membership Variable

In Chapter 5, we compared the mean FAA IRA exam score of students who prepared for the exam using BIM's software to the population mean. The focus of the study was

DOI: 10.4324/9781003308300-12

to determine the effectiveness of BIM's software by examining how well BIM-prepped students performed on the IRA exam relative to the national average. In this setting, we have one group (BIM-prepped students) and one measure (mean IRA exam scores).

Instead of focusing on one group, though, a more practical study would involve comparing BIM-prepped students' IRA exam scores to those of students who used a competing test preparation software package. We will call this competitor GWS. When considered from this perspective, we now have two groups (BIM- and GWS-prepped students), one measure (mean IRA exam scores), and we are examining the effectiveness of competing products by comparing the two groups' respective mean IRA scores. Because the respective means are from two different sets of IRA exam scores, which come from two separate groups of flight students, we refer to the two groups as independent samples, which are selected randomly from two different respective populations.

When a study involves more than one group, it automatically includes a group membership variable, which is measured on a nominal scale. For the IRA exam study, the group membership variable represents two distinct groups that differ on the test preparation software used to prepare for the exam. A group membership variable also is referred to as a single factor with n levels, where n is equal to the number of groups. For the IRA exam study, we have a single factor with two levels: The single factor is "type of software package used," and the two levels are the BIM and GWS groups.

Example 8.1: Group Membership

In the Chapter 8 Overview, we presented an example of a group membership variable relative to a study involving time to PPL. Specify the group membership variable and express it from the perspective of a "single factor with n levels," and identify what is being measured.

Solution

- The group membership variable is "Type of Flight Training Program," which is a single factor, and the two levels are Part 141 and Part 61.
- The measurement involves comparing Part 141 students' mean time to PPL vs. Part 61 students' mean time to PPL.

When considered from a research design perspective, studies that consist of a group membership variable are called between-group designs because two (or more) separate sets of data and their respective group means are being compared between groups. Both experimental and causal-comparative studies consist of a group membership variable, and hence they are between-group designs. Although between-group design studies often involve more than two groups and place no restriction on the data type of the dependent variable, we restrict our discussion in this chapter to studies involving two independent groups with a continuous dependent variable.

The t Statistic for Independent Samples

The statistical strategy for examining differences in group means between two groups is the independent-samples *t* test, which is an extension of the single-sample *t* test from Chapter 5. The corresponding test statistic that will be used for confidence intervals and hypothesis testing can be derived from the single-sample *t* statistic

$$t = \frac{M - \mu}{SE_M}$$

where

- *M* is the mean of the sample
- μ is the corresponding population mean
- SE_M is the standard error of the mean, which is equal to $\dfrac{SD}{\sqrt{n}}$

Because the independent-samples *t* test involves two independent samples and examines the difference in group means, we modify the structure of the single-sample *t* statistic formula accordingly:

$$t = \frac{(M_1 - M_2) - (\mu_1 - \mu_2)}{SE_{(M_1 - M_2)}}$$

where

- M_1 and M_2 are the respective means of the two groups
- μ_1 and μ_2 are the respective population means
- μ is the estimated standard error of the difference in group means

Focusing on the standard error for a moment, recognize from Chapter 3 in our discussion of sampling distributions that, in general, standard error represents the average difference between a sample statistic and the corresponding population parameter. When applied to the independent-samples *t* test, the sample statistic is the difference in sample means (M_1 – M_2) and the population parameter is the difference in population means (μ_1– μ_2). So, the estimated standard error is an index that tells us how far off, on average, the *difference* between the two sample means is from the actual difference between the two corresponding population means.

When applying the *t* test statistic to an independent-samples analysis, there are different formulas for calculating the independent-samples *t* statistic. One is called the *pooled variances t test* and the second is called the *separate variances t test*. The difference between these formulas is with respect to the calculation of standard error, and their use is contingent on whether the data set is compliant with the *homogeneity of variance assumption*, which we discuss next. Table 8.1 contains a summary of the two formulas.

Assumptions of the t Statistic for Independent Samples

Recall from Chapter 5 that the two primary assumptions of the single-sample *t* statistic are *independent observations*, which are met if the data are randomly sampled,

Table 8.1 Pooled and Separate Variances Formulas for the Independent-Samples *t* Test

Type of t Test	Formula	Comments
Pooled Variances t Statistic	$t = \dfrac{M_1 - M_2}{SE_{(M_1 - M_2)}}$ where $SE_{(M_1 - M_2)} = \sqrt{\dfrac{SD_P^2}{n_1} + \dfrac{SD_P^2}{n_2}}$	• The standard error term, $SE_{(M_1 - M_2)}$, uses what is known as the *pooled variance*, denoted SD_P^2, and is obtained by averaging (i.e., "pooling") the variances (SD^2) of the two samples into a single overall estimate. • The formula for the pooled variance is $$SD_P^2 = \frac{SD_1^2(n_1 - 1) + SD_2^2(n_2 - 1)}{(n_1 - 1) + (n_2 - 1)}$$ where SD_1^2 and SD_2^2 represent the variance associated with sample 1 and sample 2, respectively. • The formula for the pooled variance may be re-expressed as $SD_P^2 = \dfrac{SS_1 + SS_2}{df_1 + df_2}$ • If the two samples are of equal size ($n_1 = n_2$), then the pooled variance formula is algebraically reduced to $$SD_P^2 = \frac{SD_1^2 + SD_2^2}{2}.$$ • The overall *df* of the pooled variances *t* statistic is $df = n_1 + n_2 - 2$.
Separate Variances t Statistic	$t = \dfrac{M_1 - M_2}{SE_{(M_1 - M_2)}}$ where $SE_{(M_1 - M_2)} = \sqrt{\dfrac{SD_1^2}{n_1} + \dfrac{SD_2^2}{n_2}}$	• The difference between this formula and the pooled variances formula is with the standard error term. Instead of using pooled variances, the standard error is with respect to the variances of each sample. • The *df* for this statistic is complex and can be a decimal, but it always will be less than $n_1 + n_2 - 2$.

and *normality*, which is met if the parent population is either normally distributed or sufficiently large to apply the central limit theorem. These assumptions also apply to the independent-samples *t* statistic but are modified to reflect the concept of independent samples:

- *Independent observations.* The scores in the first sample must be independent, and the scores in the second sample must be independent. This assumption is satisfied if a random sample of size n_1 is selected from the first population, and a random sample of size n_2 is selected from the second population.
- *Normality.* The respective parent populations must be normally distributed. If not, then the size of each sample must be sufficiently large for the central limit theorem to apply. From a general perspective, this means that $n_1 \geq 30$ and $n_2 \geq 30$. To test for normality, we examine the results from the Shapiro–Wilk Goodness of Fit test separately for each sample as we did in Chapter 7.

In addition to these assumptions, the independent-samples *t* test also has one additional assumption.

- *Homogeneity of variances.* Because we now have two independent samples drawn from two separate populations, we must ensure that the variances in these two populations are statistically equivalent. More concretely:

  The variances of the scores must be equal, or homogeneous, across the two populations that correspond to the samples that are being compared.

  This is known as the homogeneity of variances assumption (or more simply, *equal variances*). Noncompliance to this assumption is acceptable if the samples are of equal size (i.e., $n_1 = n_2$). However, in the presence of disparate sample sizes, if this assumption is not met, then the result of a hypothesis test for the difference between two group means will be biased because the alpha level could deviate considerably from its preset threshold level. This could lead to either an underestimated significance level that could result in incorrectly rejecting the null hypothesis or an overestimated significance level that could result in a lower power of the test.

To check for the equal variances assumption, we use the *Levene test*, which is based on a hypothesis test of the two population variances:

$$H_0 : \sigma_1^2 = \sigma_1^2 \text{ (Population variances are statistically equal)}$$

$$H_1 : \sigma_1^2 \neq \sigma_1^2 \text{ (Population variances are not statistically equal)}$$

If the result is *fail to reject* H_0, then the data set is compliant with the equal variances assumption; otherwise, it is not compliant. The Levene test result is displayed by most statistics programs as part of the output for an independent-samples t test. Because most statistics programs use the p-value approach to significance, if p is greater than the preset alpha level ($p > \alpha$), then the equal variance assumption is satisfied. If, however, $p < \alpha$, then the assumption is not satisfied.

In addition to indicating if the homogeneity of variances assumption is met, the result of the Levene test also determines which independent-samples t test formula is used. If the data set is compliant with the equal variances assumption, then we use the pooled variances t statistic; otherwise, we use the separate variances t statistic. It also is worth noting that many statistical programs provide a third option independent of the pooled and separate variances t statistic. One such option is called *Welch's t statistic*, which does not assume equal population variances and hence can be used regardless of the outcome of the Levene test.

8.3 Statistical Inferences Involving Independent Samples

Statistical inference can be approached either directly via confidence intervals or indirectly via hypothesis testing. When applied to the independent-samples t test, the population parameter is the mean, μ, and the focus is with respect to the difference in group means, $\mu_1 - \mu_2$, where μ_1 = the mean of the first population and μ_2 = the mean of the second population. We now examine CIs and hypothesis testing with respect to independent samples.

Confidence Interval for the Difference in Means

The general form of a CI is:

$$\text{Population Parameter} = \text{Point Estimate} \pm (SE)(\text{Critical } t \text{ Score})$$

When applied to an independent-samples study:

- The point estimate is the difference in the sample means, $M_1 - M_2$.
- SE is the standard error of the difference in the sample means, $SE_{(M_1 - M_2)}$.
- The critical t score is acquired from Table 2/Appendix A for $df = n_1 + n_2 - 2$.

As a result:

$$95\% \text{ CI for } (\mu_1 - \mu_2) = [a, b]$$

where

- a is the lower bound and is equal to $(M_1 - M_2) - (SE_{(M_1 - M_2)})(t_{critical})$
- b is the upper limit and is equal to $(M_1 - M_2) + (SE_{(M_1 - M_2)})(t_{critical})$

When calculating standard error, there are two different standard error formulas for an independent-samples t test (see Table 8.1): one for pooled variances when the equal variances assumption is met, and one for separate variances when the equal variances assumption is not met.

To illustrate the construction of the 95% CI for the difference in sample means, we will use the data set given in the file "Ch_8 Time to PPL Data." This data set consists of 20 randomly selected flight students' times to PPL (in hours) acquired from a local flight school. Of these 20 students, 10 were trained using a "traditional" approach (Group 1), which involved the CFI and the flight student, and 10 were trained using a "cohort" approach (Group 2), which involved the CFI and two flight students who alternated as pilot and observer during instruction. Thus, the dependent variable is time to PPL in hours, and the independent variable is group membership—Type of Flight Instruction—and the two levels are Traditional and Cohort.

In practice, we do not manually calculate the 95% CI but instead report it from our statistics program. However, for instructional purposes, we will demonstrate how the CI is constructed. To facilitate this construction, we prepared Table 8.2, which contains the raw data for each group along with key summary statistics. From Table 8.2, note the following:

- We partitioned the data from the given data set into two separate samples (Traditional and Cohort).
- The mean times to PPL for the Traditional and Cohort groups are $M_1 = 56.35$ hours and $M_2 = 44.68$ hours, respectively. Therefore, the point estimate is $(M_1 - M_2) = 56.35 - 44.68 = 11.67$.
- The sample size of each group is the same, $n_1 = n_2 = 10$. Therefore, $df = n_1 + n_2 - 2 = 10 + 10 - 2 = 20 - 2 = 18$.
- The critical t value from Table 2/Appendix A for $df = 18$ is 2.101.

Table 8.2 Time to PPL Data Used for Constructing the 95% CI for the Difference in Sample Means

Type of Flight Instruction (X)	Time to PPL (Y)	$Y - M_Y$	$(Y - M_Y)^2$	Type of Flight Instruction (X)	Time to PPL (Y)	$Y - M_Y$	$(Y - M_Y)^2$
Traditional	49.4	−6.95	48.303	Cohort	36.4	−8.28	68.558
Traditional	40.8	−15.55	241.8	Cohort	39.5	−5.18	26.832
Traditional	69.2	12.85	165.12	Cohort	42.8	−1.88	3.5344
Traditional	56.7	0.35	0.1225	Cohort	43.4	−1.28	1.6384
Traditional	52.5	−3.85	14.823	Cohort	43	−1.68	2.8224
Traditional	54.1	−2.25	5.0625	Cohort	41	−3.68	13.542
Traditional	55.4	−0.95	0.9025	Cohort	59.5	14.82	219.63
Traditional	79.3	22.95	526.7	Cohort	43.3	−1.38	1.9044
Traditional	45.7	−10.65	113.42	Cohort	53.7	9.02	81.36
Traditional	60.4	4.05	16.402	Cohort	44.2	−0.48	0.2304
Sum of Group 1	563.5	SS_1	1132.6555	Sum of Group 2	446.8	SS_2	420.052
n_1	10			n_2	10		
M_1	56.35			M_2	44.68		
SD_1	11.22			SD_2	6.83		

We now focus on standard error. Referencing Table 8.1, note that if the homogeneity of variance assumption is satisfied, then to calculate standard error, we first need to calculate the pooled variance. Further note in the Comments section for the pooled variance t that the second, third, and fourth bullet items contain different formulas for calculating the pooled variance. Now look at the separate variances row of Table 8.1. Here, there is only one formula for standard error. It is important to recognize that regardless of which formula is used to calculate standard error (pooled or separate variances), the result will be the same. This is illustrated below using the data from Table 8.2, which contains all the needed information to perform these calculations.

Calculating Pooled Variances $\left(SD_p^2\right)$ Using the Three Different Formulas

- $$\frac{SD_1^2(n_1-1)+SD_2^2(n_2-1)}{(n_1-1)+(n_2-1)} = \frac{11.22^2(9)+6.83^2(9)}{9+9} = \frac{1133+420}{18} = \frac{1553}{18} = 86.28$$

- $$\frac{SS_1+SS_2}{df_1+df_2} = \frac{1132.6555+420.052}{9+9} = \frac{1552.7075}{18} = 86.26$$

- $$\frac{SD_1^2+SD_2^2}{2} = \frac{11.22^2+6.83^2}{2} = \frac{125.8884+46.6489}{2} = \frac{172.5373}{2} = 86.27$$

Thus, the pooled variance is $SD_p^2 = 86.3$.

Calculating Standard Error $\left(SE_{(M_1 - M_2)}\right)$

- *Pooled Variances Formula with* $SD_p^2 = 86.3$

$$\sqrt{\frac{SD_P^2}{n_1} + \frac{SD_P^2}{n_2}}$$

$$= \sqrt{\frac{86.3}{10} + \frac{86.3}{10}}$$

$$= \sqrt{8.63 + 8.63}$$

$$= \sqrt{17.26}$$

$$= 4.15$$

- *Separate Variances Formula*

$$\sqrt{\frac{SD_1^2}{n_1} + \frac{SD_2^2}{n_2}}$$

$$= \sqrt{\frac{11.22^2}{10} + \frac{6.83^2}{10}}$$

$$= \sqrt{\frac{125.8884}{10} + \frac{46.6489}{10}}$$

$$= \sqrt{12.5884 + 4.66489}$$

$$= \sqrt{17.25329}$$

$$= 4.15$$

Thus, the standard error is 4.15.

We now calculate the 95% CI where $(M_1 - M_2) = 11.67$, $SE_{(M_1 - M_2)} = 4.15$, and $t_{Critical} = 2.101$.

- Lower bound: $11.67 - (4.15)(2.101) = 11.67 - 8.72 = 2.95$
- Upper bound: $11.67 + (4.15)(2.101) = 11.67 + 8.72 = 20.39$

Thus, the 95% CI for $(\mu_1 - \mu_2) = [2.95, 20.39]$.

Interpreting this CI, we would report that with 95% certainty, we can expect that the true difference between the groups' mean time to PPL will lie within the interval [2.95, 20.39]. In other words:

95% of the time we expect the Traditional group's mean time to PPL will be between 2.95 and 20.39 more hours than the Cohort group's mean time to PPL.

Alternatively:

If we were to randomly select 100 samples of size 10 separately from the Traditional and Cohort flight instruction populations, then in 95 of these samples, the mean difference in time to PPL between the two groups will be between 2.95 and 20.39 hours.

In short, it appears that the Cohort instructional approach is more effective because it takes, on average, between 3 and 20 *fewer* hours to complete PPL training compared to the Traditional approach. Furthermore, because 0 is not within this interval, we would reject the corresponding null hypothesis, which would posit that the difference between the groups' mean time to PPL is zero. Because the width of this CI is relatively large (it's a 17-hour range), the corresponding accuracy in parameter estimation (AIPE) is not very good, which suggests that the sample data are not robust to accurately estimate the true difference between group means in the respective populations.

Hypothesis Testing for the Difference in Means

To conduct a hypothesis test using independent-samples t test, we follow the same four-step procedure we presented in our previous work except now the context is with respect to the difference between two group means, $M_1 - M_2$. As a result, the following changes are noted:

- The population parameter will be μ and will focus on mean differences, $\mu_1 - \mu_2$.
- The purpose of the test is to determine if the difference in sample means reflects more than a chance deviation from a hypothesized difference in the corresponding population means. If it does, then we can infer that the difference in sample means also exists in the population.
- Because we are examining the difference between two sample means, the null hypothesis will claim that in the population there is no significant difference between the two means. This infers that the difference is zero, or equivalently, that they are equal. Thus, the null hypothesis is: H_0: $\mu_1 - \mu_2 = 0$, which is algebraically equivalent to H_0: $\mu_1 = \mu_2$.
- The test statistic in Step 2, which establishes the critical boundary value, is t.

We now conduct a hypothesis test for the difference in group means using the time to PPL data given in Table 8.2 as well as in the file Ch_8 Time to PPL Data.

Step 1. Identify the targeted parameter, and formulate the null and alternative hypotheses. The targeted parameter is the mean, and the null and alternative hypotheses are:

H_0: $\mu_1 - \mu_2 = 0$. There is no significant difference in time to PPL between flight students trained via a Traditional approach vs. flight students trained via a Cohort approach.

H_1: $\mu_1 - \mu_2 \neq 0$. There is a significant difference in time to PPL between flight students trained via a Traditional approach vs. flight students trained via a Cohort approach. Because the direction is unknown, we apply a two-tailed test.

Step 2. Determine the test criteria. The test criteria involve three components:

Level of Significance (α). We will set alpha to $\alpha = .05$.

Test Statistic. The test statistic is t.

$$t^* = \frac{(M_1 - M_2) - (\mu_1 - \mu_2)}{SE_{(M_1 - M_2)}}$$

The Critical Boundary. The critical t value is acquired from Table 2/Appendix A. Because $n_1 = n_2 = 10$, $df = 10 + 10 - 2 = 18$, and the corresponding critical t value for a two-tailed test for $\alpha = .05$ is $t(18) = \pm 2.101$. This is shown in Figure 8.1(a).

Step 3. Collect data, check assumptions, run the analysis, and report the results.

Collect Data. The data are in the file Ch_8 Time to PPL Data.

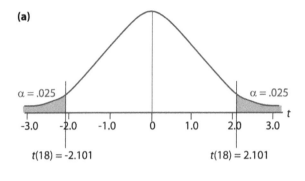

(a)

$\alpha = .025$ $\alpha = .025$

-3.0   -2.0   -1.0   0   1.0   2.0   3.0 t

$t(18) = -2.101$ $t(18) = 2.101$

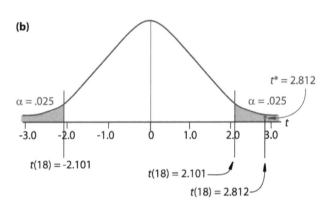

(b)

$t^* = 2.812$

$\alpha = .025$ $\alpha = .025$

-3.0   -2.0   -1.0   0   1.0   2.0   3.0 t

$t(18) = -2.101$

$t(18) = 2.101$ ⟋

$t(18) = 2.812$ ⟋

Figure 8.1 Illustration of the critical *t* values for the time to PPL data given in Table 8.2 (a), and the location of the calculated *t* statistic with respect to the critical *t* values (b).

Check Assumptions. There are three assumptions:
- Independence. The data for each sample were randomly selected from their respective populations. Therefore, this assumption is met.
- Normality. We test for this assumption via the Shapiro–Wilk Goodness of Fit test. For the current study, $p = .6287$ for the Traditional Group and $p = .0355$ for the Cohort group. Thus, the Traditional group's data set is compliant with this assumption, but the Cohort group's data set is not. However, the Cohort group has an outlier (59.5 hours). If removed, the Cohort data set is compliant with the normality assumption. We will report this and continue with the analysis.
- Homogeneity of variances. The Levene test result reported from our statistics program is $F(1, 18) = 1.42, p = .2490$, which confirms that the variances in the two populations from which the samples were selected are statistically equal. The Levene test also holds if the outlier from the Cohort group is removed from the analysis.

Run the Analysis and Report the Results. In practice, we will run the analysis using our statistics program and record the results. However, for instructional purposes, we manually calculate the *t* value. Because the homogeneity of variances assumption is met, we use the pooled variances *t* test:

$$t = \frac{(M_1 - M_2) - (\mu_1 - \mu_2)}{SE_{(M_1 - M_2)}}$$

Note from the formula that $\mu_1 - \mu_2$ will be 0 because the null hypothesis claims there will be no significant difference between the two groups in the population. Further note that M_1, M_2, and standard error were calculated earlier as part of our discussion on confidence intervals. Therefore, this leads to

$$t^* = \frac{(M_1 - M_2)}{SE_{(M_1 - M_2)}} = \frac{(56.35 - 44.68)}{4.15} = \frac{11.67}{4.15} = 2.812$$

This result is depicted in Figure 8.1(b).

Step 4. **Decide whether to reject or fail to reject the null hypothesis, and make a concluding statement relative to the RQ.**

Decision. The calculated t^* is $t(18) = 2.812$, which is greater than $t_{critical} = 2.101$ and lies in the critical region. Therefore, the decision is to *reject the null hypothesis.*

Concluding Statement. Relative to the preset significance level of $\alpha = .05$, the sample data provide sufficient evidence that the mean time to PPL for the Traditional group was significantly higher than that of the Cohort group. More specifically, the Traditional group averaged 11.67 *more* hours to PPL than the Cohort group, and this 11.67-hour difference was statistically significant, $t(18) = 2.101$, $p = .0116$. Thus, based on sample data, the Cohort approach to flight instruction is more effective than the Traditional approach.

8.4 Using the Independent-Samples *t* Test in Research: A Guided Example

We now provide a guided example of a research study that demonstrates the use of the independent-samples *t* test. This study is an extension of the one given in Chapter 5's Guided Example, which examined the effect of using BIM's test preparation software package to prepare for the 60-item FAA IRA exam relative to the national average. In this extension, we introduce a competing package, GWS, and will compare BIM-prepped students' mean score to GWS-prepped students' mean score to see which software package is more effective in preparing students to take the IRA exam. The data for this guided example are provided in the file Ch_8 Guided Example Data. We will structure this example into three distinct parts: (a) pre-data analysis, (b) data analysis, and (c) post-data analysis.

Pre-Data Analysis

Research Question and Operational Definitions

The overriding research question is: "What is the difference in mean FAA IRA exam scores between students who prepare for the exam using BIM's software and students who prepare for the exam using GWS's software?" In the context of the current study, IRA exam scores refer to the number of items students answer correctly on the 60-item, dichotomously-scored FAA IRA exam. Thus, scores could range from 0 to 60.

Research Hypothesis

The difference in group means will not be zero. More specifically, we hypothesize that the mean score of one group either will be higher or lower than the mean score of the second group. We are uncertain of the direction for any specific package, though.

Research Methodology

The research methodology that would best answer this question is *true experimental*. We will select a sample of flight students and randomly assign them to one of two groups. Group 1 will prepare for the FAA IRA exam using BIM's software, and Group 2 will prepare for the exam using GWS' software. After a specific study period, both groups will then take the IRA exam at the same time, and we will record their scores. (*Note:* This study also could be conducted using an ex post facto design.)

Sample Size Planning (Power Analysis)

To determine the minimum sample size needed, we consult Statistics Kingdom's (2022a) online calculator and enter the following:

- Tails: Two (H_1: $\mu = \mu_0$)
- Distribution: Student's t
- Significance level (α): .05
- Effect: Medium
- Effect size: 0.5
- Digits: 4
- Sample: Two samples
- Power: .8
- Effect type: Cohen's d
- Standard deviations: Equal σ

The last entry assumes equal variances, which means we believe the data set will be compliant with the homogeneity of variances assumption. These entries return a minimum sample size of $n = 64$ per group for power $= .80146$. This means to have an 80.15% probability of correctly rejecting the null hypothesis, where the effect observed in the sample also exists in the population, we need at least 128 participants with 64 in each group. Because the given data set has $n = 100$ participants per group, this suggests we will have a greater than 80.15% of correctly rejecting H_0, assuming a medium effect size.

Data Analysis

We now analyze the given data set relative to a hypothesis test of the difference in two group means.

Step 1. Identify the targeted parameter, and formulate null/alternative hypotheses. The targeted parameter is the mean, and the null/alternative hypotheses are: H_0: $\mu_1 = \mu_2$. There will be no significant mean difference in exam scores between students who prepare for the FAA IRA exam using BIM's software (Group 1) and students who prepare for the exam using GWS's software (Group 2). (The difference will be zero.)

$H_1: \mu_1 \neq \mu_2$. There will be a significant mean difference in exam scores between students who prepare for the FAA IRA exam using BIM's software (Group 1) and students who prepare for the exam using GWS's software (Group 2). One group's mean will be significantly higher (or lower) than the other group's mean.

Step 2. Determine the test criteria. The test criteria involve three components:

Level of Significance (α). We will set alpha to $\alpha = .05$.

Test Statistic. The test statistic is t.

The Critical Boundary. With $n_1 = n_2 = 100$, $df = n_1 + n_2 - 2 = 198$, and a two-tailed test for $\alpha = .05$, there is no entry in Table 2/Appendix A. Therefore, we use Omni Calculator's (n.d.) online critical t-value calculator, which yields a critical t value of $t(198) = \pm 1.972$. This is illustrated in Figure 8.2(a).

Step 3. Collect data, check assumptions, run the analysis, and report the results.

Collect Data. The data are in the file Ch_8 Guided Example Data.

Check Assumptions. There are three assumptions for the independent-samples t test.
- Independence. We assume the scores in each sample are independent because the data were collected randomly.
- Normality. The results of the Shapiro–Wilk Test for Goodness of Fit are $p = .0051$ for the BIM group and $p < .0001$ for the GWS group. Thus, neither

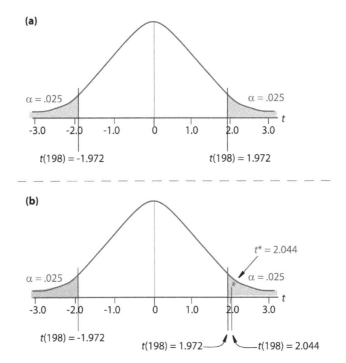

Figure 8.2 Illustration of critical t values for Chapter 8's Guided Example (a), and the location of the calculated t value with respect to the critical t values (b).

data set is compliant with this assumption. However, because each sample size is $n = 100$, we can apply the Central Limit Theorem. Therefore, we conclude that the data set is compliant with this assumption.

- Homogeneity of Variances. The Levene test result reported from our statistics program is $F(1, 198) = 1.59, p = .2088$, which confirms that the variances in the respective parent populations are statistically equal.

All assumptions have been met.

Run the Analysis and Report the Results. Because the homogeneity of variances assumption has been met, we will run a pooled variances t test and review the results, which are summarized below.

- $M_{BIM} = 43.66, M_{GWS} = 40.89$.
- The difference in group means is $43.66 - 40.89 = 2.77$ in favor of BIM: The BIM group averaged 2.77 points higher on the FAA IRA exam than the GWS group.
- The calculated t statistic is $t* = 2.043967$, which falls in the critical region as illustrated in Figure 8.2(b).
- The corresponding p value is $p = .0423$.

Step 4. Decide whether to reject or fail to reject the null hypothesis, and make a concluding statement relative to the RQ.

Decision. Reject H_0 because $t* = 2.044$ is in the critical region.

Concluding Remark. Flight students who prepared for the FAA IRA exam using BIM's software averaged 2.77 points higher on the exam than flight students who prepared for the exam using GWS's software. This 2.77-unit difference was statistically significant, $t(198) = 2.04, p = .0423$.

Post-Data Analysis

We now examine the effect size, 95% CI, post hoc power, and plausible explanations.

Effect Size

As discussed in Chapter 5, there are two ways to measure the effect size involving a t test: Cohen's estimated d, which is with respect to standard deviation, and r^2, which is with respect to explained variance.

Cohen's estimated d. When applied to a single-sample t test, Cohen's estimated d reflects the comparison between a sample mean and the population mean:

$$\text{Estimated } d = \frac{M - \mu}{SD} \text{ (for single-sample } t \text{ test)}$$

However, when we apply this to an independent-samples t test, the numerator now reflects the difference in the two sample means, and the denominator reflects the "pooled" standard deviation of both samples combined, denoted SD_P:

$$\text{Estimated } d = \frac{M_1 - M_2}{SD_P} \text{ (for independent-samples } t \text{ test)}$$

To determine the pooled standard deviation, we take the *square root of the pooled variance* (recall that standard deviation is equal to the square root of variance). This pooled variance is provided in the Analysis of Variance (ANOVA) table that is part of the output provided by our statistics program. We will discuss ANOVA tables in Chapter 9. For the moment, though, SD_p is the numerical value associated with the "Mean Square" column of the "Error" row as highlighted in Table 8.3. As a result:

$$\text{Estimated } d = \frac{M_1 - M_2}{SD_p} = \frac{43.66 - 40.89}{\sqrt{91.829}} = \frac{2.77}{9.58} = 0.289$$

Thus, students who used BIM's software to prepare for the FAA IRA exam averaged about three-tenths of a standard deviation higher than students who used GWS's software. This is a relatively small treatment effect.

Explained variance (r^2). To determine how much of the variability in the IRA exam scores is being explained by the group membership variable (i.e., the comparison between BIM's and GWS's software packages), we use the r^2 formula presented in Chapter 5 as follows:

$$r^2 = \frac{t^2}{t^2 + df} = \frac{2.04^2}{2.04^2 + 198} = \frac{4.1616}{4.1616 + 198} = \frac{4.1616}{202.1616} = .02$$

Thus, 2% of the variability in the IRA exam scores is being explained by the group membership variable, which means 98% of the variability is being explained by other factors. Based on Cohen's metric-free units, $r^2 = .02$ is considered a small effect.

The 95% Confidence Interval

The lower and upper bounds of the 95% CI reported by our statistics program are 0.09751 and 5.44249, respectively, which we round to two decimal places:

$$95\% \text{ CI}:(\mu_1 - \mu_2) = [0.10,\ 5.44]$$

This tells us that 95% of the time we can expect the true mean difference between the two groups' mean exam scores to lie within the interval [0.1, 5.44]. In other words, 95% of the time we can expect BIM's mean exam score to range from one-tenth of a point higher to as much as 5.44 points higher than GWS's mean exam score. Another way of looking at this is to say that if we were to randomly select two samples of size 100 each—one sample from the population using BIM's software, and one sample from the population using GWS's software—then in 95 of these samples, BIM's mean exam score would range between 0.10 and 5.44 units higher than GWS's mean exam score. Furthermore, because 0 is not within this interval, we would reject the corresponding null hypothesis, which claims that the difference between the groups' mean score is zero. Because the width of this CI is

Table 8.3 Analysis of Variance (ANOVA) Table for Independent-Samples *t* Test Guided Example

Source	df	Sum of Squares (SS)	Mean Square (MS)	F	p
Test Prep Software	1	383.645	383.645	4.1778	.0423
Error	198	18182.230	91.829		
C. Total	199	18565.875			

comparatively small (it's a little more than a 5-point difference with respect to a 60-point scale), the corresponding AIPE is relatively good. This suggests that the sample data provide a reasonable estimate of the true difference between group means in the population.

Power (Post Hoc)

To determine the power of the study, we consult Statistics Kingdom's (2022b) online power calculator and enter the following inputs:

- Tails: Two (H_1: $\mu = \mu_0$)
- Distribution: T
- Significance level (α): .05
- Effect type: Standardized effect size
- Standard deviations: Equal σ
- n_2: 100

- Digits: 4
- Sample: Two samples
- Effect: Small
- Effect size: 0.289
- n_1: 100

This returns a power of .5294, which indicates there is a little more than a 50% probability that the three-tenths standard deviation effect found in the sample truly exists in the population. In other words, there is a 52.94% likelihood that we made a strong, correct decision to reject the null hypothesis and conclude that the 2.77-point difference in mean exam scores between the two groups also exists in the population. This is not very strong power and infers that we could have probably obtained a similar result simply by flipping a coin, which has a 50% probability, to decide if we should reject H_0.

Plausible Explanations for the Results

One plausible explanation is perhaps the BIM group was more studious than the GWS group. Without any information on how many hours each group studied on average this explanation is reasonable. A second plausible explanation is perhaps the GWS group students were more stressed or fatigued when they took the exam. Without capturing any information about students' stress or fatigue levels, this too seems reasonable. Lastly, with respect to the low power of 53%, recall from sample size planning there was a greater than 80% probability to detect a medium effect of $d = 0.50$. The results of the current study, though, yielded a small effect ($d = 0.289$), which means the sample size was insufficient to detect this effect. If we return to Statistics Kingdom's (2022a) sample size calculator and enter a "small" effect with an effect size of 0.289, a sample size of $n = 189$ per group would be needed for power = .80.

Chapter Summary

1. Research studies that involve comparing group means from separate groups are referred to as between-group designs, and the groups are represented by a group membership variable.
2. An independent-samples *t* test is a statistical strategy used to compare the means of two separate groups, which have been randomly selected from two separate populations. There are two forms of this *t* statistic. The pooled variances *t* is used if the equal

variances assumption is met, and the separate variances t is used if this assumption is not met. The primary difference between these formulas is with respect to the calculation of the standard error of the difference in group means. Regardless of the approach, though, the standard error will be the same.

3. To test for equal variances, known as the homogeneity of variances assumption, we use the Levene test and observe the reported p value. If $p > \alpha$, then the assumption is met. Independent of this assumption, Welch's t statistic can be used.

4. A hypothesis test involving the independent-samples t test focuses on the difference in population means and has null hypothesis H_0: $\mu_1 - \mu_2 = 0$, which is equivalent to $\mu_1 = \mu_2$.

Vocabulary Check

Between-group designs	Pooled variances t test
Group membership variable	Separate variances t test
Homogeneity of variances assumption	Single factor with n levels
Independent samples	t test statistic for independent samples
Levene test	Welch's t statistic

Review Exercises

A. Check Your Understanding

In 1–10, choose the best answer among the choices provided.

1. Which of the following research objectives describes a between-groups design?
   a. To determine the relationship between risk perception and number of hours as PIC on a cross-country flight.
   b. To determine the difference in risk perception between students with aviation experience and those without aviation experience.
   c. To determine if the relationship between risk perception and hours of actual instrument flight is the same as the relationship between risk perception and hours of simulated instrument flight for a group of GA pilots.
   d. To determine if the most common reason for choosing to fly for a regional airline is the same as the most common reason for choosing to fly for a major airline for a group of Part 121 pilots who have flown for both types of airlines.

2. The FAA's mandatory retirement age, which was increased from age 60 to age 65 for Part 121 pilots, became effective December 13, 2007. Let's assume that you endeavor to conduct a study to determine if there is any difference in piloting skills between 64-year-old Part 121 pilots and their 60-year-old counterparts, where piloting skills will be measured in a simulator and defined as the mean deviation from the glidepath in degrees. In this study, the group membership variable is _____ and the groups are _____.
   a. counterparts, Part 121 pilots
   b. 60- and 64-year-old pilots, difference in mean deviation from the glidepath
   c. pilot age, 60- and 64-year-old pilots
   d. Part 121 pilots, differences in piloting skills

3. Given H_0: $\mu_1 - \mu_2 = 0$ and H_1: $\mu_1 - \mu_2 \neq 0$. If the 95% CI ($\mu_1 - \mu_2$): [2, 7], then which of the following is true?
   a. The 95% CI for the difference in the means equals 95% of the distance from 2 to 7.
   b. The two means differ significantly with respect to $\alpha = .05$.
   c. The two means do not differ significantly with respect to $\alpha = .05$.
   d. The sample means differ by exactly 5 points.
4. The homogeneity of variances assumption states that
   a. the variances of the samples are equal.
   b. the sample means come from populations with equal variances.
   c. the samples come from the same population.
   d. the variance must stay constant for each participant in the experiment.
5. An aviation researcher compared the job-related mental health, or *well-being*, of flight attendants from a full-service carrier (FSC) and a low-cost carrier (LCC). Flight attendants' mental health was measured using Lukat et al.'s (2016) Positive Mental Health (PMH) instrument, which is a 9-item, 4-point Likert-type scale where scores range from 0 (do not agree) to 3 (agree). Thus, scores could range from 0 to 27, with higher scores reflecting higher PMH, which translates to how happy, healthy, or successful a person feels. The results of an independent-samples t test were as follows: $t(21) = 0.39$, $M_{FSC} = 18.5$, $M_{LCC} = 19.2$, $SE_{Difference} = 0.94$, 95% CI = [−1.6, 2.3]. Based on these results, what can we conclude?
   a. There was no significant difference in PMH scores between the two groups.
   b. The pooled variances t test was used to analyze the data.
   c. The overall standard deviation of the sample independent of group membership was 0.94.
   d. Either something is wrong with the data or there was an error in the calculations because flight attendants at a FSC should have a higher PMH than those at an LCC.
6. A study was done that compared the FAA IRA exam scores with respect to biological sex (male or female) for a particular year. The researcher acquired exam scores from a random sample of $n_1 = 79$ males and a random sample of $n_2 = 71$ females. The null and alternative hypotheses were H_0: $\mu_1 = \mu_2$, H_1: $\mu_1 > \mu_2$, and the results of an independent-samples t test with respect to $\alpha = .05$ were $t(148) = -3.0$, $M_1 = 90.9$, $M_2 = 93.1$, $SE_{Difference} = 0.72$, and the Levene test had $p = .3045$. Based on these results, what can we conclude?
   a. There was no significant difference between the groups because t^* was negative and H_1 was a right-tailed test.
   b. There was no significant difference between the groups because $p > \alpha$.
   c. The standard deviations of each group were statistically equal.
   d. The 95% CI included 0 within the interval.
7. A recruiter for naval aviators was interested in comparing the levels of physical fitness of students attending the local 2-year college and those attending the local 4-year university. He selected a random sample of 100 students from the 2-year college and a random sample of 100 students from the 4-year university. The descriptive statistics of students' fitness scores were $M = 95$ ($SD = 10$) and $M = 92$ ($SD = 13$) for the 2-year college and 4-year university, respectively. Based on $\alpha = .05$, the result of an independent-samples t test was $t(198) = 3.27$, $p = .0264$. Which of the following is an appropriate conclusion for this study?
   a. The population means are not significantly different.
   b. The mean fitness score for the second group is significantly higher than the mean fitness score of the first group.

    c.   The difference in mean fitness scores between the groups is statistically significant.

    d.   It is not appropriate to compare groups that come from different populations.

8. Let's say an aviation researcher conducted a one-tailed test of the hypothesis that avia-tion maintenance technicians (AMTs) who listen to music during their shift will be more productive than AMTs who do not listen to music during their shift. Using a research-er-developed measure of productivity where 0 indicates "not productive" and 10 indi-cates "highly productive," the researcher reported that the mean score for the sample of 40 AMTs who listened to music during their shift was 3, the mean score for the sam-ple of 50 AMTs who did not listen to music during their shift was 6, and the standard error of the difference in the means was 0.5. What should you conclude from this study?

    a.   Listening to music during their shift caused AMTs to be less productive.

    b.   There was no significant difference in productivity between AMTs who listened to or did not listen to music during their shift.

    c.   AMTs who listened to music during their shift had significantly lower productiv-ity scores than AMTs who did not listen to music during their shift.

    d.   AMTs who listened to music during their shift had significantly higher productiv-ity than AMTs who did not listen to music during their shift.

9. Let's assume a researcher is examining the difference in sample means from two in-dependent samples. Let's further assume that instead of randomly selecting each sample, she decides to match the participants with respect to key demographics so that every participant in the first sample has a matched counterpart in the second sample. This process of matching participants

    a.   increases the observed mean difference.

    b.   increases the probability of failing to reject a true null hypothesis.

    c.   decreases variability.

    d.   decreases the size of the calculated t value.

10. An aviation researcher examined the effectiveness of a new method for teaching the phonetic alphabet used by pilots and air traffic controllers (ATCs) to new flight stu-dents. She randomly selected 24 flight students from a local flight school to receive either the old phonics method ($n_1 = 12$) or the new phonics method ($n_2 = 12$). At the conclusion of the lesson, she administered a 20-item dichotomously scored exam and the results were $M_1 = 14.75$ ($SD_1 = 2.29$), $M_2 = 15.16$ ($SD_2 = 2.44$), and $SD_P = 6.425$. Based on these results, what could the researcher conclude?

    a.   There is evidence to suggest there is a significant difference between the old and new lesson designs.

    b.   There is evidence to suggest there is no significant difference between the old and new lesson designs.

    c.   The estimated effect size of the difference in group means was $d = 0.06$.

    d.   There is not enough information to answer this question.

B. Apply Your Knowledge

Use the following research description and the corresponding data set to conduct the activities given in Parts A–C (see also Section 8.4).

An airline's passenger load factor is a measure of the use of aircraft capacity that rep-resents the ratio (expressed as a percentage) of the total number of seats that are filled for an airline—called Revenue Passenger-Kilometers (RPKs)—to the total number of seats available—called Available Seat-Kilometers (ASKs). For example, consider the very simple case where an airline has 500 planes, and each plane has a 500-passenger seat capacity (ASK). If the airline's systemwide passenger load factor is 80%, then this would mean that the airline's average RPK is 400 passengers per flight. The file Ch_8 Exercises Part B Data contains the monthly passenger load factors from October 2002 through July 2014, for all U.S. and foreign carriers for both domestic and international routes (Bureau of Transportation Statistics, n.d.). Your assignment is to import this data set into your statistics program and compare the load factors between domestic and international routes relative to the assumption that domestic routes have a greater load factor (and hence are more profitable) than international routes.

A. Pre-Data Analysis

1. Specify the research question and corresponding operational definitions.
2. Specify the research hypothesis.
3. Determine the appropriate research methodology/design and explain why it is appropriate.
4. Conduct an a priori power analysis to determine the minimum sample size needed. Compare this result to the size of the given data set and explain what impact the size of the given sample will have on the results relative to the minimum size needed.

B. Data Analysis

1. Conduct a hypothesis test of independent samples (difference in two means) by applying all four steps as presented in Section 8.4.

C. Post-Data Analysis

1. Determine and interpret the effect size from both standard deviation units via Cohen's estimated d and explained variance perspectives (r^2).
2. Determine and interpret the 95% confidence interval, including its precision and AIPE.
3. Determine and interpret the power of the study from a post hoc perspective.
4. Present at least two plausible explanations for the results.
5. Now conduct an outlier analysis using Jackknife distances and rerun the analysis (there should be 11 outliers). Summarize the key differences in the results between outliers present and outliers absent in Table 8.4 and explain which findings you would be more comfortable supporting: those with outliers present or those with outliers absent.

Table 8.4 **Summary Chart for Chapter 8 Exercises, Part B–Section C**

Findings	Outliers Present	Outliers Absent
1. Pooled or separate variances?	1. _____	1. _____
2. Overall mean difference?	2. _____	2. _____
3. Mean load factor for Domestic?	3. _____	3. _____
4. Mean load factor for International?	4. _____	4. _____
5. Is mean difference significant?	5. _____	5. _____
6. What is the 95% CI?	6. _____	6. _____
7. What is the effect size (d and r^2)?	7. _____	7. _____
8. What is the power of the study?	8. _____	8. _____

References

Bureau of Transportation Statistics (n.d.). *Passengers: All carriers—All airports*. https://www.transtats.bts.gov/Data_Elements.aspx?Data=2

Lukat, J., Margraf, J., Lutz, R., van der Veld, W. M., & Becker, E. S. (2016). Psychometric properties of the positive mental health scale (PMH-scale). *BMC Psychology, 4*(8). https://doi.org/10.1186/s40359-016-0111-x

Omni Calculator (n.d.). *The critical value calculator*. https://www.omnicalculator.com/statistics/critical-value

Statistics Kingdom (2022a). *Z-test and T-test sample size calculator*. https://www.statskingdom.com/sample_size_t_z.html

Statistics Kingdom (2022b). *Normal, T—Statistical power calculators*. https://www.statskingdom.com/32test_power_t_z.html

9 Single-Factor ANOVA

Student Learning Outcomes

After studying this chapter, you will be able to do the following with respect to analyzing data using a single-factor ANOVA:

1. Determine the number of pairwise comparisons to be made, calculate the difference between the testwise and experimentwise alpha levels, and distinguish between the two primary strategies for controlling the experimentwise error rate.
2. Contrast between-group variability vs. within-group variability and describe the F ratio and corresponding F distribution.
3. Prepare and interpret an ANOVA summary table.
4. Conduct and interpret the results of a hypothesis test involving a single-factor ANOVA.
5. Engage in pre-data analysis, data analysis, and post-data analysis activities as described in Section 9.5.

9.1 Chapter Overview

In this chapter, which is an extension of Chapter 8, we build on our discussion of the independent-samples t test by presenting the single-factor analysis of variance (ANOVA), a statistical strategy for comparing the means from three or more groups. We first introduce the concept of ANOVA, including its rationale and logic, and then present the F ratio, which is the test statistic for ANOVA, and its corresponding sampling distribution, called the F distribution, which is an extension of the t distribution. We also discuss the construction and interpretation of the summary table that results from performing an ANOVA as well as how to conduct a hypothesis test using ANOVA. We conclude this chapter with a guided example that demonstrates the application of the single-factor ANOVA to a research study that involves comparing seven group means.

9.2 The Concept of ANOVA

Limitations of the Independent-Samples t Test

When comparing two group means using the independent-samples t test as we did in Chapter 8, we are making a *pairwise comparison* because we are comparing a pair of groups selected from two separate populations, relative to some dependent measure.

DOI: 10.4324/9781003308300-13

In the FAA IRA exam study, the DV was the mean and we compared BIM's group mean to GWS's group mean.

Let's extend the IRA exam study to include two additional hypothetical companies: JAE and DSC. Doing so now involves comparing the means of four groups—BIM, GWS, JAE, and DSC—and the corresponding null hypothesis would claim there is no significant difference in mean scores between any of the groups. Unlike the two-group case, though, there will be more than one pairwise comparison. More specifically, these four groups lead to *six* pairwise comparisons: BIM vs. GWS, BIM vs. JAE, BIM vs. DSC, GWS vs. JAE, GWS vs. DSC, and JAE vs. DSC. Generalizing, the total number of pairwise comparisons is equal to $\frac{k(k-1)}{2}$, where k = the total number of groups involved (we divide by 2 to remove redundancies). This also implies we must perform six separate independent-samples t tests to test the corresponding null hypotheses:

- $H_0: \mu_{BIM} = \mu_{GWS}$
- $H_0: \mu_{BIM} = \mu_{JAE}$

- $H_0: \mu_{BIM} = \mu_{DSC}$
- $H_0: \mu_{GWS} = \mu_{JAE}$

- $H_0: \mu_{GWS} = \mu_{DSC}$
- $H_0: \mu_{JAE} = \mu_{DSC}$

The problem with conducting multiple independent-samples t tests involving the same factor is we risk *inflating the preset alpha level*. For example, in Step 2 of hypothesis testing, the level of significance we specify is the preset alpha level, which represents the probability, or level of risk, we are willing to accept in making a Type I error (a "false positive"). By convention, this preset alpha level is set to $\alpha = .05$ and indicates there is a 5% chance that the effect found in the sample was a fluke and really does not exist in the population. In other words, there is a 5% chance of incorrectly concluding—based on sample data—that the null hypothesis is false and therefore is to be rejected. This preset alpha level also is called the testwise alpha because it is the level at which we are testing the null hypothesis.

If several hypothesis tests are conducted for the same investigation, we get a *pyramiding effect* of Type I errors. For example, if we were to perform six separate independent-samples t tests for the IRA exam study involving BIM, GWS, JAE, and DSC, and each test was based on a preset alpha of $\alpha = .05$, then there is a 5% risk of a Type I error for *each* test. As a result, the risk of committing a Type I error associated with these six separate tests is no longer 5% but instead accumulates to produce a much higher risk. This accumulated Type I error is called the experimentwise, or inflated alpha, and is equal to

$$1-(1-\alpha)^k$$

where k = the total number of pairwise comparisons. For the running example where we have four group means, if the testwise $\alpha = .05$ and we perform six separate independent-samples t tests, then the inflated alpha level is

$$1-(1-\alpha)^k$$
$$=1-(1-.05)^6$$
$$=1-(.95)^6$$
$$=1-.735$$
$$=.265$$

Thus, the probability of committing a Type I error associated with performing a "family" of six pairwise comparisons is 26.5%, not 5%.

As a result, the independent-samples t test is appropriate for studies that compare exactly two groups. If, however, there are more than two groups, then $\frac{k(k-1)}{2}$ tests will be needed to examine all the corresponding pairwise comparisons. In such situations, the independent-samples t test is not appropriate and should not be used because of the risk of inflated alpha levels. The more tests we do, the greater the risk, and by the time we finish, we will have a much higher probability of committing a Type I error than the initial preset α.

Strategies for Dealing with Inflated Alpha Levels

Bonferroni Procedure

One way to limit the risk of inflated Type I error is to apply the Bonferroni procedure, which reduces the testwise alpha level by dividing it by the number of comparisons. For the running example, we would test each pairwise hypothesis against an alpha level of $\frac{\alpha}{6} = .0083$ instead of .05. To illustrate the effect of this procedure, as shown in Figure 9.1(a), the critical t values for a two-tailed independent-samples t test with $\alpha = .05$ and $n = 128$ (64 per group) involving all six pairwise comparisons are $t(126) = \pm 1.979$. As shown in Figure 9.1(b), when we apply the Bonferroni procedure, the alpha level is $\alpha = .0083$, and the corresponding t critical values are $t(126) = \pm 2.682$ (Omni Calculator, n.d.), which is located at the more extreme ends of each tail than the critical t for $\alpha = .05$. This means that stronger and more compelling evidence is needed to reject H_0 for $\alpha = .0083$ than for $\alpha = .05$. As a result, the Bonferroni procedure protects against inflated alpha levels by making the preset alpha much more stringent.

To apply the Bonferroni procedure in practice, we establish the critical t value relative to $\frac{\alpha}{k}$ and compare the calculated t (i.e., t^*) from the statistics program's output to the Bonferroni critical t. Alternatively, if our statistics program provides the flexibility to change α, then we would change it to $\frac{\alpha}{k}$. Lastly, for those who profess a p-value approach to significance, you would compare the p value to the Bonferroni alpha level: If p is less than the Bonferroni alpha, then you would claim statistical significance and reject H_0.

Omnibus Approach via ANOVA

A second approach to dealing with inflated alpha levels resulting from conducting multiple independent-samples t tests is to perform a single omnibus test, which examines all the comparisons in the study collectively. One such omnibus test is ANOVA, which may be thought of as a generalization of the independent-samples t test. ANOVA is an omnibus test because it treats all the groups as a single entity and performs a single analysis at the preset alpha level. If the overall ANOVA model is significant, then we can proceed with pairwise comparisons without incurring an inflated alpha level.

(a)

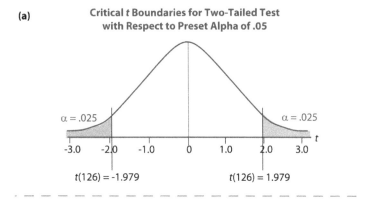

(b)

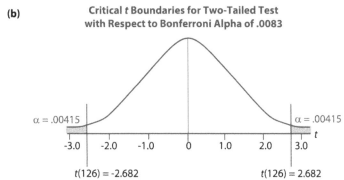

Figure 9.1 Differences in critical region boundary values between a preset alpha level of $\alpha =$ .05 vs. the Bonferroni alpha level of $\alpha = .05/6 = .0083$, which is based on six pairwise comparisons.

For the running IRA exam example, instead of performing six separate independent-samples t tests at an experimentwise alpha of $\alpha = .265$, or examining the results relative to a Bonferroni alpha level of $\alpha = .0083$, we conduct an ANOVA. We perform a single test based on the presumption that there is no significant difference among the four mean test scores in the population, and by doing so, we keep the chance of committing a Type I error at the initial preset alpha level. If the result of this single analysis is significant, then we can perform the respective six pairwise comparisons via separate independent-samples t tests without inflating the alpha level. We present examples of a hypothesis test using ANOVA in Sections 9.4 and 9.5.

The Logic of Analysis of Variance

As implied by its name, the focus of ANOVA is on *variances*, and the overall question is, "Is the variability among the sample means due to treatment or chance?" To understand what we mean by this question, consider the file Ch_9 IRA Exam Scores Data for ANOVA Concept, which contains hypothetical IRA exam score data for the four test preparation software packages. A summary of the group means for the first column of this data set is given in Table 9.1, and a similar summary for the second column is given in Table 9.2. Note

Table 9.1 Small Within-Groups Variability but Large Between-Groups Variability

Sample 1 (BIM)	Sample 2 (GWS)	Sample 3 (JAE)	Sample 4 (DSC)	
48	50	53	55	
47	51	54	56	
48	49	52	54	
49	50	53	55	
$T_1 = 192$	$T_2 = 200$	$T_3 = 212$	$T_4 = 220$	
$M_1 = 48$	$M_2 = 50$	$M_3 = 53$	$M_4 = 55$	$M_T = 51.5$

Table 9.2 Large Within-Groups Variability but Small-Between Groups Variability

Sample 1 (BIM)	Sample 2 (GWS)	Sample 3 (JAE)	Sample 4 (DSC)	
48	52	53	58	
55	48	45	47	
52	46	52	44	
49	54	50	55	
$T_1 = 204$	$T_2 = 200$	$T_3 = 200$	$T_4 = 204$	
$M_1 = 51$	$M_2 = 50$	$M_3 = 50$	$M_4 = 51$	$M_T = 50.5$

that the group means in Table 9.1 vary from 48 to 55, but vary from 50 to 51 in Table 9.2. Why is there so much variability in the group means in Table 9.1 but not in Table 9.2?

ANOVA examines variability by partitioning the total variability of the entire data set into two components: between-groups variability, which refers to the *mean differences* between *individual groups*, and within-groups variability, which refers to *differences among the individual scores* within *each group*. Focusing on Table 9.1, first observe that the differences among the raw scores *within each group* are small, indicating there is little *within-groups* variability. However, when we examine the respective mean scores across the four groups, there is a considerable amount of *between-groups* variability, particularly with respect to BIM ($M_1 = 48$) and DSC ($M_4 = 55$). Thus, the data in Table 9.1 reflect small within-groups variability but relatively large between-groups variability. There are three reasons for between-groups variability:

- *Individual differences*. The variability in mean scores between groups is the result of participants' different backgrounds, abilities, attitudes, etc.
- *Experimental error*. The variability in mean scores between groups is due to unexplained or uncontrolled events such as from equipment, lack of participants' attention, unpredictable changes in the way the study was conducted, etc.
- *Treatment effect*. The variability in mean scores between groups is related to the actual treatment. In other words, it was the treatment that caused the means to be different.

Now look at Table 9.2, where we have the reverse situation. There is considerable variability in the raw scores *within each group*, but the difference in group means across the four groups is very small, which indicates little between-groups variability. There are two reasons for this: *individual differences* and e*xperimental error.*

When using samples acquired from different populations, we expect large between-group differences but little within-group differences. This expectation is reasonable because if each sample represents a treatment, then everyone *within* a particular group is receiving the same treatment, but each group is receiving a different treatment. In the context of our IRA exam study as illustrated in Table 9.1, we would expect students *within* each respective group to have similar exam scores because they all pre-pared for the IRA exam using the same materials associated with that group. However, we also would expect students *between* these groups to have different scores because they are using different preparation materials.

The logic behind partitioning the total variability of a data set into between- and with-in-groups variance is that it enables us to examine their ratio:

$$\frac{\text{Between-Groups Variance}}{\text{Within-Groups Variance}}$$

Furthermore, because the variability between groups can be explained by individual differences, experimental error, and treatment effect, whereas the within-group variability can only be explained by individual differences and experimental error, this ratio may be re-expressed as

$$\frac{\text{Treatment Effect} + \text{Individual Differences} + \text{Experimental Error}}{\text{Individual Differences} + \text{Experimental Error}}$$

This ratio is called the *F ratio*, and it is the test statistic for ANOVA. In Section 9.3 we will discuss how this ratio is constructed along with the corresponding *F* distribution. For the moment, though, our focus is conceptual, and we observe the following:

- If the treatment effect is 0, then the individual differences and experimental error in the numerator and denominator cancel each other out, which results in an *F* ratio of $F = 1$.
- If there is a nonzero treatment effect, then $F > 1$. If the treatment effect is "small," then $F \rightarrow 1$, and the effect most likely will not be statistically significant. If the treatment effect is "large," then $F \rightarrow \infty$, and the effect most likely will be statistically significant. Thus, the *F* ratio always will be greater than or equal to 1 ($F \geq 1$) depending on the magnitude of the corresponding treatment effect.
- The denominator of the *F* ratio is called the error term, which is a measure of the variance due to random, unsystematic differences.

Applying this concept to the running IRA exam study, the data in Table 9.1 suggest that the *F* ratio will be $F > 1$ because there is considerable between-groups variability:

- 2-point difference for BIM vs. GWS and JAE vs. DSC
- 3-point difference for JAE vs. DSC
- 5-point difference for BIM vs. JAE and GWS vs. DSC
- 7-point difference for BIM vs. DSC

Similarly, the data in Table 9.2 suggest that $F \to 1$ because there is very little between-groups variability:

- 0-point difference for GWS vs. JAE and BIM vs. DSC
- 1-point difference for BIM vs. GWS, GWS vs. DSC, BIM vs. JAE, and JAE vs. DSC.

Example 9.1: *The Concept of the Single-Factor ANOVA*

To test the assumption that sleep deprivation impacts aggression, a group of volunteer flight attendants were randomly assigned to sleep deprivation periods of 0, 6, 12, and 18 hours, and then tested for aggressive behaviors in a controlled work environment involving pilots and passengers. The researchers operationally defined aggressive behaviors as verbal "put downs," getting into an argument, or verbal interruptions. The researchers then observed the flight attendants' behavior as they interacted with pilots and passengers during the study period and recorded the total number of aggressive behaviors. A copy of the corresponding data set is given in the file Ch_9 Example 9.1 Data. (a) Write the overall research question. (b) Determine the overall factor and corresponding levels. (c) Explain why the study's data should be analyzed using ANOVA. (d) Determine the group means, assess if the *F* ratio from the ANOVA results will be statistically significant, and explain the reason(s) for your assessment.

Solution

(a) What is the effect of sleep deprivation on flight attendants' aggressive behavior? An alternative RQ is: What is the difference in the number of aggressive behaviors among flight attendants with respect to sleep deprivation periods of 0, 6, 12, and 18 hours?

(b) The factor is "sleep deprivation," and the levels of the factor are the four groups of 0, 6, 12, and 18 hours of sleep deprivation.

(c) There are four groups and hence six different pairwise comparisons. If performed using multiple independent-samples *t* tests, there will be an inflated alpha level of .265. By performing an ANOVA as an omnibus test, this will protect the analysis from inflated alpha levels.

(d) $M_0 = 1.625$, $M_6 = 3.125$, $M_{12} = 5.0$, $M_{18} = 7.5$. There is considerable between-groups variability for three of the six pairwise comparisons:
- $M_{18} - M_0 = 7.5 - 1.625 = 5.875$
- $M_{18} - M_6 = 7.5 - 3.125 = 4.375$
- $M_{12} - M_0 = 5.0 - 1.625 = 3.375$

This suggests that the *F* ratio will be $F > 1$, and therefore, the overall ANOVA probably will be statistically significant.

9.3 The ANOVA Summary Table and the *F* Distribution

The ANOVA Summary Table

Before we continue with our discussion of ANOVA, a review of standard deviation and variance is in order. Recall from Chapter 2 that we used the simple data set of {4, 6, 8, 12,

15} to introduce the concept of standard deviation, which represents the "average distance" scores in a distribution are from the mean. We indicated that the best way to determine this "average distance" was to first calculate how much each score deviated from the mean, and then find the average deviation by adding these deviation scores and dividing by N. Although this was a reasonable approach, the problem with this logic was that the overall sum of the deviations from the mean is 0, and hence the corresponding average deviation from the mean also is 0. This was illustrated in Table 2.7, which is replicated in Table 9.3 as a convenience to the reader.

To fix this problem, we squared each deviation from the mean and then added the squared deviations as shown in Table 9.3. When applied to our data set, the *sum of the squared deviations from the mean*, denoted SS for "sum of squares," is 80. However, because we wanted the "average distance" scores are from the mean, we divided SS by the sample size, N, which yielded the *mean of the sum of the squared deviations*. Thus, as shown in Table 9.3, the mean of the sum of the squared deviations is 80/5 = 16, which we called *variance*. Finally, to calculate standard deviation, we took the square root of the variance to "undo" the squaring. We also made a distinction between the population vs. sample variance and standard deviation, which was based on whether we divided SS by N or by N − 1. This is summarized in Table 9.3.

Let's now focus on variance. Recognize that variance is a measure of dispersion that represents the "average *squared* deviations" scores are from the mean because we are dividing the sum of the squared deviations by the total number of scores (N or N − 1). Furthermore, because this is the *arithmetic* average, variance reflects the mean of the squared deviations, and is denoted as MS, which stands for *mean squares*. Because we generally work with samples and not populations, when we put this all together, we have:

$$\text{Variance} = MS = \frac{SS}{df}$$

Let's now apply this to our ANOVA discussion. As indicated earlier, ANOVA partitions the entire data set into two components: between-group variability and within-group variability. This partitioning serves as the basis for the F ratio, which is the test statistic for ANOVA. In Figure 9.2, when we apply our review of variance to ANOVA and the F ratio, we can get a better idea of the structure of ANOVA.

Table 9.3 Replica of Table 2.7 from Chapter 2 for Calculating Standard Deviation

	Score (X)	Deviations from Mean	Squared Deviations	Calculations
	4	$4 - 9 = -5$	$(-5)^2 = 25$	**Population**
	6	$6 - 9 = -3$	$(-3)^2 = 9$	• Variance = Mean of $SS = \dfrac{SS}{N} = \dfrac{80}{5} = 16$
	8	$8 - 9 = -1$	$(-1)^2 = 1$	
	12	$12 - 9 = 3$	$(3)^2 = 9$	• Standard Deviation = $\sqrt{\text{Variance}} = \sqrt{16} = 4$
	15	$15 - 9 = 6$	$(6)^2 = 36$	**Sample**
Sums	45	0	80	• Variance = Mean of $SS = \dfrac{SS}{N-1} = \dfrac{80}{4} = 20$
				• Standard Deviation = $\sqrt{\text{Variance}} = \sqrt{20} = 4.47$

Note. The sum of the squared deviations, denoted SS, is 80.

The sum of the squared deviations of the dataset is partitioned into the sum of squared deviations relative to between groups and within groups. Similarly, the degrees of freedom for the dataset is partitioned into the degrees of freedom relative to between groups and within groups.

When the sum of the squared deviations is divided by the degrees of freedom, we get the "average" sum of squared deviations, or "mean squares" (*MS*), which is variance. In ANOVA, variance is partitioned into between-groups variance and within-groups variance.

$$MS_{Between} = \text{Between-Groups Variance} = \frac{SS_{Between}}{df_{Between}} = \frac{②}{⑤} \Bigg\} \; ⑦$$

$$MS_{Within} = \text{Within-Groups Variance} = \frac{SS_{Within}}{df_{Within}} = \frac{③}{⑥} \Bigg\} \; ⑧$$

The *F* ratio is the proportion of the between-groups variance and the within-groups variance.

$$F = \frac{MS_{Between}}{MS_{Within}} = \frac{\text{Between-Groups Variance}}{\text{Within-Groups Variance}} = \frac{\frac{SS_{Between}}{df_{Between}}}{\frac{SS_{Within}}{df_{Within}}} = \frac{⑦}{⑧} \Bigg\} \; ⑨$$

Figure 9.2 The structure of ANOVA and its nine components as denoted by the circled numbers in bold. (Adapted from Gravetter & Wallnau, 2017, p. 377.)

Table 9.4 Basic Structure of the ANOVA Summary Table

Source	SS	df	MS	F
Between	a	d	$g = \dfrac{a}{d}$	$F = \dfrac{MS_{Between}}{MS_{Within}} = \dfrac{g}{h}$
Within (Error)	b	e	$h = \dfrac{b}{e}$	
Total	c	f		

The structure of ANOVA consists of nine separate components: SS_{Total}, $SS_{Between}$, SS_{Within}, df_{Total}, $df_{Between}$, df_{Within}, $MS_{Between}$, MS_{Within}, and the *F* ratio. These nine components are placed in an ANOVA summary table, which is shown in Table 9.4. Observe from Table 9.4 that:

- $SS_{Between} + SS_{Within} = SS_{Total}$ $(a + b = c)$
- $df_{Between} + df_{Within} = df_{Total}$ $(d + e = f)$
- The between-groups variance ("g") is the ratio of $SS_{Between}$ and $df_{Between}$

$$MS_{Between} = \frac{SS_{Between}}{df_{Between}} \left(\frac{a}{d} \right)$$

- The within-groups variance ("h") is the ratio of SS_{Within} and df_{Within}

$$MS_{Within} = \frac{SS_{Within}}{df_{Within}} \left(\frac{b}{e} \right)$$

- F is the ratio of the between-groups variance and within-groups variance

$$F = \frac{MS_{Between}}{MS_{Within}} \left(\frac{g}{h} \right)$$

We commonly refer to the within-groups variance (MS_{Within}) as the error variance because, as noted earlier, the denominator is a function of random error, unsystematic differences, and measurement error.

From Raw Data to the ANOVA Summary Table: An Example

In practice, we do not perform an ANOVA using hand calculations but instead rely on the output from our statistics program. However, to facilitate an appreciation for and an understanding of the ANOVA process, we demonstrate how to perform an ANOVA using the hypothetical IRA exam scores given in Table 9.1. These data are also given in the first column of the file Ch_9 IRA Exam Scores Data for ANOVA Concept. At the end of each stage of the ANOVA process, we will append a letter (*a–h*) that corresponds to the cells in the ANOVA summary table given in Table 9.4.

Stage 1: The Between-Groups Sum of Squared Deviations ($SS_{Between}$)

$SS_{Between}$ examines the difference between group means. To calculate $SS_{Between}$, we compute the deviation of each group's mean from the overall mean ($M_i - M_T$), square this difference, and then add the squared deviations. We also must weight each calculation by each group's sample size (n_i). If, however, the sample size of each group is the same, which is the case here, then we may simply multiply the result by the common sample size by factoring out the common factor.

$$SS_{Between} = n_1 (M_1 - M_T)^2 + n_2 (M_2 - M_T)^2 + n_3 (M_3 - M_T)^2 + n_4 (M_4 - M_T)^2$$

$$= 4(48 - 51.5)^2 + 4(50 - 51.5)^2 + 4(53 - 51.5)^2 + 4(55 - 51.5)^2$$

$$= 4 \times \left[(48 - 51.5)^2 + (50 - 51.5)^2 + (53 - 51.5)^2 + (55 - 51.5)^2 \right]$$

$$= 4 \times \left[(-3.5)^2 + (-1.5)^2 + (1.5)^2 + (3.5)^2 \right]$$

$$= 4 \times [12.25 + 2.25 + 2.25 + 12.25]$$

$$= 4 \times 29$$

$$= 116 (a)$$

Stage 2: The Within-Groups Sum of Squared Deviations (SS_{Within})

SS_{Within} examines the differences among the raw scores in each group separately. To calculate SS_{Within}, we compute the deviation each score within a group is from that group's mean ($x_i - M_i$), square each deviation, and then add the squared deviations. We then add each group's SS.

- SS_{Within} for BIM: $(48 - 48)^2 + (47 - 48)^2 + (48 - 48)^2 + (49 - 48)^2 = 2$
- SS_{Within} for GWS: $(50 - 50)^2 + (51 - 50)^2 + (49 - 50)^2 + (50 - 50)^2 = 2$
- SS_{Within} for JAE: $(53 - 53)^2 + (54 - 53)^2 + (52 - 53)^2 + (53 - 53)^2 = 2$
- SS_{Within} for DSC: $(55 - 55)^2 + (56 - 55)^2 + (54 - 55)^2 + (55 - 55)^2 = 2$
- SS_{Within} overall $= 2 + 2 + 2 + 2 = 8 (b)$

Stage 3: The Total Sum of Squared Deviations (SS_{Total})

This is the sum of $SS_{Between}$ and SS_{Within}: $116 + 8 = 124 (c)$

Stage 4: The Degrees of Freedom

The degrees of freedom are as follows:

- $df_{Between} = $ (Number of groups $-$ 1) $= (G - 1) = (4 - 1) = 3$ (*d*)
- $df_{Within} = $ (Total sample size $-$ Number of groups) $= (N - G) = (16 - 4) = 12$ (*e*)
- $df_{Total} = $ (Total sample size $-$ 1) $= (N - 1) = (16 - 1) = 15$ (*f*)

Stage 5: The Between- and Within-Groups Variances

Because variance is equal to the *mean of the squared deviations*, we divide the respective *SS* by the corresponding *df*.

- $MS_{Between} = \dfrac{SS_{Between}}{df_{Between}} = \dfrac{116}{3} \approx 38.667$ rounded to 3 decimal places (*g*)

- $MS_{Within} = \dfrac{SS_{Within}}{df_{Within}} = \dfrac{8}{12} \approx 0.667$ rounded to 3 decimal places (*h*)

Stage 6: The F ratio

F is the ratio of the between- and within-groups variances:

$$F = \frac{MS_{Between}}{MS_{Within}} = \frac{38.667}{0.667} \approx 58$$

If we use fractions instead of converting to decimals, then *F* is *exactly* equal to 58:

$$F = \frac{\dfrac{116}{3}}{\dfrac{8}{12}} = \left(\frac{116}{3} \times \frac{12}{8}\right) = \left(\frac{116}{1} \times \frac{4}{8}\right) = \left(\frac{116}{1} \times \frac{1}{2}\right) = \left(\frac{116}{2}\right) = 58$$

Stage 7: The ANOVA Summary Table

The final ANOVA summary table based on the results from the previous six stages is provided in Table 9.5.

The F Distribution

Before we introduce the *F* distribution, let's review the structure of the *F* ratio. First recall that the *F* ratio is a fraction where the numerator represents the between-groups variance, and the denominator represents the within-groups variance. Next recall that from

Table 9.5 ANOVA Summary Table for IRA Exam Scores Example

Source	SS	df	MS	F
Between	116	3	38.667	58.00
Within (Error)	8	12	0.667	
Total	124	15		

a mathematical perspective, variance is equal to the sum of the squared deviations (*SS*) divided by *df*. Thus, the *F* ratio may be constructed as a complex fraction as follows:

$$F = \frac{\text{Variance between groups}}{\text{Variance within groups}}$$

$$= \frac{SD^2_{Between}}{SD^2_{Within}}$$

$$= \frac{\dfrac{SS_{Between}}{df_{Between}}}{\dfrac{SS_{Within}}{df_{Within}}}$$

$$= \frac{SS_{Between}}{df_{Between}} \times \frac{df_{Within}}{SS_{Within}}$$

Given this representation, the *F* ratio consists of two *dfs*: one for the numerator ($df_{Between}$) and one for the denominator (df_{Within}). This means that unlike the *t* test statistic, *F* will have two *dfs*, and the degrees of freedom are always ($df_{Between}$, df_{Within}) or ($df_{Numerator}$, $df_{Denominator}$). A word of caution: Although the actual calculation of the *F* ratio involves multiplying the fraction in the numerator by the inverse of the fraction in the denominator (as shown above), it appears that $df_{Between}$ is in the denominator. This is *not* the case because $df_{Between}$ is in the fraction that is in the numerator of the initial complex fraction.

When using the *F* ratio as the test statistic for hypothesis testing involving ANOVA, consider the following:

- If the null hypothesis is "true" (i.e., there is no treatment effect), then the numerator and denominator of the *F* ratio are measuring the same variance, and therefore, *F* should be around 1.00. The ratio will not be exactly equal to 1, though, because of sampling error. This implies that *F* rarely ever will be between 0 and 1.
- If the null hypothesis is "false" (i.e., there is a treatment effect), then the numerator of the *F* ratio will be greater in magnitude than the denominator. This implies that *F* will be greater than 1. How much greater than 1 will be a function of the difference in magnitude between the numerator and denominator.

Given this background information, we are now ready to introduce the *F* distribution.

From Figure 5.1/Chapter 5, the shape of the *t* distribution is a function of sample size (*df*). For small *df*, the shape is relatively flat, but as *df* increases, the shape of the *t* distribution begins to approximate the shape of the normal distribution, and for *df* ≥ 30, the *t* distribution closely approximates the normal distribution. Applying this structure to the *F* distribution, the shape and variation of the *F* distribution also depend on sample sizes relative to the corresponding variances. In other words, the shape of the *F* distribution is a function of $df_{Between}$ and df_{Within}. Thus, there will be different *F* distributions for each pair of *df* associated with the respective sample variances. This is illustrated in Figure 9.3 for various pairs of *df*. Notice that the shape of the *F* distribution is positively skewed but approaches a normal distribution when the pairs of *df* become larger.

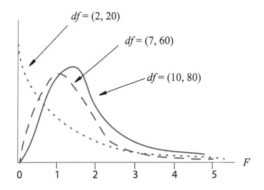

Figure 9.3 Various F distributions for different degrees of freedom.

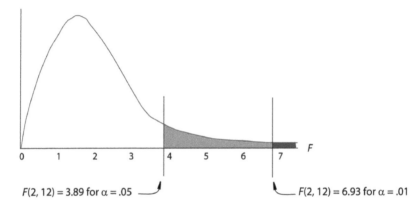

$F(2, 12) = 3.89$ for $\alpha = .05$

$F(2, 12) = 6.93$ for $\alpha = .01$

Figure 9.4 The critical regions for $F(2, 12)$ with respect to $\alpha = .05$ and $\alpha = .01$ from Table 5/Appendix A.

The critical values associated with the F distribution are given in Table 5/Appendix A for $\alpha = .05$ and $\alpha = .01$. When reading the F table, df for the numerator are listed across the top of the table and df for the denominator are listed along the left side of the table. For example, as illustrated in Figure 9.4, $F(2, 12) = 3.89$ for $\alpha = .05$ and 6.93 for $\alpha = .01$. If df are not listed in the F table, we use Soper's (2022b) online critical F-value calculator.

The Relationship between F and t

To understand the relationship between F and t, let's revisit the BIM vs. GWS IRA exam study from Chapter 8's Guided Example. Following is a summary of the results from the independent-samples t test we conducted:

- $N = 200$ and $df = 198$.
- $M_{BIM} = 43.66$, $M_{GWS} = 40.89$.
- The difference in group means was $43.66 - 40.89 = 2.77$ in favor of BIM.
- The calculated t statistic and corresponding p value were $t(198) = 2.04397$, $p = .0423$.

- The final decision was to "reject H_0" and conclude that students who prepared for the FAA IRA exam using BIM's software scored significantly higher on the exam than students who used GWS's software.

Let's now run this same analysis using ANOVA. The corresponding ANOVA summary table, which is generated from our statistics software, is given in Table 9.6. Note that $F(1, 198) = 4.1778$. Comparing t and F, observe the following:

- $t = 2.04397$ for $df = 198$, and $F = 4.1778$ for $df = (1, 198)$.
- $t^2 = (2.04397)^2 = 4.1778$, which is equal to F. Alternatively, $t = \sqrt{F}$.
- The degrees of freedom for t are equal to the degrees of freedom associated with the within-groups variance of F.

In general, the t distribution with $df = (n - 2)$ is equal to the square root of the F distribution with $df = (1, n - 2)$. Thus, if we performed an ANOVA to compare the means of exactly two groups, we get the same results as those generated by an independent-samples t test. The reason is because the normal distribution, t distribution, and F distribution are all related to each other. Of course, the utility of ANOVA is far greater than an independent-samples t test because ANOVA provides a single omnibus test that enables us to compare the means of more than two groups without having to conduct multiple independent-samples t tests and inflating the alpha level. This example, though, demonstrates why ANOVA is considered an extension, or generalization, of the independent-samples t test.

9.4 Statistical Inferences Involving the Single-Factor ANOVA: Hypothesis Testing

Statistical inference can be approached either directly via confidence intervals or indirectly via hypothesis testing. Because a single-factor ANOVA is considered an omnibus test, it does not yield any confidence intervals *directly*. Instead, if the overall ANOVA model is significant, then CIs are computed for each of the respective follow-up pairwise comparisons. Because the calculation and interpretation of these corresponding CIs are the same as what we presented in Section 8.3/Chapter 8, we focus exclusively on hypothesis testing in the current section.

A hypothesis test involving a single-factor ANOVA is performed the same as an independent-samples t test with three differences. The first is with respect to how we formulate the null and alternative hypotheses, the second is with respect to how we check the normality assumption, and the third is with respect to follow-up pairwise comparisons.

Table 9.6 ANOVA Summary Table for BIM-GWS IRA Exam Study from Ch_8 Guided Example Data

Source	SS	df	MS	F
Between	383.645	1	383.645	4.1778
Within (Error)	18182.230	198	91.829	
Total	18565.875	199		

Null and Alternative Hypotheses

Null Hypothesis

Because a single-factor ANOVA compares three or more group means simultaneously, the null hypothesis is extended to reflect three or more groups. So, if k = the number of groups under discussion, then the corresponding null hypothesis for the single-factor ANOVA is

$$H_0: \mu_1 = \mu_2 = \mu_3 = \ldots = \mu_{k-1} = \mu_k$$

In the IRA exam study involving the four groups BIM, GWS, JAE, and DSC, the null hypothesis is written as

$$H_0: \mu_{BIM} = \mu_{GWS} = \mu_{JAE} = \mu_{DSC}$$

This notation signifies that the means of each group are identical, and effectively states there is "no treatment effect."

Alternative Hypothesis

The alternative hypothesis of a single-factor ANOVA focuses on the various pairwise comparisons where researchers specify which of the group means they believe will be different. Recall, though, that with k groups there are $\dfrac{k(k-1)}{2}$ comparisons. So, a research study with k = 5 groups will have 10 pairwise comparisons, and if the researcher believes there will be significant differences among the 10 comparisons, then all 10 must be listed as alternative hypotheses, which can be cumbersome. Because of the many different possible conditions, the alternative hypothesis is stated more generally as

$$H_1: \text{At least one group mean is different from another}$$

This representation of the alternative hypothesis implies there is a treatment effect and is preferred over listing all the specific alternatives.

Testing for Normality

In our discussion of the independent-samples t test in Chapter 8, we tested for normality by examining the results from the Shapiro–Wilk Goodness of Fit test *separately* for each of the two groups. With ANOVA, though, we have three or more groups. So, for the IRA exam study with four groups, we must conduct four separate tests for normality, one for each group. There is, however, a more efficient approach to testing for normality in ANOVA. This approach involves examining the residuals of the dependent variable for normality. This will report the results of the corresponding Shapiro–Wilk test, which then can be applied to the overall ANOVA model. This is the approach we will use because it involves a single test instead of many separate tests.

Post Hoc Comparisons

Because the single-factor ANOVA is considered an omnibus test that protects against inflated alpha levels, we do not conduct any pairwise comparisons unless the overall ANOVA model is significant. If the model is significant, then we proceed with what are called *post hoc tests* or *post hoc comparisons* to determine which group means are significantly different. This is generally done in Step 4 of hypothesis testing. Of the various types of post hoc tests available, we use Tukey's HSD, which stands for "honestly significant difference."

We now present an example of a hypothesis test involving the single-factor ANOVA. This example is an extension of the FAA IRA exam study we presented in Chapter 8's Guided Example (Section 8.4). Instead of working with two groups, though, we include two additional groups: JAE and DSC. A copy of the hypothetical data set is provided in the file Ch_9 IRA Exam Data for ANOVA Hypothesis Test Example.

Step 1. Identify the targeted parameter, and formulate the null/alternative hypotheses. The targeted parameter is the mean, and the null and alternative hypotheses are:

H_0: $\mu_{BIM} = \mu_{GWS} = \mu_{JAE} = \mu_{DSC}$. There is no significant difference in mean IRA exam scores among the four groups.

H_1: At least one group mean is significantly different from the others.

Step 2. Determine the test criteria. The test criteria involve three components:

Level of Significance (α). We will set alpha to $\alpha = .05$.

Test Statistic. The test statistic is F.

The Critical Boundary. The critical F value is acquired from Table 5/Appendix A or from Soper's (2022b) online critical F-value calculator. To determine the degrees of freedom, recall that $df_{Between}$ is equal to the number of groups minus one $(G - 1)$, and df_{Within} is equal to the total sample size minus the number of groups $(N - G)$. Given four groups with a total sample size of $n = 400$, $df_{Between} = 4 - 1 = 3$ and $df_{Within} = 400 - 4 = 396$. As a result, the corresponding critical F value for $\alpha = .05$ obtained from Soper is $F(3, 396) = 2.627$. The critical region is depicted in Figure 9.5(a).

Step 3. Collect data, check assumptions, run the analysis, and report the results.

Collect Data. The data are in the file Ch_9 IRA Exam Data for ANOVA Hypothesis Test Example.

Check Assumptions. There are three primary assumptions for ANOVA:
- Independence. We assume that the data for each sample were randomly generated, and the samples (and their respective populations) are independent of each other. Therefore, this assumption is met.
- Normality. The p value for the Shapiro--Wilk Goodness of Fit Test for the residuals of the IRA exam scores is $p < .0001$, which means that this assumption is *not* met for the given data set. However, the central limit theorem can be applied because of the large sample sizes ($n_1 = n_2 = n_3 = n_4 = 100$). Therefore, we will claim that this assumption is met.
- Homogeneity of variances. The Levene test result from our statistics program is $F(3, 396) = 2.534$, $p = .057$, which confirms that the variances in the four respective populations are statistically equivalent.

Run the Analysis and Report the Results. As shown in the ANOVA summary table in Table 9.7, the results of this analysis are $F(3, 396) = 4.218$, $p = .006$.

Step 4. Decide whether to reject or fail to reject the null hypothesis, and conduct post hoc comparisons if appropriate, make a concluding statement relative to the RQ.

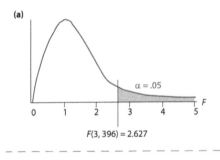

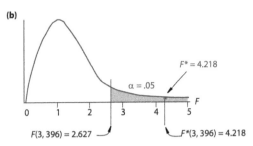

Figure 9.5 Illustration of the critical F value for the data set from Ch_9 IRA Exam Data for ANOVA Hypothesis Test Example (a), and the location of the calculated F value with respect to the critical F value (b).

Table 9.7   ANOVA Summary Table for Ch_9 IRA Exam Data Used in ANOVA Hypothesis Test Example

Source	SS	df	MS	F	Prob > F
Between	1324.620	3	441.540	4.218	.006
Within (Error)	41454.140	396	104.682		
Total	42778.760	399			

Decision. The calculated $F^* = 4.218$ is greater than the critical $F = 2.627$ and lies in the critical region as shown in Figure 9.5(b). Therefore, the decision is to *reject H_0*. Because the omnibus test is significant, we now conduct Tukey's HSD post hoc tests to determine which group means are significantly different. As summarized in Table 9.8, two of the six pairwise comparisons are statistically significant: BIM vs. DSC, $M_{Difference} = 4.50$, $t(396) = 3.11$, 95% CI = [0.77, 8.23], $p = .011$; and BIM vs. JAE, $M_{Difference} = 4.01$, $t(396) = 2.77$, 95% CI = [0.28, 7.74], $p = .03$. Although the mean differences of the four remaining pairwise comparisons are nonzero—for example, there is a 2.85-point difference between GWS and DSC in favor of GWS—these mean differences are not statistically significant.

Concluding Statement. Given a statistically significant omnibus, $F(3, 396) = 4.218$, $p = .006$, follow-up Tukey's HSD post hoc tests revealed two significant findings:
- Students who prepared for the IRA exam using BIM's software scored on average 4.5 points higher than students who prepared for the IRA exam using DSC's software.

Table 9.8 Summary of Tukey's HSD Post Hoc Comparisons
for Ch_9 IRA Exam Data Used in ANOVA
Hypothesis Test Example

Group Comparisons	M_{Diff}	$t(396)$	95% CI	p
BIM – DSC	4.50	3.11	[0.77, 8.23]	.011
BIM – GWS	1.65	1.14	[−2.08, 5.38]	.665
BIM – JAE	4.01	2.77	[0.28, 7.74]	.030
GWS – DSC	2.85	1.97	[−0.88, 6.58]	.201
GWS – JAE	2.36	1.63	[−1.37, 6.09]	.362
JAE – DSC	0.49	0.34	[−3.24, 4.22]	.987

Note. SE = 1.447.

- Students who prepared for the IRA exam using BIM's software scored on average 4 points higher than students who prepared for the IRA exam using JAE's software.

9.5 Using the Single-Factor ANOVA in Research: A Guided Example

We now provide a guided example of a research study that involves the use of the single-factor ANOVA. The data for this guided example were acquired from undergraduate flight students' records who completed a first course in aviation meteorology during the 4-year period 2018–2021. Students were administered the final examination online and given the flexibility to determine what day (Sunday–Saturday) of final exams week to take the exam. A pool of N = 399 students' exam scores was divided into seven samples each of n = 57 reflecting each day of the week. A copy of the data set is given in the file Ch_9 Guided Example Data. We will structure this example into three distinct parts: (a) pre-data analysis, (b) data analysis, and (c) post-data analysis.

Pre-Data Analysis

Research Question and Operational Definitions

The overriding research question is: "What is the difference in mean final exam scores among the seven separate groups of students who took the final exam, respectively, on Sunday, Monday, Tuesday, Wednesday, Thursday, Friday, and Saturday?" In the context of the current study, final exam scores refer to students' derived scores expressed as a percentage (0–100) from a teacher-prepared final examination for a first course in aviation meteorology. (This operational definition should include the course name and course description but is presented in general terms to maintain anonymity.)

Research Hypothesis

The mean exam score of one group will be different from that of at least one other group.

Research Methodology

The research methodology that best answers the RQ is *causal comparative*. This is because there is a group membership variable (Day of the Week), the groups are pre-existing, and students previously completed the final exam during past semesters.

Sample Size Planning (Power Analysis)

To determine the minimum sample size, we consult *G*Power* (Faul et al., 2007, 2009), a free-to-use software for conducting a power analysis, with the following inputs:

- Test family: *F* tests
- Statistical test: ANOVA: Fixed effects, omnibus, one-way.
- Type of power analysis: A priori: Compute required sample size—given α, power, and effect size.
- Input parameters: Effect size $f = 0.25$ (medium effect), α error prob = .05, Power = .80, and Number of groups = 7.

The minimum sample size needed is $n = 231$. Thus, we will need at least 231 scores (33 per group if we intend to maintain equal sample sizes across the seven groups) to have an 80% probability of finding a medium effect and correctly rejecting the H_0. The given data set consists of $n = 399$ cases (57 per group), so we meet this minimum threshold.

Data Analysis

We now analyze the given data set via a hypothesis test involving the single-factor ANOVA.

Step 1. Identify the targeted parameter and formulate null/alternative hypotheses. The targeted parameter is the mean, and the null/alternative hypotheses are:

H_0: $\mu_{Sunday} = \mu_{Monday} = \mu_{Tuesday} = \mu_{Wednesday} = \mu_{Thursday} = \mu_{Friday} = \mu_{Saturday}$.
There will be no significant mean difference in final exam scores among the seven groups regardless of the day of the week students took the exam.
H_1: At least one group mean will be significantly different from the others.

Step 2. Determine the test criteria. The test criteria involve three components:

Level of Significance (α). We will set alpha to α = .05.

Test Statistic. The test statistic is *F* with $df_{Between} = (G - 1) = (7 - 1) = 6$, and $df_{Within} = (N - G) = (399 - 7) = 392$. Thus, $df = (6, 392)$.

The Critical Boundary. There is no table entry for $df = (6, 392)$ in Table 5/Appendix A, so we will use Soper's (2022) online critical *F* calculator, which yields a critical *F* value of 2.12 as illustrated in Figure 9.6(a).

Step 3. Collect data, check assumptions, run the analysis, and report the results.

Collect Data. The data are in the file Ch_9 Guided Example Data.

Check Assumptions. There are three assumptions for the single-factor ANOVA.
- Independence. We assume the scores in each of the seven samples are independent because no student took the exam on more than one day.

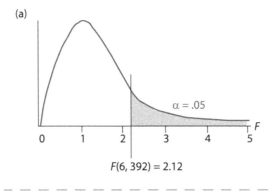

$F(6, 392) = 2.12$

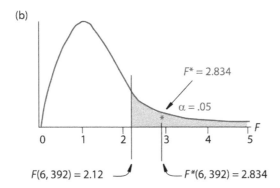

$F(6, 392) = 2.12$ $F*(6, 392) = 2.834$

Figure 9.6 Illustration of the critical F value for Chapter 9's Guided Example (a), and the location of the calculated F value with respect to the critical F value.

- Normality. The p value for the Shapiro–Wilk Goodness of Fit Test for the residuals of the final exam scores is $p = .0019$, which means that this assumption is *not* met for the given data set. However, the central limit theorem can be applied because of the relatively large sample sizes where each sample is of size $n = 57$. Therefore, we will claim that this assumption is met.
- Equal Variances. The Levene test reported from our statistics program is $F(6, 392) = 1.099, p = .362$, which confirms that the variances from the seven populations are statistically equal.

All assumptions have been met.

Run the Analysis and Report the Results. The results of this analysis as reported by our statistics program are $R^2 = .042, F(6, 392) = 2.834, p = .010$. A copy of the corresponding ANOVA summary table is given in Table 9.9.

Step 4. Decide whether to reject or fail to reject the null hypothesis, conduct post hoc comparisons if appropriate, and make a concluding statement relative to the RQ.

Decision. The calculated $F* = 2.834$ is greater than the critical $F = 2.12$ and lies in the critical region as shown in Figure 9.6(b). Therefore, the decision is to *reject* H_0. Because the omnibus test is significant, we now conduct Tukey's HSD post hoc comparisons to determine which group means are significantly different.

Table 9.9 ANOVA Summary Table for Chapter 9's Guided Example

Source	SS	df	MS	F	Prob > F
Day of Week	3717.920	6	619.653	2.834	.010
Error	85716.491	392	218.665		
Total	89434.411	398			

According to our statistics program, 2 of the 21 pairwise comparisons are statistically significant:

- Wednesday vs. Thursday, $M_{Difference} = 9.67$, $t(392) = 3.49$, 95% CI = [1.46, 17.88], $p = .010$.
- Saturday vs. Thursday, $M_{Difference} = 8.86$, $t(392) = 3.20$, 95% CI = [0.65, 17.07], $p = .025$.

Concluding Statement. Given a statistically significant omnibus, $F(6, 392) = 2.834$, $p = .010$, follow-up Tukey's HSD post hoc tests revealed two significant findings:

- Students who took the exam on Wednesday averaged 9.67 points higher than students who took the exam on Thursday.
- Students who took the exam on Saturday averaged 8.86 points higher than students who took the exam on Thursday.

Post-Data Analysis

We now examine the effect size, 95% CI, post hoc power, and plausible explanations.

Effect Size

The effect size of a single-factor ANOVA can be presented from an explained variance perspective or with respect to standard deviation.

Effect Size as a Measure of Explained Variance (Eta Squared). A common measure of effect size for the single-factor ANOVA is eta squared, denoted η^2. From a regression perspective, eta squared is equivalent to the coefficient of determination, R^2, and therefore measures the proportion of the total variance in a dependent variable that is associated with or explained by the group membership variable, which is the independent variable. Although the results of the ANOVA reported R^2, we also could calculate eta squared directly from the output of the ANOVA summary table as the ratio of the between-group variance and total variance:

$$\eta^2 = \frac{SS_{Between}}{SS_{Total}} = \frac{3717.92}{89434.411} = .04$$

Thus, 4% of the variance in the final exam scores is being explained by group membership, which represents the comparisons of the seven groups. Alternatively, if we know what day of the week a student took the final exam, then we have 4% of the information needed to perfectly predict this student's final exam score.

Some statistics programs do not report eta squared (η^2) but instead report the *partial eta squared* (η_p^2):

$$\eta_P^2 = \frac{SS_{Between}}{SS_{Between} + SS_{Within}}$$

η_P^2 is used as a measure of effect size in more complex ANOVAs, but for single-factor ANOVA (one variable with three or more independent groups, or levels), there is no difference between η^2 and η_P^2.

Effect Size as a Measure of Standard Deviation (Cohen's f). As presented in previous chapters, an effect size in inferential statistics is an index that reflects the amount of difference between two groups in standard deviation units. Because ANOVA involves comparing the means of more than two groups, we need a measure of effect size that represents the multiple pairwise comparisons of all the groups involved in the analysis. This measure is Cohen's effect size index, *f*, denoted as

$$f = \sqrt{\frac{\eta^2}{1-\eta^2}}$$

Cohen's *f*, which assumes equal sample sizes, is defined as the *standard deviation of standardized means*. In other words, if we were to standardize the scores in each separate group by converting them to *z* scores and then calculate the mean for each group, then *f* is the standard deviation of these mean *z* scores. Thus, *f* is a measure of dispersion that reflects the average distance each group's mean is from the overall mean of the group means. If *f* = 0, then this indicates that the group means are equal because they do not deviate from the mean of the means. As the disparity among the group means increases, though, *f* also will increase.

Cohen (1988) operationally defined *f* = 0.10, 0.25, and 0.40 as small, medium, and large effect sizes, respectively. When applied to the current study

$$f = \sqrt{\frac{\eta^2}{1-\eta^2}} = \sqrt{\frac{.04}{1-.04}} = \sqrt{\frac{.04}{.96}} = \sqrt{.04167} = .20$$

This is a little smaller than Cohen's operational definition of a medium effect. This effect size is representative of the *overall* ANOVA model. The effect size relative to any subsequent pairwise comparisons derived from a significant omnibus would be computed separately by Cohen's *d* for the independent-samples *t* test. (*Note:* G*Power, which was used for sample size planning earlier, uses Cohen's *f* for power analysis.)

The 95% Confidence Interval

For a single-factor ANOVA, and assuming that the omnibus is significant, we report the 95% CIs as part of the post hoc comparisons, but only interpret those associated with the significant pairwise comparisons.

Wednesday vs. Thursday. The 95% CI for the difference in means = [1.46, 17.88]. Thus, 95% of the time we can expect in the population students who take the final exam on Wednesday will score on average anywhere between 1.5 points to about 18 points higher than students who take the exam on Thursday. We further observe that 0 is not in this interval, and therefore, this difference in mean scores is significant. The accuracy in parameter estimation is relatively poor, though, because the interval is relatively wide.

Saturday vs. Thursday. The 95% CI for the difference in means = [0.65, 17.07]. Thus, 95% of the time we can expect in the population students who take the final exam on Saturday will score on average anywhere between about one-half point to approximately 17 points higher than students who take the exam on Thursday. We further observe that 0 is not in this interval. Therefore, the difference in mean scores is significant, but because the interval is fairly wide, the accuracy in parameter estimation is relatively poor.

The Power of the Study

To determine the power of the study we use *G*Power* (Faul et al., 2007, 2009) as we did for sample size planning with the following inputs:

- Test family: *F* tests
- Statistical test: ANOVA: Fixed effects, omnibus, one-way.
- Type of power analysis: "Post hoc: Compute achieved power—given α, sample size, and effect size."
- Input parameters: Effect size $f = 0.20$, α error prob = .05, Total sample size = 399, and Number of groups = 7.

This results in a power of .86. Thus, there is an 86% probability that we made a strong, correct decision to reject the null hypothesis, and that the effect size of $f = 0.20$ found in the sample also exists in the population.

Plausible Explanations for the Results

It appears that the best day to take the final exam is either Wednesday or Saturday. One plausible explanation for this result is that by waiting until midweek or the end of the week provides more time to study and prepare for the exam. A second plausible explanation is sample size. Recall from sample size planning, we needed a minimum sample size of $n = 231$ to have a power of .80 to detect an effect size of $f = 0.25$. Although we detected a smaller effect size, we also had a larger sample size, which increases power.

Chapter Summary

1. Conducting multiple independent-samples *t* tests of the same factor inflates the preset alpha level, called the testwise alpha, and leads to an experimentwise alpha, which increases the risk of committing a Type I error. The Bonferroni procedure controls the experimentwise error rate by dividing the testwise alpha by the number of pairwise comparisons, which yields a much more stringent testwise alpha.

2. An ANOVA is an omnibus test that examines multiple pairwise comparisons collectively at one time without inflating the preset alpha level. One ANOVA design is the

single-factor ANOVA, which involves one factor with three or more levels (i.e., groups). It compares group means by examining the amount of variability in the scores between groups and comparing this to the amount of variability in the scores within groups.

3. The test statistic for ANOVA is F, which is defined as the ratio of the between-groups variance and within-groups variance. The sampling distribution of F is an extension of the t distribution and is a function of the sample sizes upon which the targeted variances are based. The shape of the F distribution is positively skewed but approaches a normal distribution as sample sizes increase. The results of an ANOVA are presented in an ANOVA summary table, with the last step reflecting the corresponding F value.

4. The null hypothesis of the single-factor ANOVA is H_0: $\mu_1 = \mu_2 = \mu_3 = \ldots = \mu_{k-1} = \mu_k$, where k is the number of targeted groups, and the alternative hypothesis is H_1: At least one group mean is different from another. If the overall ANOVA is significant, then post hoc pairwise comparisons are conducted using Tukey's HSD test. Two measures of effect size are Cohen's f, which is expressed in standard deviation units, and eta-squared, η^2, which is a measure of explained variance.

Vocabulary Check

Analysis of variance (ANOVA)	Individual differences
ANOVA assumptions	Inflated alpha
ANOVA summary table	Mean of squared deviations (MS)
Between-groups variability	Mean squares
Bonferroni procedure	Omnibus test
Cohen's f	Pairwise comparisons
Error term	Post hoc comparisons
Error variance	Pyramiding effect of Type I errors
Eta-squared (η^2)	Standard deviation of standardized means
Experimental error	Testwise alpha
Experimentwise alpha	Treatment effect
F distribution	Tukey's HSD
F ratio	Within-groups variability

Review Exercises

A. Check Your Understanding

In 1–10, choose the best answer among the choices provided.

1. Consider the following research question: "What is the difference in IQ between flight students and non-flight students?" Which statistical test is the best one to use to answer this question?
   a. t test
   b. Pearson r
   c. F ratio
   d. Regression

2. In an *F* test, the individual differences among subjects that are not related to the independent variables are expressed as
   a. difference between groups.
   b. variance within groups.
   c. variance between groups.
   d. degrees of freedom.
3. The results of an ANOVA yielded an *F* statistic equal to −0.012. This means that
   a. it is unlikely that the ANOVA is statistically significant.
   b. the coefficient of determination is also negative.
   c. the *F* statistic was calculated incorrectly.
   d. the *SS* must be equal to zero.
4. Let's assume you endeavor to compare the effectiveness of sleep aids taken by pilots who are assigned international flights relative to how restful they feel, which is a self-reported response that can vary from 1 = "not very restful" to 5 = "extremely restful." Let's further assume that your primary focus is comparing over-the-counter aids such as melatonin to prescription drugs such as Lunesta as well as to a control group of pilots who do not use any type of sleep aid. Which statistical strategy is appropriate for this study?
   a. Single-sample *t* test because you are comparing a sample mean to a control group.
   b. Independent-samples *t* test because you are comparing two groups.
   c. Single-factor ANOVA because you are comparing three groups.
   d. Regression because you want to be able to predict pilots' level of "restfulness."
5. Three groups—A, B, and C—were selected randomly to participate in a research study. At the end of the study, an exam was administered and the scores were analyzed via ANOVA, which resulted in a calculated *F** value of $F(2, 9) = 15.9$. If the critical *F* values are 4.26 for $\alpha = .05$ and 8.02 for $\alpha = .01$, respectively, the researcher should report that
   a. Group C is superior to Groups A and B at the .05 level of significance.
   b. Group B is superior to Group A at the .01 level of significance.
   c. the differences among the group means could easily have occurred by chance.
   d. the differences among the group means are unlikely to have occurred by chance.
6. The following partial ANOVA summary table resulted from a study that involved four groups of 12 each. Which of the following statements is NOT true?

Source	SS	df	MS	F
Between	9			
Within				
Total	20	47		

   a. The overall sample size is $N = 48$.
   b. The overall sample size is $N = 47$.
   c. $SS_{Within} = 11$.
   d. $MS_{Between} = 3$.

7. Refer to the information given in Problem 6. If the ANOVA summary table is completed correctly, then $df_{Between} =$ _____, $MS_{Within} =$ _____, and the F ratio is _____.
   a. 3; 4; $F(3, 11) = 0.75$
   b. 3; 11; $F(3, 11) = 0.25$
   c. 3; 44; $F(3, 44) = 0.068$
   d. 3; 0.25; $F(3, 0.25) = 12$

8. A study examined the effect a person's level of education (high school diploma, 2-year college degree, 4-year college degree, and graduate/advanced degree) has on his or her career happiness. The study involved a random selection of airline industry employees, including airline ticket agents, transportation security officers, baggage handlers, and flight crew personnel. A person's career happiness was defined as scores on the Career Happiness Scale, which could range from 40 to 200 with higher scores reflecting higher levels of happiness. In this study, "level of education" is _____ and "career happiness" is _____.
   a. a categorical dependent variable; a continuous independent variable
   b. a categorical dependent variable; a categorical independent variable
   c. a single-factor categorical independent variable with four levels; a continuous dependent variable
   d. a four-factor categorical independent variable; a continuous dependent variable

9. Refer to the research description given in Problem 8. Let's assume that each group's mean Career Happiness score is $M_{HS} = 128.2$, $M_{2\text{-year}} = 64.0$, $M_{4\text{-year}} = 80.8$, and $M_{Grad/Adv} = 57.0$, and that the results are $F(3, 16) = 10.87$, $p = .0004$. Which of the following statements *most likely* is true based on these results?
   a. People working in the aviation industry who have a 2-year college degree are significantly happier in their career than people working in the aviation industry who have a graduate/advanced degree.
   b. People working in the aviation industry who have a high school diploma are significantly happier in their career than people working in the aviation industry who have a 2-year college, 4-year college, or graduate/advanced degree.
   c. People working in the aviation industry who have a 2-year college degree are significantly happier in their career than people working in the aviation industry who have a 4-year college degree.
   d. For people working in the aviation industry, the less educated they are based on the type of degree they have, the happier they are in their career.

10. An aviation researcher compared the job-related mental health, or *well-being*, of flight attendants from four different airlines: Delta, United, JetBlue, and Southwest. Flight attendants' mental health was measured using the self-administered Positive Mental Health (PMH) instrument, which is a 25-item, 6-point Likert-type response scale. Thus, scores could range from 25 to 150, with higher scores reflecting higher job-related well-being. The results of a single-factor ANOVA were as follows: $R^2 = .32$, $F(3, 32) = 5.10$, and $p = .0053$. Based on these results, what can we conclude about the sample?
    a. The overall sample size was $N = 36$ and each group was of equal size.
    b. The overall sample size was $N = 36$ and each group was of unequal size.
    c. The overall sample size was $N = 36$, but there is not enough information to determine the size of each group.
    d. The overall sample size was $N = 35$, but there is not enough information to determine the size of each group.

B. Apply Your Knowledge

Use the following research description and the corresponding data set to conduct the activities given in Parts A–C (see also Section 9.5).

As part of a capstone project, a student examined the time (in hours) to earning a private pilot's license (PPL) among different age groups of flight students. The data were acquired from a random sample of $N = 258$ flight students who completed their PPL at a local Part 141 program, and the data are provided in the file, Ch_9 Exercises Part B Data. Your assignment is to import this data set into your statistics program and compare the time to PPL among the targeted age groups. Keep in mind that fewer hours to PPL is "better" than more hours.

A. Pre-Data Analysis

1. Specify the research question and corresponding operational definitions.
2. Specify the research hypothesis.
3. Determine the appropriate research methodology/design and explain why it is appropriate.
4. Conduct an a priori power analysis to determine the minimum sample size needed. Compare this result to the size of the given data set and explain what impact the size of the given sample will have on the results relative to the minimum size needed.

B. Data Analysis

1. Conduct a hypothesis test for the single-factor ANOVA by applying all four steps as presented in Sections 9.4 and 9.5.

C. Post-Data Analysis

1. Report and interpret the two effect size indexes, η^2 and Cohen's *f*.
2. Determine and interpret the 95% confidence intervals, including their precision and AIPE, for all significant Tukey HSD post hoc comparisons, if appropriate.
3. Determine and interpret the power of the study.
4. Present at least two plausible explanations for the results.

References

Cohen, J. (1988). *Statistical power analysis for the behavioral sciences* (2nd ed.). Lawrence Erlbaum Associates.

Faul, F., Erdfelder, E., Buchner, A., & Lang, A.-G. (2009). Statistical power analyses using G*Power 3.1: Tests for correlation and regression analyses. *Behavior Research Methods, 41*, 1149–1160. [Download software at http://www.gpower.hhu.de]. https://www.psychologie.hhu.de/fileadmin/redaktion/Fakultaeten/Mathematisch-Naturwissenschaftliche_Fakultaet/Psychologie/AAP/gpower/GPower31-BRM-Paper.pdf

Faul, F., Erdfelder, E., Lang, A.-G., & Buchner, A. (2007). G*Power 3: A flexible statistical power analysis program for the social, behavioral, and biomedical sciences. *Behavior Research Methods*, *39*, 175–191. [Download software at http://www.gpower.hhu.de]. https://www.psychologie.hhu.de/fileadmin/redaktion/Fakultaeten/Mathematisch-Naturwissenschaftliche_Fakultaet/Psychologie/AAP/gpower/GPower3-BRM-Paper.pdf

Gravetter, F. J., & Wallnau, L. B. (2017). *Statistics for the behavioral sciences* (10th ed.). Cengage Learning.

Omni Calculator (n.d.). *The critical value calculator.* https://www.omnicalculator.com/statistics/critical-value

Soper, D. (2022). *Calculator: Critical F-value.* https://www.danielsoper.com/statcalc/calculator.aspx?id=4

10 Factorial ANOVA

Student Learning Outcomes

After studying this chapter, you will be able to do the following with respect to analyzing data using factorial ANOVA:

1. Prepare a group-means table, and compute and interpret the numerical row, column, and interaction effects.
2. Construct an interaction plot, identify the type of interaction illustrated, and interpret the interaction.
3. Prepare and interpret a factorial ANOVA summary table.
4. Conduct and interpret the results of a hypothesis test involving factorial ANOVA.
5. Engage in pre-data analysis, data analysis, and post-data analysis activities as described in Section 10.5.

10.1 Chapter Overview

This chapter extends our discussion of single-factor ANOVA from Chapter 9 to include two-factor ANOVA, also known as factorial ANOVA. We begin by introducing the concept of factorial designs, including the concept of interactions, which are an inherent component of factorial designs. We then introduce factorial ANOVA, which is a statistical strategy for analyzing data from a factorial design. We include in this discussion numerical and graphical representations of factorial designs, and then demonstrate how factorial ANOVA is used as an omnibus test to examine three separate effects: two main effects and an interaction effect. We follow this with an example of a hypothesis test involving factorial ANOVA, and conclude with a guided example that demonstrates the application of factorial ANOVA to a research study based on a factorial design.

10.2 The Concept of Factorial Designs

We introduced the concept of ANOVA in Chapter 9 by examining a single factor with multiple levels. The IRA exam study, for example, consisted of the single-factor, IRA exam test preparation packages and had four levels (or groups): BIM, GWS, JAE, and DSC. In many research studies, though, a single-factor design is limited in scope with respect to the information it provides. A more informative approach would be to include two (or more) factors with two or more levels.

DOI: 10.4324/9781003308300-14

For example, consider a research study that examines differences in GA pilots' understanding of basic weather knowledge based on their flight training program (Part 61 vs. Part 141) and the source of their CFIs' training (Civilian-based vs. Military-based). In this study, there are two factors, Flight Training Program and Source of CFIs' Training, and each factor has two levels. When a research study involves two or more factors with multiple levels, it is called a *factorial design*, and the combination of two factors and their respective levels is called a *two-factor research design*. Factorial designs are described by the number of levels of each factor separated by the symbol, ×, which is read as "by." Thus, the weather knowledge study is a *2 × 2 factorial design*.

Matrix Representation of Factorial Designs

Two-factor factorial designs are commonly represented numerically as a rectangular array, or *matrix*, and follow the convention of a matrix, namely, *row-by-column*. Thus, the rows of a factorial design represent the levels of the first factor, and the columns represent the levels of the second factor. This structure is illustrated in Figure 10.1 for the 2 × 2 weather knowledge study presented above. Observe the following from Figure 10.1:

- The levels of Factor *A*, "Flight Training Program," represent the rows.
- The levels of Factor *B*, "Source of CFI Training," represent the columns.
- Each cell represents a specific group that reflects the "cross" of the level of one factor with the level of the second factor. For example, Group 1 represents GA pilots who received flight training via a Part 61 program and whose flight instructor received civilian-based CFI training.
- We use a matrix naming convention of "row–column" to identify each cell. For example, Cell 3 (Group 3) is identified by Row 2–Column 1.
- The number of cells is equal to the product of the number of levels of each factor. Because this is a 2 × 2 factorial design, the total number of cells is 2 × 2 = 4.

Example 10.1: *The Structure of a Two-Factor Factorial Design*

Prepare the matrix representation for the two-factor IRA exam study where the first factor is Type of Preparation Software and has four levels (BIM, GWS, JAE, and DSC), and the second factor is biological sex and has two factors (Female and Male).

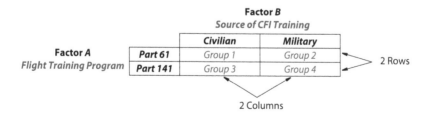

Figure 10.1 Matrix structure of a 2 × 2 factorial design.

Solution

Because Factor *A* has four levels and Factor *B* has two levels, the factorial design will be presented as a 4 × 2 matrix with four rows and two columns as shown in Figure 10.2. Notice there are 4 × 2 = 8 cells (or groups).

Let's return to the 2 × 2 factorial design of the weather knowledge study from Figure 10.1 and complete the cells of the matrix, which represent the respective group means. To do this, we will use the data set given in the file Ch_10 Weather Knowledge Study Data. These data were collected as part of a former student's unpublished study that partially replicated Burian (2002). Overall scores could range from 10 to 65, with higher scores reflecting higher basic weather knowledge. The respective group means are provided in Figure 10.3. For example, the mean for the first cell is $M_1 = 42.6$, which is the mean weather knowledge score for the sample that consisted of GA pilots who received flight training under a Part 61 program and whose flight instructor received civilian-based CFI training. We refer to this structure as a *group-means table*.

In addition to individual group means, we also can report the mean for each level of the two factors, as well as the differences in means between the levels of each factor. This is illustrated in Figure 10.4.

Observe the following:

- The means for the levels of Factor *A* are the *row means*, and the difference in row means is the *row effect*. For example, going across each row and independent of the columns, GA pilots who received flight training via a Part 61 program averaged 6.6 points *lower* on the weather knowledge test than GA pilots who received flight training via a Part 141 program:

$$M_{\text{Part 61}} - M_{\text{Part 141}} = 43.75 - 50.35 = -6.60 \ (\text{Row effect})$$

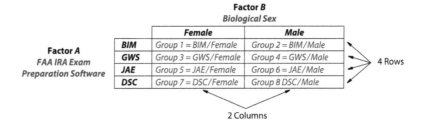

Factor B
Biological Sex

		Female	Male	
Factor A	**BIM**	Group 1 = BIM/Female	Group 2 = BIM/Male	
FAA IRA Exam	**GWS**	Group 3 = GWS/Female	Group 4 = GWS/Male	4 Rows
Preparation Software	**JAE**	Group 5 = JAE/Female	Group 6 = JAE/Male	
	DSC	Group 7 = DSC/Female	Group 8 DSC/Male	

2 Columns

Figure 10.2 Matrix structure of a 4 × 2 factorial design for Example 10.1.

Factor B
Source of CFI Training

		Civilian	Military
Factor A	**Part 61**	$M_1 = 42.6$	$M_2 = 44.9$
Flight Training Program	**Part 141**	$M_3 = 52.1$	$M_4 = 48.6$

Figure 10.3 The cells of the 2 × 2 factorial design for the weather knowledge study are completed with group means using the data from the file Ch_10 Weather Knowledge Data.

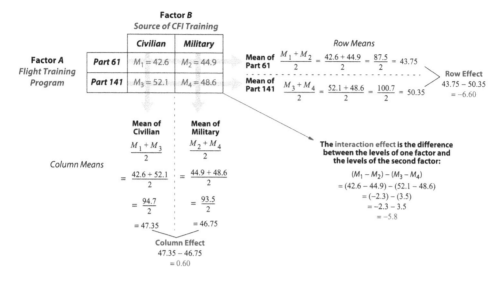

Figure 10.4 Descriptive illustration of the three effects of the 2 × 2 factorial design related to the weather knowledge study (see also Figure 10.3).

- The means for the levels of Factor B are the *column means*, and the difference in column means is the *column effect*. For example, going down each column and independent of the type of flight program, GA pilots whose flight instructors received civilian-based CFI training averaged 0.6 points higher on the weather knowledge test than GA pilots whose flight instructors received their CFI training in the military:

$$M_{\text{Civilian}} - M_{\text{Military}} = 47.35 - 46.75 = 0.60 \ (\text{Column effect})$$

- In addition to assessing the row and column effects, factorial designs also enable researchers to examine *interactions* between factors with respect to the DV. Numerically, an interaction is the difference in means between the levels of the factors:

$$(M_1 - M_2) - (M_3 - M_4)$$
$$= (42.6 - 44.9) - (52.1 - 48.6)$$
$$= (-2.3) - (3.5)$$
$$= -2.3 - 3.5$$
$$= -5.8 \ (\text{Interaction effect})$$

As a result, a two-factor factorial design enables us to simultaneously examine three separate effects: row, column, and interaction. In Sections 10.3 and 10.4, we introduce factorial ANOVA and demonstrate via hypothesis testing how to determine if any of these effects are statistically significant as well as how to interpret their results. Before doing so, though, we need to first understand the concept of interactions.

The Concept of Interactions

In a factorial design, an interaction occurs when *the levels of one factor operate differently under the levels of a second factor.* To understand what we mean by this statement, it is helpful to visualize an interaction graphically. To do this, we present the weather knowledge study from three different hypothetical situations, which are depicted in Figures 10.5(a), 10.5(b), and 10.5(c), respectively. As we examine these plots, observe that the endpoints of each line segment—shown as closed circles—represent the mean score for each group. For example, the left endpoint for the Military line segment is the mean weather knowledge score for the group of GA pilots who received flight training via a Part 61 program and whose CFIs received CFI training in the military.

Zero Interaction

In Figure 10.5(a), the lines are parallel. This indicates that regardless of the type of flight training program, GA pilots whose CFIs received military-based CFI training always have higher weather knowledge than GA pilots whose CFIs received civilian-based CFI training. As a result, the interaction is zero, which indicates that the levels of one factor are operating the same under the levels of the second factor. Also recall that from a mathematical perspective, parallel lines have equal slopes. Thus, if we were to subtract the slopes of these lines, we get 0.

When the lines of an interaction plot are parallel, the numerical representation of an interaction also will be zero—that is, $(M_1 - M_2) - (M_3 - M_4) = 0$. For example, Figure 10.6 presents a group-means table involving hypothetical data and the corresponding interaction plot. The graphical representation of the interaction reveals parallel lines, and the numerical representation confirms this: $(6 - 4) - (10 - 8) = (2) - (2) = 0$.

Disordinal Interaction

In Figure 10.5(b), the lines cross. This indicates that the levels of one factor are operating *differently* under the levels of the second factor. For example, GA pilots who received flight training from a Part 61 program have higher weather knowledge if their CFIs received military-based CFI training, but it is just the opposite with respect to Part 141. GA pilots who received flight training from a Part 141 program have higher weather knowledge if their CFIs received civilian-based CFI training. As a result, we have an interaction between the levels of Flight Training Program and the levels of Source of CFI Training.

This situation is called a disordinal interaction and demonstrates a "one size does not fit all" concept. In other words, and in the context of the given setting, neither the Part 61 nor the Part 141 flight training program is appropriate for GA pilots with respect to learning about weather. GA pilots from a Part 61 program are better suited to learn about weather from CFIs who received their training in the military, but GA pilots from a Part 141 program are better suited to learn about weather from CFIs who received civilian-based CFI training.

Ordinal Interaction

In Figure 10.5(c), the lines are neither parallel nor cross. This outcome is similar to the zero interaction situation in Figure 10.5(a) because GA pilots whose CFIs received military-based CFI training always have higher weather knowledge than GA pilots whose CFIs received civilian-based CFI training. Although graphically both situations reflect "no interaction," there is a difference in their numerical representations. When the lines are parallel, the numerical representation is equal to zero, but in all other cases, the numerical representation is not zero. As a result, we refer to this type of interaction plot as an ordinal interaction, which distinguishes it from both zero and disordinal interactions.

(a)

Zero Interaction

This plot shows that regardless of the Flight Training Program (Part 61 or Part 141), GA pilots whose CFIs received Military-based CFI training *always* have higher weather knowledge than GA pilots whose CFIs received Civilian-based CFI training. As a result:

> The levels of one factor (Flight Training Program) are operating the same under the levels of the second factor (Source of CFI Training) with respect to the DV (Weather Knowledge).

(b)

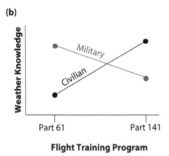

Disordinal Interaction

This plot shows that GA pilots from a Part 61 flight training program have higher weather knowledge if their CFIs received Military-based CFI training, whereas GA pilots from a Part 141 flight training program have higher weather knowledge if their CFIs received Civilian-based CFI training. Unlike the "No Interaction" plot, the lines are not parallel but instead "cross," resulting in a disordinal interaction:

> The levels of one factor (Flight Training Program) are operating differently under the levels of the second factor (Source of CFI Training) with respect to the DV (Weather Knowledge).

(c)

Ordinal Interaction

This plot is similar to the Zero Interaction plot because regardless of the Flight Training Program (Part 61 or Part 141), GA pilots whose CFIs received Military-based CFI training *always* have higher weather knowledge than GA pilots whose CFIs received Civilian-based CFI training. Thus, *the levels of one factor are operating the same under the levels of the second factor with respect to the DV.* However, because the lines are not parallel, we cannot claim "zero interaction," and because the lines do not cross, we cannot claim a disordinal interaction. As a result, this situation is called an *ordinal interaction.*

Figure 10.5 Graphical illustrations of the three different situations involving interactions.

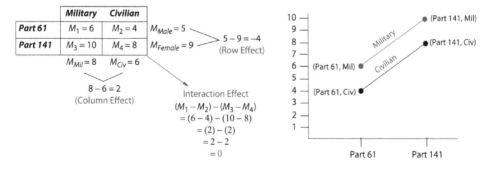

Figure 10.6 Numerical and graphical illustrations of a zero interaction.

Example 10.2: *Graphing and Interpreting Interactions*

Construct an interaction plot using the data from the file Ch_10 Weather Knowl-edge Study Data. These data were summarized in the group-means table in Figure 10.3.

Solution

Graphing the Interaction Plot

The endpoints of the line segments representing an interaction plot are the mean scores for each group. Thus, to construct the interaction plot, we plot the cell means and connect the points with a line. There is a protocol we follow, though: The key to plotting these means correctly is to work *vertically* from the table— that is, column-wise—and always place Factor A, which is the "row" factor, on the horizontal axis ("rows" are horizontal). Working with the group-means table from Figure 10.3:

- First plot $M_1 = 42.6$, which is the Part 61 mean under Civilian, and $M_3 = 52.1$, which is the Part 141 mean under Civilian, and connect them with a line segment as shown in Figure 10.7(a). This line segment represents the Civilian group.
- Next plot $M_2 = 44.9$, which is the Part 61 mean under Military, and $M_4 = 48.6$, which is the Part 141 mean under Military, and connect them with a line segment as shown in Figure 10.7(b). This line segment represents the Military group.

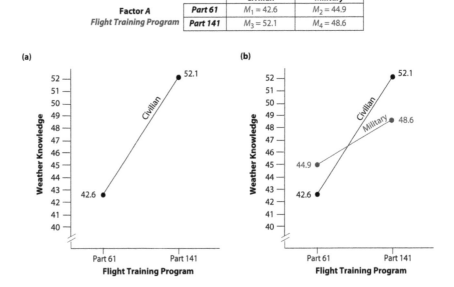

		Factor B	
		Source of CFI Training	
		Civilian	Military
Factor A	Part 61	$M_1 = 42.6$	$M_2 = 44.9$
Flight Training Program	Part 141	$M_3 = 52.1$	$M_4 = 48.6$

Figure 10.7 Constructing an interaction plot from the group-means table given in Figure 10.3 and replicated here for the convenience of the reader.

Interpreting the Interaction Plot

The interaction plot reveals a disordinal interaction (the lines cross). The two levels of Flight Training Program are operating differently under the two levels of Source of CFI Training: GA pilots in a Part 61 flight training program have higher weather knowledge when their CFIs received military-based CFI training ($M = 44.9$), whereas GA pilots in a Part 141 flight training program have higher weather knowledge when their CFIs received civilian-based CFI training ($M = 52.1$).

10.3 The Logic and Structure of Factorial ANOVA

We now turn our attention to the concept of factorial ANOVA, also called two-way ANOVA, which is a statistical strategy used to analyze data from a factorial design. Similar to single-factor ANOVA, factorial ANOVA is an omnibus test that examines the simultaneous effects of two different factors on a dependent variable via a single analysis. For example, from the weather knowledge study presented earlier, we now know that a two-factor factorial design consists of three separate effects: (a) the *row effect*, which is the difference in mean scores across the rows, independent of the columns; (b) the *column effect*, which is the difference in mean scores among the columns, independent of the rows; and (c) an *interaction effect*, which examines if the levels of one factor are operating the same or differently under the levels of the second factor. Thus, instead of examining the effect Flight Training Program has on weather knowledge in one study, and then examining the effect Source of CFI Training has on weather knowledge in a second study, and then examining the interaction effect in a third study, factorial ANOVA examines all of these effects in a single analysis.

Recall that a single-factor ANOVA partitions the overall variance into between-groups variance ($MS_{Between}$) and within-groups variance (MS_{Within}), which also is called the error term. Also recall that $MS_{Between}$ represents between-groups differences, individual differences, and experimental error; MS_{Within} represents only individual differences and experimental error; and the F ratio reflects the comparison between these two variances.

$$F = \frac{\text{Treatment Effect} + \text{Individual Differences} + \text{Experimental Error}}{\text{Individual Differences} + \text{Experimental Error}}$$

$$= \frac{\text{Between-Groups Variance}}{\text{Within-Groups Variance}}$$

$$= \frac{SD^2_{Between}}{SD^2_{Within}}$$

$$= \frac{\dfrac{SS_{Between}}{df_{Between}}}{\dfrac{SS_{Within}}{df_{Within}}}$$

$$= \frac{MS_{Between}}{MS_{Within}}$$

In factorial ANOVA, the within-groups variance (MS_{Within}) of the F ratio remains the same, but the between-groups variance ($MS_{Between}$) is partitioned into three separate components: the variability associated with the row factor (Factor A); the variability associated with the second factor (Factor B); and the variability associated with the interaction, denoted A × B. This partitioning of the between-groups variance is illustrated in Figure 10.8. As a result, factorial ANOVA produces three separate F ratios: one each for the row and column effects, which are collectively referred to as the main effects, and one for the interaction effect. The structure of these F ratios is the same as they are for a single-factor ANOVA except now the numerators reflect a specific effect (row, column, or interaction) as presented in the bullet list below. In all three F ratios, though, the denominators remain the same, namely, they reflect the variance within groups, which is the error term.

- $F \text{ ratio for Factor } A(\text{row effect}) = \dfrac{\text{Between-Groups Variance}}{\text{Within-Groups Variance}} = \dfrac{\dfrac{SS_A}{df_A}}{\dfrac{SS_{\text{Within}}}{df_{\text{Within}}}} = \dfrac{MS_A}{MS_{\text{Within}}}$

- $F \text{ ratio for Factor } B(\text{column effect}) = \dfrac{\text{Between-Groups Variance}}{\text{Within-Groups Variance}} = \dfrac{\dfrac{SS_B}{df_B}}{\dfrac{SS_{\text{Within}}}{df_{\text{Within}}}} = \dfrac{MS_B}{MS_{\text{Within}}}$

- $F \text{ ratio for Interaction effect} = \dfrac{\text{Between-Groups Variance}}{\text{Within-Groups Variance}} = \dfrac{\dfrac{SS_{A\times B}}{df_{A\times B}}}{\dfrac{SS_{\text{Within}}}{df_{\text{Within}}}} = \dfrac{MS_{A\times B}}{MS_{\text{Within}}}$

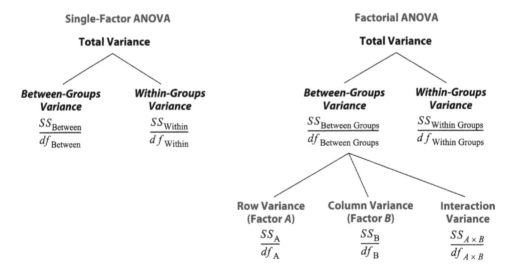

Figure 10.8 Comparing the partitioning of total variance between single-factor and factorial ANOVAs. (Adapted from Gravetter & Wallnau, 2017, p. 377.)

Note the following about the degrees of freedom:

- $df_{Between} = G - 1$, where G = number of groups (same as single-factor ANOVA)
- $df_A = a - 1$, where a = the number of levels of Factor A
- $df_B = b - 1$, where b = the number of levels of Factor B
- $df_{A \times B} = (a - 1)(b - 1)$
- $df_{Within} = N - G$ where N = the total sample size (same as single-factor ANOVA)

Putting this together leads to the general structure of the factorial ANOVA summary table. This structure is first presented in generic form in Table 10.1 using letters of the alphabet to represent each cell, and then again in Table 10.2, which replaces these letters with the proper notation. You should recognize that the structure is exactly the same as that of the single-factor ANOVA summary table (see Chapter 9/Table 9.4), except for the partitioning of the between-groups variance as highlighted in the tables.

Rather than manually calculate these cell entries as we did in Chapter 9/Section 9 for the single-factor ANOVA, let's analyze a completed factorial ANOVA summary table. The data for this table came from a former student's research study that examined the effect Biological Sex (Female vs. Male) and Flight Hours had on the annual salary of a random group

Table 10.1 Generic Structure of the Factorial ANOVA Summary Table Using Letters of the Alphabet to Represent Cells

Source	SS	df	MS	F
Between Groups	a	g	$m = a/g$	$F = m/q$
• Factor A (Row)	b	h	$n = b/h$	$F = n/q$
• Factor B (Column)	c	i	$o = c/i$	$F = o/q$
• Interaction ($A \times B$)	d	j	$p = d/j$	$F = p/q$
Within Groups	e	k	$q = e/k$	
Total	f	l		

Table 10.2 General Structure of the Factorial ANOVA Summary Table with Proper Notations

Source	SS	df	MS	F
Between Groups	$SS_{Between}$	$(G-1)$	$\dfrac{SS_{Between}}{(G-1)}$	$\dfrac{MS_{Between}}{MS_{Within}}$
• Factor A (Row)	SS_A	$(a-1)$	$\dfrac{SS_A}{(a-1)}$	$\dfrac{MS_A}{MS_{Within}}$
• Factor B (Column)	SS_B	$(b-1)$	$\dfrac{SS_B}{(b-1)}$	$\dfrac{MS_B}{MS_{Within}}$
• Interaction ($A \times B$)	$SS_{A \times B}$	$(a-1)(b-1)$	$\dfrac{SS_{A\times B}}{(a-1)(b-1)}$	$\dfrac{MS_{A\times B}}{MS_{Within}}$
Within Groups	SS_{Within}	$(N-G)$	$\dfrac{SS_{Within}}{(N-G)}$	
Total	SS_{Total}	$(N-1)$		

of pilots employed by a regional airline. Flight Hours were defined as Low (fewer than 1500 hours) and High (1500 or more hours), and annual salary was in U.S. dollars. A copy of the data is provided in the file Ch_10 Salary Example Data, the corresponding group-means table is given in Figure 10.9, and the factorial ANOVA summary table is given in Table 10.3.

With respect to Table 10.3, ignoring the shaded region and focusing solely on the first and last two rows, the structure and calculations of the corresponding cell entries for factorial ANOVA follow those of the single-factor ANOVA summary table presented in Chapter 9/Section 9.3. Notice further that the F ratio for the overall model is $F(3, 58) = 3.9385$, which is statistically significant for $\alpha = .05$ where the critical F value is $F(3, 58) = 2.76$. Thus, we have a significant omnibus and may now examine the main and interaction effects, which are highlighted in the shaded region:

- The Factor A variance examines the significance of the row effect. As illustrated in Figure 10.9, this effect is $52613 - 55944 = -3331$ and represents the difference in mean salaries between the two levels of Biological Sex independent of Factor B. Thus, Female pilots' mean annual salary was $3,331 less than Male pilots' mean annual salary. Based on the corresponding F value in Table 10.3, $F(1, 58) = 1.7677$. Thus, the $3,331 difference is *not* statistically significant for $\alpha = .05$ with critical $F(1, 58) = 4.00$.
- The Factor B variance is examining the significance of the column effect. As illustrated in Figure 10.9, this effect is $51477 - 57925 = -6448$ and represents the difference in salaries between the two levels of Flight Hours independent of Factor A. Thus, the mean annual salary of pilots with Low flight hours was $6,448 less than the mean annual salary of pilots with High flight hours. Based on the corresponding F value in Table 10.3, $F(1, 58) = 7.0205$. Thus, the $6,448 difference is statistically significant for $\alpha = .05$ with critical $F(1, 58) = 4.00$.
- The $A \times B$ variance examines the significance of the interaction effect. From Figure 10.9, this effect is $5,060$, which is not 0, and therefore, the lines are not parallel. As shown in Figure 10.10, the corresponding interaction plot reveals an ordinal interaction: Regardless of whether you are a female or male pilot flying for a regional airline, pilots with High flight hours always have higher annual salaries than pilots with Low flight hours. When compared to critical $F(1, 58) = 4.00$, the calculated F^* value as reported in Table 10.3 is $F(1, 58) = 1.3368$. Thus, this interaction is not statistically significant for $\alpha = .05$.

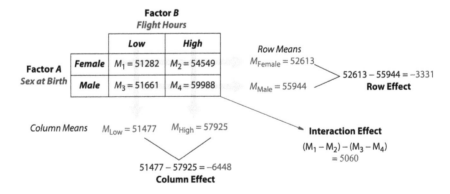

Figure 10.9 Group-means table for the regional airline pilots' salary study.

Table 10.3 Factorial ANOVA Summary Table for the Salary Example

Source	SS	df	MS	F
Between Groups	844955591	3	281651864	3.9385
• A = Sex at Birth	126414871	1	126414871	1.7677
• B = Flight Hours	502054447	1	502054447	7.0205
• Interaction (A × B)	95598437	1	95598437	1.3368
Within Groups	4147722942	58	71512465	
Total	4992678533	61		

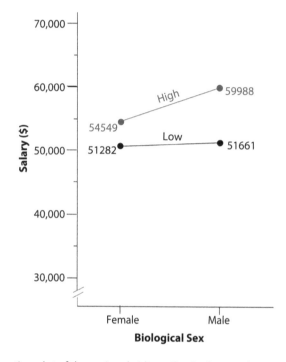

Figure 10.10 Interaction plot of the regional airline pilots' salary study.

This example demonstrates the utility of factorial ANOVA, namely, it enables us to conduct—under a single (omnibus) analysis—three separate F tests that are independent of each other. Furthermore, because main effects and interaction effects are independent of each other, this can lead to one of five possible outcomes:

(a) One significant main effect but no significant second main effect or interaction effect (this is what resulted in the regional pilots' annual salary study).
(b) Two significant main effects but no significant interaction effect.
(c) No significant main effects, but a significant interaction effect.
(d) One significant main effect and a significant interaction effect.
(e) Two significant main effects and a significant interaction effect.

We will discuss these possible outcomes further in Section 10.4 as part of our presentation on post hoc tests. However, for the moment, it is important to recognize that if the interaction is significant—cases (c), (d), and (e) above—then we should *not* interpret the results of the row and column effects. This is because a significant interaction means that the levels of one factor are *not* operating the same under the levels of the second factor. In other words, the levels of one factor *depend* on the levels of the second factor, which implies *there is no consistent main effect because the effect on the dependent variable changes from one level to the next.* This can be visualized in a disordinal interaction plot where the crossed lines show that the levels of one factor are not operating consistently under the levels of the second factor. Although there is nothing to preclude us from reporting the results of the two main effects in the presence of a significant interaction, it is prudent to include the caveat that the findings relative to the main effects might be spurious.

10.4 Statistical Inferences Involving Factorial ANOVA: Hypothesis Testing

Statistical inference can be approached either directly via confidence intervals or indirectly via hypothesis testing. Because factorial ANOVA is considered an omnibus test, it does not yield any specific confidence intervals. Instead, CIs are computed for any applicable post hoc tests. Because the calculation and interpretation of these post hoc related CIs are the same as what we presented in Chapter 8/Section 8.3, we refer the reader to this section and focus exclusively on hypothesis testing in the current section.

A hypothesis test involving factorial ANOVA is performed the same as a single-factor ANOVA with one exception: Instead of one set of null and alternative hypotheses, we will have three separate sets: one each for two main effects, and one for the interaction effect. It's also possible that, depending on the statistics program being used, you might have to "build" the final ANOVA summary table by assembling the results from the overall analysis with the results of the effect tests for the main and interaction effects. Some statistics programs keep these results separate. For our discussion, we will simply report the final summary table that shows how the between-groups variance is partitioned into row, column, and interaction variances.

We now present an example of a hypothesis test for a factorial ANOVA using the data from the weather knowledge study. We will not perform any manual calculations but instead work directly from the group means summary table given in Figure 10.4. The respective research questions are as follows:

RQ 1: What is the effect of a flight training program (Part 61 vs. Part 141) on GA pilots' understanding of basic weather knowledge?
RQ 2: What is the effect of CFIs' training source (Civilian vs. Military) on GA pilots' understanding of basic weather knowledge?
RQ 3: What is the interaction effect between the levels of Factor *A* and the levels of Factor *B* with respect to GA pilots' understanding of basic weather knowledge?

Step 1. Identify the targeted parameter, and formulate the null/alternative hypotheses. The targeted parameter is the mean, and the null and alternative hypotheses are:
Factor A (Row Effect)
$H1_0$: $\mu_{Part61} = \mu_{Part141}$ There is no significant difference in mean weather knowledge scores between flight training programs.
$H1_1$: $\mu_{Part61} \neq \mu_{Part141}$ The different flight training programs will have significantly different mean weather knowledge scores.

Factor B (Column Effect)

$H2_0$: $\mu_{Civilian} = \mu_{Military}$ There is no significant difference in weather knowledge scores between the CFI training sources.

$H2_1$: $\mu_{Civilian} \neq \mu_{Military}$ The different CFI training sources will have significantly different mean weather knowledge scores.

A × B (Interaction Effect)

$H3_0$: There is no significant interaction effect. The two levels of Flight Training Program operate the same under the two levels of Source of CFI Training.

$H3_1$: There is a significant interaction effect. The two levels of Flight Training Program operate significantly differently under the two levels of Source of CFI Training.

Step 2. Determine the test criteria. There are four separate test criteria: one for the omnibus and one each for the three effects (row, column, interaction). The test statistic and level of significance for each is F and $\alpha = .05$, respectively. What will be different are the degrees of freedom for each F value. Using the presentation on the degrees of freedom from Section 9.3, the entries given in Table 10.2, and Soper's (2022) online critical F calculator, we get the following critical F values for $N = 180$, which is the size of the given sample:

$F_{Critical}$ **for Omnibus**

$$F(G - 1, N - G) = F(3 - 1, 180 - 4) = F(2, 176) = 3.05$$

$F_{Critical}$ **for Row Effect (Factor A)**

$$F(a - 1, N - G) = F(2 - 1, 180 - 4) = F(1, 176) = 3.89$$

$F_{Critical}$ **for Column Effect (Factor B)**

$$F(b - 1, N - G) = F(2 - 1, 180 - 4) = F(1, 176) = 3.89$$

$F_{Critical}$ **for Interaction Effect (A × B)**

$$F[(a - 1)(b - 1), N - G] = F[(2 - 1)(2 - 1), 180 - 4)] = F(1, 176) = 3.89$$

Note that the critical region for each effect is the same, namely, $F(1, 176) = 3.89$. This is because we are working with a 2×2 factorial ANOVA. The degrees of freedom and corresponding critical values will not be the same when the number of levels or the number of factors is different.

Step 3. Collect data, check assumptions, run the analysis, and report the results.

Collect Data. The data are in the file Ch_10 Weather Knowledge Study Data.

Check Assumptions. There are three primary assumptions for ANOVA:
- Independence. We assume that the data for each sample were randomly generated, and the samples (and their respective populations) are independent of each other. Therefore, this assumption is met.
- Normality. The p value for the Shapiro–Wilk Goodness of Fit Test for the residuals of the weather knowledge scores is $p = .1103$, which means that this assumption is met for the given data set.
- Homogeneity of variances. The Levene test result reported from our statistics program is $F(3, 176) = 2.376$, $p = .072$, which is not significant ($p > \alpha$), and therefore, the variances from the four populations are statistically equivalent.

Run the Analysis and Report the Results. As shown in the ANOVA summary table in Table 10.4, the results are as follows:
- Omnibus: $F(3, 176) = 11.3490$, $p < .0001$

Table 10.4 Factorial ANOVA Summary Table for the Weather Knowledge Study Data

Source	SS	df	MS	F	p
Between Groups	2331.444	3	777.148	11.3490	< .0001
• A = Flight Training Program	11933.8889	1	1983.8889	28.2413	< .0001
• B = Source of CFI Training	16.2000	1	16.200	0.2366	.6273
• Interaction (A × B)	381.3556	1	381.356	5.5691	.0194
Within Groups	12052.000	176	68.477		
Total	14383.444	179			

- Factor A (Row effect): $F(1, 176) = 28.2413, p < .0001$
- Factor B (Column effect): $F(1, 176) = 0.2366, p = .6273$
- A × B (Interaction effect): $F(1, 176) = 5.5691, p = .0194$

Step 4. Decide whether to reject or fail to reject each of the three null hypotheses, conduct post hoc tests if appropriate, and make a concluding statement relative to each corresponding RQ.

Decisions and Concluding Statements. The overall model is significant, $F(3, 176) = 11.35, p < .0001$. Because the omnibus is significant, we may now examine the results of the three hypothesis tests. Before doing so, though, it is important to also recognize that the findings from the three hypothesis tests yielded one significant main effect with respect to Factor A, and a significant interaction effect. Recall from our discussion in Section 10.3 that we must include a caveat when interpreting main effects in the presence of a significant interaction effect because the levels of one factor *depend* on the levels of the other factor.

- Factor A = Flight Training Program (Row Effect). The main effect for Flight Training Program is significant, $F(1, 176) = 28.24, p < .0001$. Therefore, reject $H1_0$ and conclude there is a significant difference in weather knowledge scores between those who attended a Part 61 school vs. those who attended a Part 141 school. More specifically, as summarized in the group-means table of Figure 10.4, Part 61 participants averaged 6.6 points *lower* on the weather knowledge test than Part 141 participants, and this 6.6-unit difference is statistically significant. The reader is cautioned that although this 6.6-unit difference is significant, it is in the presence of a statistically significant interaction effect and therefore might be spurious.
- Factor B = Source of CFI Training (Column Effect). The main effect for Source of CFI Training is not significant, $F(1, 176) = 0.24, p = .6273$. Therefore, fail to reject $H2_0$ and conclude there is no significant difference in weather knowledge scores between participants whose flight instructors received civilian-based CFI training vs. participants whose flight instructors received military-based CFI training. Once again, the reader is cautioned that this interpretation is in the presence of a significant interaction effect and therefore might be spurious.
- A × B (Interaction Effect). The interaction is significant, $F(1, 176) = 5.5691, p = .0194$. Therefore, reject $H3_0$ and conclude that the two levels of Flight Training Program are operating significantly differently under the two levels of Source of CFI Training with respect to GA pilots' understanding

of basic weather knowledge. Furthermore, as previously depicted in Figure 10.7 as part of Example 10.2, the interaction plot is disordinal: GA pilots from a Part 61 program had higher weather knowledge when taught by CFIs who received military-based CFI training, whereas GA pilots from a Part 141 program had higher weather knowledge when taught by CFIs who received civilian-based CFI training.

Post Hoc Tests. As presented in Section 10.3, a factorial ANOVA can lead to five possible outcomes. Two involve a nonsignificant interaction, and three involve a significant interaction. Let's examine these cases relative to post hoc tests.

- No significant interaction effect, but at least one significant main effect. If one or both main effects is significant, but the interaction is not significant, then pairwise comparisons are appropriate. For factors with two levels, no additional work is necessary because the pairwise comparisons are provided in the corresponding group-means table, which is the case for the current example (see Figure 10.4). If, however, the significant factor has three or more levels, then we conduct post hoc tests using Tukey's HSD exactly as we did in Chapter 9 for a single-factor ANOVA.

- Significant interaction effect. One post hoc analysis strategy for a significant interaction effect, regardless of the significance of the main effects, is to conduct a *test of simple main effects*, which compares the effects of one factor at a single level of the other factor. For example, in the current study, this would involve comparing (a) Civilian–Part 61 to Civilian–Part 141, (b) Military–Part 61 to Military–Part 141, (c) Part 61–Civilian to Part 61–Military, and (d) Part 141–Civilian to Part 141–Military. This effectively involves separating the two-factor study into a series of single-factor studies, which we will *not* do. Instead, we graph the interaction and interpret it in the context of the given research setting exactly as we did in the current example. Furthermore, if at least one main effect is significant, then we report this finding, but we do *not* perform any pairwise comparisons. Once again, this is because a significant interaction indicates that the effect of one factor depends on the levels of the second factor, which means the first factor's effect is not uniformly consistent across the levels of the second factor.

10.5 Using Factorial ANOVA in Research: A Guided Example

We now provide a guided example of a research study that was conducted by a former student that examined the effect of two factors on students' academic achievement. The first factor was Student Type, which had two levels (Flight vs. Nonflight), and the second factor was Aerobic Activity, which had three levels (bicycling, jogging, and swimming). Flight students were enrolled in a Part 141 flight training program at a 4-year university, nonflight students were attending a 4-year university but not enrolled in a flight training program, and the aerobic activities were performed at least 3 days per week for at least 30 minutes per day. Academic achievement was defined as students' cumulative grade-point average (GPA) at the end of their first academic year. A copy of the data set is given in the file Ch_10 Guided Example Data. We will structure this example into three distinct parts: (a) pre-data analysis, (b) data analysis, and (c) post-data analysis.

Pre-Data Analysis

Research Questions and Operational Definitions

The study is a 2×3 factorial design and hence there are three separate research questions: one each for the two main effects, and one for the interaction. Operational definitions were provided in the research description.

RQ 1. What is the effect of Student Type on academic achievement? Alternatively, "What is the difference in mean GPAs between flight and nonflight students?"

RQ 2. What is the effect of Aerobic Activity on academic achievement? Alternatively, "What is the difference in mean GPAs among the three aerobic activities of bicycling, jogging, and swimming?"

RQ 3. What is the *interaction effect* between the two levels of Student Type and the three levels of Aerobic Activity relative to academic achievement? Alternatively, "Do the levels of Student Type operate the same or differently under the levels of Aerobic Activity with respect to academic achievement?"

Research Hypothesis

The corresponding research hypotheses are as follows:

Hyp 1. Flight and nonflight students will differ in their academic achievement.

Hyp 2. The academic achievement related to at least one aerobic activity will be different from the other groups.

Hyp 3. There will be an interaction effect: The two levels of Student Type will operate differently under the three levels of Aerobic Activity with respect to academic achievement.

Research Methodology

The research methodology that best answers the RQs is *causal-comparative*. This is because the groups formed from crossing the two factors are intact and could not be manipulated. For example, we could not assign participants to be a flight student who jogs or a nonflight student who swims.

Sample Size Planning (Power Analysis)

For factorial ANOVA, in addition to the omnibus, we are testing three separate effects and therefore must perform four separate power analyses. To do this, we consult *G*Power* (*Faul et al., 2007, 2009) and apply it four different times: once for the omnibus, and once each for the three different effects (row, column, and interaction). For each analysis the following inputs are the same. What differs are the $df_{Numerator}$ and number of groups:

- Test family: *F* tests
- Statistical test = ANOVA: Fixed effects, special, main effects, and interactions
- Type of power analysis: A priori: Compute required sample size—given α, power, and effect size.

- Input parameters: Effect size $f = 0.25$ (medium effect), α error prob $= .05$, and Power $= .80$.

  Omnibus. Numerator $df = 5$ and Number of groups $= 6$, which yield a minimum size of $N = 211$.

  Factor A (Row Effect). Numerator $df = 1$ and Number of groups $= 2$, which yield a minimum sample size of $N = 128$. Thus, to have an 80% chance of finding a significant difference in GPAs between flight and nonflight students, independent of Factor B, requires at least 128 participants.

  Factor B (Column Effect). Numerator $df = 2$ and Number of groups $= 3$, which yield a minimum sample size of $N = 158$. Thus, to have an 80% chance of finding the mean GPA associated with one aerobic activity to be significantly different from the mean GPAs of the other aerobic activities, independent of Factor A, requires at least 158 participants.

  $A \times B$ *(Interaction Effect).* Numerator $df = 2$ and Number of groups $= 6$, which yield a minimum sample size of $N = 158$. Thus, to have an 80% chance of finding a significant interaction effect requires at least 158 participants.

Data Analysis

We now analyze the data from the given data set via a hypothesis test involving factorial ANOVA.

Step 1. **Identify the targeted parameter, and formulate null/alternative hypotheses.** The targeted parameter is the mean, and the three sets of null/alternative hypotheses are:

Factor A (Row Effect)

$H1_0$: $\mu_{\text{Flight}} = \mu_{\text{Nonflight}}$ There is no significant difference in academic achievement between flight and nonflight students.

$H1_1$: $\mu_{\text{Flight}} \neq \mu_{\text{Nonflight}}$ There will be a significant difference in academic achievement between flight and nonflight students.

Factor B (Column Effect)

$H2_0$: $\mu_{\text{Bicycling}} = \mu_{\text{Jogging}} = \mu_{\text{Swimming}}$ There is no significant difference in academic achievement among the three aerobic activities.

$H2_1$: At least one group mean will be different.

A $\times$ B (Interaction Effect)

$H3_0$: There is no significant interaction effect. The two levels of Student Type operate the same under the three levels of Aerobic Activity.

$H3_1$: There is a significant interaction effect. The two levels of Student Type operate significantly differently under the three levels of Aerobic Activity with respect to academic achievement.

Step 2. **Determine the test criteria.** There are four separate test criteria: one for the omnibus and one each for the three effects (row, column, interaction). The test statistic and level of significance for each is F and $\alpha = .05$, respectively. What will be different are the degrees of freedom for each F value. To determine the appropriate degrees of freedom, we consult Soper's (2022) online critical F calculator.

Recall that a = number of levels of Factor A, b = number of levels of Factor B, N = sample size, and G = total number of groups. All critical F values are with respect to the sample size of the given data set ($N = 60$).

$F_{Critical}$ *for the Omnibus*

$$F(G - 1, N - G) = F(6 - 1, 60 - 6) = F(5, 54) = 2.39$$

$F_{Critical}$ *for Factor A (Row Effect)*

$$F(a - 1, N - G) = F(2 - 1, 60 - 6) = F(1, 54) = 4.02$$

$F_{Critical}$ *for Factor B (Column Effect)*

$$F(b - 1, N - G) = F(3 - 1, 158 - 6) = F(2, 54) = 3.17$$

$F_{Critical}$ *for A $\times$ B (Interaction Effect)*

$$F[(a - 1)(b - 1), N - G]$$

$$= F[(2 - 1)(3 - 1), (60 - 6)] = F[(1)(2), (54)] = F(2, 54) = 3.17$$

Step 3. Collect data, check assumptions, run the analysis, and report the results.

Collect Data. The data are in the file Ch_10 Guided Example Data.

Check Assumptions. There are three primary assumptions for ANOVA:
- Independence. We assume the scores are independent and that each participant either is a flight or nonflight student and is engaged in only one of the targeted aerobic activities.
- Normality. The p value for the Shapiro–Wilk Goodness of Fit Test for the residuals of the GPAs is $p = .4456$, which means that this assumption is met for the given data set.
- Equal Variances. The Levene test result reported from our statistics program is $F(5, 54) = 2.041$, $p = .087$, which is not significant ($p > \alpha$), and therefore, the variances from the six populations are statistically equal.

Run the Analysis and Report the Results. As shown in the ANOVA summary table in Table 10.5, the results are:
- Omnibus: $F(5, 54) = 73.60$, $p < .0001$
- Factor A (Row effect): $F(1, 54) = 18.48$, $p < .0001$
- Factor B (Column effect): $F(2, 54) = 34.47$, $p < .0001$
- A $\times$ B (Interaction effect): $F(2, 54) = 140.28$, $p < .0001$

A copy of the group-means table is provided in Figure 10.11.

Table 10.5 Factorial ANOVA Summary Table for the Guided Example

Source	SS	df	MS	F	p
Between Groups	56.08	5	11.22	73.60	< .0001
• A = Student Type	2.82	1	2.82	18.48	< .0001
• B = Aerobic Activity	10.51	2	5.25	34.47	< .0001
• Interaction (A $\times$ B)	42.76	2	21.38	140.28	< .0001
Within Groups (Error)	8.23	54	0.15		
Total	64.31	59			

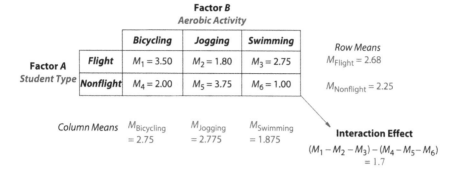

Figure 10.11 Group-means table for Chapter 10's Guided Example.

Step 4. Decide whether to reject or fail to reject each of the three null hypotheses, conduct post hoc tests if appropriate, and make a concluding statement relative to each corresponding RQ.

Decisions and Concluding Statements

- Omnibus. The calculated F^* for the omnibus is $F(5, 54) = 73.60$, which is greater than $F_{critical} = 2.39$. Because the overall model is significant, we may examine the significance of the variances associated with the three effects (row, column, and interaction).
- Factor A (Row effect). The calculated F^* for Factor A is $F(1, 54) = 18.48$, which is greater than $F_{critical} = 4.02$. Therefore, reject the corresponding null hypothesis $(H1_0)$ and conclude there is a significant difference in mean GPA between flight and nonflight students.
- Factor B (Column effect). The calculated F^* for Factor B is $F(2, 54) = 34.47$, which is greater than $F_{critical} = 3.17$. Therefore, reject the corresponding null hypothesis $(H2_0)$ and conclude that among the three aerobic activities of bicycling, jogging, and swimming, at least one group's mean GPA is significantly different from the other groups' mean.
- $A \times B$ (Interaction Effect). The calculated F^* for the interaction effect is $F(2, 54) = 140.28$, which is greater than $F_{critical} = 3.17$. Therefore, reject the corresponding null hypothesis $(H3_0)$ and conclude that the two levels of Student Type are operating significantly differently under the three levels of Aerobic Activity. To determine the type of interaction (disordinal or ordinal) and where the interaction occurs, we need to examine the interaction plot, which we do as part of the post hoc test discussion.

Post Hoc Tests. The factorial ANOVA summary table (Table 10.5) shows that both main effects and the interaction effect are significant. Therefore, post hoc tests are warranted. However, because the interaction is significant, our post hoc tests should be restricted to the interaction only and NOT include the significant main effects. Once again, this is because the levels of one factor are dependent on the levels of the second factor; therefore, we cannot interpret the main effects, which are independent of each other at face value. Nevertheless, we provide guidance on interpreting these post hoc tests for instructional purposes.

- Factor A = Student Type. The post hoc test for Factor A is automatically provided in the group-means table (Figure 10.11) because the factor involves two levels: M_{Flight} = 2.68, $M_{Nonflight}$ = 2.25, and the difference in GPAs is 2.68 – 2.25 = 0.43. Furthermore, from the ANOVA summary table (Table 10.5), this 0.43-unit difference is statistically significant. Thus, *independent* of Aerobic Activity, flight students' mean GPA is significantly higher than nonflight students' mean GPA. However, because this finding is in the presence of a significant interaction effect, the result might be spurious.
- Factor B = Aerobic Activity. From the group-means table in Figure 10.11, the mean GPAs for the three levels of Factor B *independent* of Student Type are $M_{Bicycling}$ = 2.75, $M_{Jogging}$ = 2.775, and $M_{Swimming}$ = 1.875. Furthermore, as reported in the ANOVA summary table in Table 10.5, this factor is statistically significant, which means that the mean GPA of least one aerobic activity is significantly different than that of another group. Because Factor B involves three levels, an appropriate post hoc test is Tukey's HSD to determine which of the three pairwise comparisons is significant: Bicycling vs. Jogging, Bicycling vs. Swimming, or Jogging vs. Swimming. However, we should neither conduct nor interpret the results of this post hoc test because of the significant interaction. For completeness, though, follow-up Tukey results showed that independent of Student Type, the mean difference between Bicycling and Swimming (M_{Diff} = 0.875) was significant, $t(54)$ = 7.088, $p < .0001$, and the mean difference between Jogging and Swimming (M_{Diff} = 0.900) was significant, $t(54)$ = 7.290, $p < .0001$. If you are instructed to report these results, then it is incumbent to caution the reader that the results might be spurious due to the significant interaction effect, which voids the independence of the two factors.
- $A \times B$ = Interaction. We examine the respective interaction plots and interpret the interactions two at a time. As depicted in Figure 10.12, there are three such interactions to interpret:

Bicycling vs. Jogging. This is a disordinal interaction. Flight students have higher GPAs if their aerobic activity is bicycling, whereas nonflight students have higher GPAs if their aerobic activity is jogging.

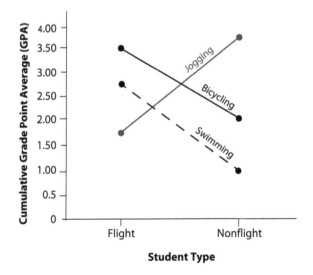

Figure 10.12 Interaction plot for Chapter 10's Guided Example.

Bicycling vs. Swimming. In this interaction plot, the lines do not cross. Further-more, when we calculate the numerical interaction from the group-means table in Figure 10.11 with respect to the cells associated with these levels we get

$$(M_1 - M_3) - (M_4 - M_6)$$
$$= (3.5 - 2.75) - (2 - 1)$$
$$= -0.25$$

Because the numerical interaction is not 0, the lines also are not parallel. As a result, this is an ordinal interaction: Regardless of Student Type, flight and non-flight students have higher GPAs if their aerobic activity is bicycling as opposed to swimming.

Jogging vs. Swimming. This is a disordinal interaction. Flight students have higher GPAs if their aerobic activity is swimming, whereas nonflight students have higher GPAs if their aerobic activity is jogging.

Post-Data Analysis

We now examine the effect sizes, 95% CIs, post hoc power, and plausible explanations.

Effect Size

The effect size for factorial ANOVA can be presented from an explained variance perspec-tive or with respect to standard deviation.

Effect Size as a Measure of Explained Variance (Partial Eta Squared). The effect size for fac-torial ANOVA, which is a bit more complex than single-factor ANOVA because it involves more than one factor, is partial eta squared (η_p^2):

$$\eta_p^2 = \frac{SS_{Between}}{SS_{Between} + SS_{Within}}$$

where $SS_{Between}$ and SS_{Within} are derived directly from the ANOVA summary table relative to the partitioning of the between-groups variance for each targeted effect. Although many statistics programs automatically report partial eta squares, we will manually calcu-late them for the current example.

- Effect Size for Factor *A* (Student Type)

$$\eta_p^2 = \frac{SS_{Between}}{SS_{Between} + SS_{Within}} = \frac{2.816667}{2.816667 + 8.23} = \frac{2.816667}{11.046667} = 0.25498$$

- Effect Size for Factor *B* (Aerobic Activity)

$$\eta_p^2 = \frac{SS_{Between}}{SS_{Between} + SS_{Within}} = \frac{10.508333}{10.508333 + 8.23} = \frac{10.508333}{18.738333} = 0.56079$$

- Effect Size for Interaction (*A* × *B*)

$$\eta_p^2 = \frac{SS_{Between}}{SS_{Between} + SS_{Within}} = \frac{42.758333}{42.758333 + 8.23} = \frac{42.758333}{50.98833} = 0.83859$$

Relative to Cohen's (1988) guidance, these are medium-to-large effect sizes, and they provide an estimate of the amount of variance in GPAs explained by the respective factor.

In addition to reporting partial eta squared, Levine and Hullett (2002) recommended researchers report η^2 and *omega squared* (ω^2), which is another effect index of explained variance:

$$\omega^2 = \frac{SS_{\text{Between}} - (df_{\text{Between}})(MS_{\text{Within}})}{SS_{\text{Total}} + MS_{\text{Within}}}$$

According to Levine and Hullet, ω^2 is preferable to η^2 when group sizes are equal. Following their recommendation, we created Table 10.6 as an extension of Table 10.5 to include entries for η^2, η_P^2, and ω^2.

Effect Size as a Measure of Standard Deviation (Cohen's f). As presented in Chapter 9, Cohen's effect size index, *f*, with respect to eta squared is

$$f = \sqrt{\frac{\eta^2}{1-\eta^2}}$$

Applying this to partial eta squared and omega squared, we get the following:

$$f_P = \sqrt{\frac{\eta_P^2}{1-\eta_P^2}} \text{ and } f_\omega \sqrt{\frac{\omega^2}{1-\omega^2}}$$

Because omega squared is an unbiased estimator of the population eta squared, and because we had equal group sizes in our guided example, we use it as the basis for calculating the respective *f* values, which we also provide in Table 10.6.

The 95% Confidence Interval

We report 95% CIs for factorial ANOVA as part of any appropriate post hoc comparisons. For example, with respect to Factor *A* in the current example, the corresponding 95% CI

Table 10.6 Factorial ANOVA Summary Table for the Guided Example Extended to Include Effects Sizes

						Effect Size Indexes [a]			
Source	*SS*	*df*	*MS*	*F*	*p*	η^2	η_P^2	ω^2	$f\omega$
Between Groups	56.08	5	11.22	73.60	< .0001	0.872	0.873	0.882	2.729
• *A* = Student Type	2.82	1	2.82	18.48	< .0001	0.044	0.253	0.041	0.208
• *B* = Aerobic Activity	10.51	2	5.25	34.47	< .0001	0.163	0.561	0.158	0.434
• Interaction (*A* × *B*)	42.76	2	21.38	140.28	< .0001	0.665	0.839	0.659	0.562
Within Groups (Error)	8.23	54	0.15						
Total	64.31	59							

Note. [a] η^2 is eta squared; η_P^2 is partial eta squared; ω^2 is omega squared where effect sizes of 0.01, 0.059, and 0.138 are considered small, medium, and large, respectively; and $f\omega$ is Cohen's *f* based on omega squared, and effect sizes of 0.10, 0.25, and 0.40 are considered small, medium, and large, respectively.

with respect to comparing flight to nonflight students' mean GPAs is [0.23, 0.64]. Thus, in the population, we can expect that 95% of the time flight students' mean GPA, independent of aerobic activity, will be anywhere from 0.23 to 0.64 units higher on average than that of nonflight students. However, remember that the interaction was significant in the current example, and hence we do not interpret any post hoc comparisons for the individual factors. Therefore, for factorial ANOVA, the only time we would report and interpret CIs for the factors is when the interaction is not significant.

With respect to the interaction, if we were to conduct a test of simple main effects as a post hoc analysis of a significant interaction, then we would report and interpret the corresponding 95% CIs. For example, from the group-means table in Figure 10.11, the difference in GPAs between flight students who swim ($M = 2.75$) and nonflight students who swim ($M = 1.00$) is 1.75. Based on the results from our statistics program, the corresponding 95% CI is [1.23, 2.27]. Thus, in the population, we can expect that 95% of the time the mean GPA of flight students who swim will be on average anywhere between 1.23 and 2.27 units higher than the mean GPA of nonflight students who swim, and because 0 is not in this interval, this difference is statistically significant. However, because we opt not to conduct any post hoc tests of a significant interaction, we do not report CIs for the interaction.

The Power of the Study

To determine the power of the study we use *G*Power* (Faul et al., 2007, 2009) as we did for sample size planning with two exceptions: we change Type of power analysis to "Post hoc: Compute achieved power—given α, sample size, and effect size"; and we enter the actual effect size using Cohen's *f* from Table 10.6 for the omnibus and each effect.

Omnibus. Numerator $df = 5$, Number of groups $= 6$, and $f = 2.729$, which yield a power of 1.00. However, we report this power as $> .99$. Thus, there is a greater than 99% probability that the significant omnibus observed from the sample involving the two targeted factors and their interaction is significant in the population.

Factor A (Row Effect). Numerator $df = 1$, Number of groups $= 2$, and $f = 0.208$, which yield a power of .35. Thus, there is a 35% probability that the differences in GPA between flight and nonflight students, independent of aerobic activity, found in the sample exists in the population. This also means there is a 35% probability that the decision to reject the corresponding null hypothesis was correct.

Factor B (Column Effect). Numerator $df = 2$, Number of groups $= 3$, and $f = 0.434$, which yield a power of .84. Thus, there is an 84% probability that the differences in GPA among the three different aerobic activities, independent of student type, found in the sample exist in the population. This also means there is an 84% probability that the decision to reject the corresponding null hypothesis was correct.

A × B (Interaction Effect). Numerator $df = 2$, Number of groups $= 6$, and $f = 0.562$, which yield a power of .97. Thus, there is a 97% probability that the interactions found in the sample exist in the population. This also means there is a 97% probability that the decision to reject the corresponding null hypothesis was correct.

Plausible Explanations for the Results

For Factor *A*, the group-means table in Figure 10.11 shows that independent of Aerobic Activity, the difference in mean GPAs between flight and nonflight students is 0.43

(M_{Flight} = 2.68 and $M_{Nonflight}$ = 2.25). Although this difference was significant, the corresponding power was low (.35). A plausible explanation for this is sample size. Recall from the sample planning stage, we needed a minimum sample size of N = 128 to have an 80% chance of finding a significant medium effect (f = 0.25). However, the sample size of the data set was N = 60, and the magnitude of the actual effect was f = 0.208. Thus, we had one-half of the sample size needed, and the effect was smaller than initially anticipated. To have at least an 80% chance of finding an effect of f = 0.208 would require a sample of size N =184.

For Factor B, the group-means table in Figure 10.11 shows that independent of Student Type, bicycling and jogging have nearly the same mean GPAs ($M_{Bicycling}$ = 2.75 and $M_{Jogging}$ = 2.775), and both are about 1 point higher than the mean GPA for swimming ($M_{Swimming}$ = 1.875). Although we did not conduct post hoc comparisons because of the significant interaction, it is reasonable to conclude that the aerobic activities of bicycling and jogging are superior to swimming with respect to academic achievement. One plausible explanation for this finding is the *selection threat* to internal validity. Because participants were not randomly assigned to a particular aerobic activity, it is possible that the academic achievement of those who comprised the swimming group was low prior to the study. This is further supported by the fact that no pre-assessments were administered to ensure group equivalency.

As for the interaction, Figure 10.12 shows two disordinal interactions: bicycling is superior to jogging for flight students, and jogging is superior to bicycling for nonflight students with respect to GPA. A plausible explanation for both findings also is the selection threat to internal validity. In the absence of any relative pre-assessments, it is conceivable that the flight-bicycling group and the nonflight-jogging group had higher academic achievement prior to participating in the study.

Chapter Summary

1. Factorial designs involve two (or more) factors each with multiple levels and are represented numerically by a group-means table, which is a rectangular array of numbers where each cell consists of that group's mean score. The group-means table follows the structure of a matrix such that the rows represent the levels of the first factor (Factor A) and the columns represent the levels of the second factor (Factor B).

2. In a factorial design, the difference in row means of Factor A, independent of the levels of Factor B, is called the row effect, and the difference in column means of Factor B, independent of the levels of Factor A, is called the column effect. Independent of row and column effects is an interaction effect, which numerically is the difference in means between the levels of Factor A and the levels of Factor B.

3. An interaction means that the levels of one factor are operating differently under the levels of the second factor. When illustrated graphically where the DV is on the vertical axis, the row factor is on the horizontal axis, and the second factor is the "overlay," three general graphs emerge: if the lines are parallel, then this denotes a zero interaction; if the lines cross, then this denotes a disordinal interaction; and if the lines are neither parallel nor cross, then this denotes an ordinal interaction.

4. Factorial ANOVA is an omnibus statistical strategy used to analyze data from a factorial design. If the omnibus is significant, then we can examine the significance of the

three separate effects: two independent main effects (row and column), and interaction. Factorial ANOVA examines the significance of these three effects simultaneously by partitioning the between-groups variance into the variance associated with Factor A, Factor B, and interaction ($A \times B$). Thus, a hypothesis test involving factorial ANOVA has three separate sets of null and alternative hypotheses: one each for the two main effects, and one for the interaction effect. The corresponding assumptions of factorial ANOVA are the same as those for single-factor ANOVA: independence, normality, and equal variances.

5. Given a significant omnibus and a nonsignificant interaction effect, it is appropriate to interpret the statistical significance of the two main effects by independently examining the difference in row means and the difference in column means, including any corresponding posthoc tests. However, if the interaction is significant, then we do not interpret the main effects because the two factors are no longer independent of each other. A significant interaction means that the effect of Factor A depends on the levels of Factor B, and the effect of Factor B depends on the levels of Factor A. As a result, we plot the interaction and interpret the plot.

Vocabulary Check

Column effect	Omega squared (ω^2)
Column means	Partial eta squared (η_p^2)
Disordinal interaction	Row effect
Factorial ANOVA	Row means
Factorial design	Single-factor design
Group-means table	Two-factor design
Interaction	Two-way ANOVA
Main effects	Zero interaction

Review Exercises

A. Check Your Understanding

In 1–10, choose the best answer among the choices provided.

1. A study examined the effects of GA pilots' Self-Efficacy (Low vs. High) and their level of Risk Perception (Low vs. High) on the number of self-reported hazardous events in which they were involved. Based on the results of a 2×2 factorial ANOVA, the omnibus was not significant, $F(3, 12) = 2.53$, Self-Efficacy was not significant, $F(1, 12) = 0.0033$, Risk Perception was significant, $F(1, 12) = 5.54$, and the interaction was not significant, $F(1, 12) = 2.06$. Which of the following is correct?
   a. $N = 15$.
   b.   The $MS_{Risk Perception}$ was larger than MS_{Within}.
   c.   There were two treatments.
   d.   There is an error. You cannot have a significant factor with a nonsignificant omnibus.

2. After completing her data analysis, a researcher concluded: "With respect to final grades, attending evening classes is an advantage for female flight students but a

disadvantage for male flight students." Which of the following research strategies did the researcher most likely use to arrive at this conclusion?
   a.   *t* test for independent sample means
   b.   *z* test for independent sample means
   c.   Single-factor ANOVA
   d.   Factorial ANOVA

3. A 2 × 6 factorial ANOVA was used to examine the effect of airline passengers' biological sex (Female or Male) and Class of Service they normally purchase with respect to their general attitudes toward the airline industry. Class of Service was based on Delta Air Lines (Basic Economy, Main Cabin, Delta Comfort +, First Class, Delta Premium Select, and Delta One), and attitudes were measured on a scale from 1 to 100 with higher scores indicating more positive attitudes. The author reported, "The results of data analysis showed that $M_{Male} = 90$, $M_{Female} = 60$, $F(1, 415) = 32.05$, $p < .001$, and the effects of Class of Service and the corresponding interactions were small and not significant." What should the researcher do next?
   a.   Conclude that male passengers have more positive attitudes than female passengers.
   b.   Conduct a post hoc test to determine if the different seat classifications are significant.
   c.   Conclude that the results are inconclusive because the interactions were not significant.
   d.   Conclude that a computation mistake was made because *F* cannot be that large.

4. One distinction between an ordinal interaction and a disordinal interaction is
   a.   the graphs of the corresponding lines of an ordinal interaction are parallel.
   b.   the slopes of the corresponding lines of a disordinal interaction are equal.
   c.   the graphs of the corresponding lines of an ordinal interaction do not cross.
   d.   a disordinal interaction is always statistically significant.

5. As part of a research study, a researcher examined the effect "professional experience" has on airport managers' level of professionalism, which was measured using Hall's Professional Inventory (HPI). Professional experience was based on two different factors. Factor *A* = Professional Affiliation had two levels: "Member of American Association of Airport Executives (AAAE)" and "Not a Member of AAAE." Factor *B* = Level of Education had three levels: "2 or fewer years of college," "4-year degree," and "graduate degree." A total of $N = 150$ airport managers participated in the study. Below is a partially completed ANOVA summary table that resulted from data analysis. Based on these results, $SS_{Between} = $ _____, $SS_{Interaction} = $ _____ and the *dfs* for the *F* ratio are (___, ___).

Source	SS
Between	?
• Professional Affiliation	120
• Level of Education	0
• Interaction	?
Within	254
Total	454

a. 200, 0, $F(6, 149)$
b. 120, 80, $F(6, 149)$
c. 200, 80, $F(5, 144)$
d. 120, 200, $F(5, 144)$

6. A study was designed to examine the effect a person's level of education (high school diploma, 2-year college degree, 4-year college degree, and graduate/advanced degree) has on his or her career happiness. The study involved a random selection of people working in the aviation industry, including airline ticket agents, security officers, baggage handlers, and pilots. A person's career happiness was defined as scores on the Career Happiness Rating Scale and could range from 40 to 200; the higher the score, the happier the individual. The statistical procedure the researcher used to measure this effect was

a. factorial ANOVA.
b. *t* test for independent sample means.
c. single-factor ANOVA.
d. correlation.

7. A group of subject matter experts (SMEs) developed a new method for teaching the phonetic alphabet used by pilots and air traffic controllers to new flight students. They randomly selected 24 flight students to receive either the old phonics method or the new phonics method. Because the students have morning and afternoon classes, the SMEs wanted to build this factor into the research design. Given below are two tables. The first summarizes the descriptive statistics and the second is a partially completed ANOVA summary table

	Morning			Afternoon		
	N	M	SD	N	M	SD
Old	6	16.33	1.03	6	13.17	3.55
New	6	13.50	2.74	6	16.83	2.14

Source	SS	df	MS	F
Between				
Time of Day	0.042			
Lesson Design	1.042			
Interaction				
Error	128.500			
Total	192.959			

If the ANOVA summary table is completed successfully, the designers of the study can conclude:

a. There is evidence to suggest a difference between the morning and afternoon classes.
b. There is evidence to suggest a difference between the old and new lesson designs.
c. There is evidence to suggest that the effect of Lesson Design in the morning is different from the effect of Lesson Design in the afternoon.
d. There is insufficient information to make a conclusion.

8. An aviation researcher examined the effect of alcohol consumption on pilot performance. The study was conducted in a flight simulator and pilot performance was defined as the measure of deviation (in degrees) while following the glide slope on an instrument landing system. Deviations were recorded over a 20-second period both prior to and after pilots drank the prescribed amount of alcohol. Because of the potential influence age could have on the results, the researcher conducted a factorial design. Factor A = Blood Alcohol Content (BAC) had three levels: None (BAC = 0), Low (BAC = .05), and Moderate (BAC = .11); and Factor B = Age had three levels: 25–34 years old, 35–44 years old, and 45 years old or older. The results of a factorial ANOVA were: R^2 = .61, $F(8, 210)$ = 41.92, $p < .0011$, and the three respective null hypotheses were rejected. Which of the following statements represents an accurate interpretation of the results?
   a. The interaction was disordinal.
   b. The interaction was ordinal.
   c. The interaction effect was not significant.
   d. The interaction was significant.

9. A study based on Smith-Jentsch et al. (1996) examined the effect of two factors involving Part 121 pilots' assertiveness: Factor A = Number of prior negative events pilots previously experienced (None vs. At Least One). Examples of prior negative events included having previously flown with a captain who used unsafe procedures, having previously experienced a potentially life-threatening incident in which crew coordination was a causal factor, and having been previously pressured to take a flight that the pilot felt uncomfortable taking because of weather or mechanical problems. Factor B = Treatment (Trained vs. Untrained) with pilots being randomly assigned to one of the two groups. The Trained group received 2 hours of assertiveness training taught by commercial pilots and designed to help pilots be more assertive in the cockpit during a conflict situation. The Untrained group did not receive any type of assertiveness training. One week after the training session, both groups participated in a simulated flight where their assertiveness skills were rated. The ratings ranged from 1 (least effective) to 5 (most effective), and the simulated flight involved conflict situations that required assertiveness during the simulation. Figure 10.13 contains the interaction plot that resulted from data analysis. Which statement reflects an accurate assessment of the interaction?
   a. There appears to be a significant interaction effect.
   b. There appears to be a significant main effect with respect to Factor A.
   c. There appears to be a significant main effect with respect to Factor B.
   d. There appears to be a significant main and interaction effect.

10. An aviation human factors study examined the relationship between military pilots' biological sex (Male vs. Female) and three different types of stressful situations: Performing mental arithmetic, pain induction, and social role-playing. For mental arithmetic, participants were presented with five arithmetic problems and given 3 minutes to solve them, social role-playing was conducted via computer simulation, and pain induction was performed via a cold pressor test. The dependent variable was heart rate (measured in beats per minute), with higher heart rates signaling higher levels of stress. Pilots were randomly assigned to one of the three groups and as they engaged in the respective activity, their heart rate was measured and recorded. The primary focus of the study was to determine which situation was least stressful for male and female pilots. Figure 10.14 contains the interaction plot that resulted from

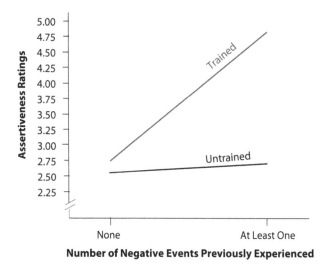

Figure 10.13 Interaction plot for Chapter 10 Review Exercises Part A9.

data analysis. If you were to make a recommendation to a military pilot's association group, which of the following statements would reflect an accurate recommendation based on the interaction plot?

a. Male pilots should not be asked to perform mental arithmetic because this activity stresses them out more than female pilots who perform the same task.

b. Female pilots would make better spies than male pilots because if they were captured female pilots would endure less stress than male pilots if they were to undergo painful torture.

c. Regardless of biological sex, social role-playing always leads to higher levels of stress than pain induction.

d. Female pilots, regardless of the situation in which they were placed, always had lower levels of stress than male pilots.

B. Apply Your Knowledge

Use the following research description and the corresponding data set to conduct the activities given in Parts A–C (see also Section 10.5).

This research study follows the description given in Problem 8 of Check Your Understanding. The study involved $N = 225$ participants who were randomly assigned to one of the nine groups. The data are given in the file Ch_10 Exercises Part B Data. Import this data set into your statistics program and perform a factorial ANOVA. Keep in mind that smaller deviation measures mean better pilot performance.

A. Pre-Data Analysis

1. Specify the research questions and corresponding operational definitions.
2. Specify the research hypotheses.

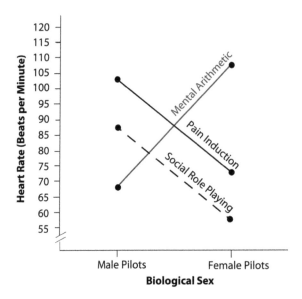

Figure 10.14 Interaction plot for Chapter 10 Review Exercises Part A10.

3. Determine the appropriate research methodology/design and explain why it is appropriate.
4. Conduct an a priori power analysis to determine the minimum sample size needed. Compare this result to the size of the given data set and explain what impact the size of the given sample will have on the results relative to the minimum size needed.

B. Data Analysis

1. Conduct a hypothesis test using factorial ANOVA by applying all four steps of hypothesis testing as presented in Sections 10.4 and 10.5.
2. Determine what post hoc test(s) to conduct based on the results of the analysis and then conduct and interpret the results.
3. Prepare and interpret the interaction plot.

C. Post-Data Analysis

1. Report and interpret the effect size indexes, η^2, η_p^2, ω^2, and Cohen's f for all three effects.
2. If appropriate, determine and interpret the 95% confidence intervals, including their precision and AIPE, for all significant post hoc comparisons.
3. Determine and interpret the power of the three effects (row, column, and interaction).
4. Present at least one plausible explanation for the results.

References

Burian, B. K. (2002, September). *General aviation pilot weather knowledge and training.* (FAA Grant #00-G-020, Final Report). Federal Aviation Administration, Washington, DC. https://www.tc.faa.gov/its/worldpac/grants/00g20a.pdf

Cohen, J. (1988). *Statistical power analysis for the behavioral sciences* (2nd ed.). Lawrence Erlbaum Associates.

Faul, F., Erdfelder, E., Buchner, A., & Lang, A.-G. (2009). Statistical power analyses using G*Power 3.1: Tests for correlation and regression analyses. *Behavior Research Methods*, 41, 1149–1160. [Download software at http://www.gpower.hhu.de]. https://www.psychologie.hhu.de/fileadmin/redaktion/Fakultaeten/Mathematisch-Naturwissenschaftliche_Fakultaet/Psychologie/AAP/gpower/GPower31-BRM-Paper.pdf

Faul, F., Erdfelder, E., Lang, A.-G., & Buchner, A. (2007). G*Power 3: A flexible statistical power analysis program for the social, behavioral, and biomedical sciences. *Behavior Research Methods*, 39, 175–191. [Download software at http://www.gpower.hhu.de]. https://www.psychologie.hhu.de/fileadmin/redaktion/Fakultaeten/Mathematisch-Naturwissenschaftliche_Fakultaet/Psychologie/AAP/gpower/GPower3-BRM-Paper.pdf

Gravetter, F. J., & Wallnau, L. B. (2017). *Statistics for the behavioral sciences* (10th ed.). Cengage Learning.

Levine, T. R., & Hullett, C. R. (2002). Eta squared, partial eta squared, and misreporting of effect size in communication research. *Human Communication Research*, 28(4), 612–625. https://doi.org/10.1111/j.1468-2958.2002.tb00828.x

Smith-Jentsch, K. A., Jentsch, F., Payne, S. C., & Salas, E. (1996). Can pretraining experiences explain individual differences in learning? *Journal of Applied Psychology*, 81(1), 110–116. https://doi.org/10.1037/0021-9010.81.1.110

Soper, D. (2022). *Calculator: Critical F-value.* https://www.danielsoper.com/statcalc/calculator.aspx?id=4

Part E

Analyzing Research Data Using a Within-Groups Design

11 Repeated-Measures *t* Test

Student Learning Outcomes

After studying this chapter, you will be able to do the following with respect to a repeated-measures *t* test:

1. Distinguish between a between-groups design and a within-groups design.
2. Perform and interpret the results of a hypothesis test for the mean difference.
3. Determine and interpret the 95% confidence interval for the mean difference.
4. Engage in pre-data analysis, data analysis, and post-data analysis activities as described in Section 11.4.

11.1 Chapter Overview

In this chapter, we introduce the concept of a repeated-measures research design, which involves a single group that is exposed to all the treatment conditions, one at a time, and then assessed on the same dependent variable at the end of each treatment administration. As part of this discussion, we present various examples of aviation-based repeated-measures research studies, as well as advantages and disadvantages of such designs. We then introduce the repeated-measures *t* test, which is used to analyze data from a repeated-measures study that involves exactly two measurements, and provide examples of confidence intervals and hypothesis testing for the repeated-measures *t* test. We conclude this chapter with a guided example that demonstrates the application of the repeated-measures *t* test to a repeated-measures research study.

11.2 The Concept of Repeated-Measures

Let's assume we would like to examine the effectiveness of a 15-hour workshop designed to reduce CFIs' level of complacency (see Example 4.3(a) and Section 4.4). To conduct this study, we could assign participants to either a treatment or control group, and when the workshop is over, administer to both groups Dunbar's (2015) 7-item CFI complacency instrument, where higher scores reflect a higher level of complacency in flight instruction. If the treatment group's mean score is significantly lower than the control group's mean score, then the workshop was effective. This approach employs a *between-groups design* because we are comparing group means between groups.

As an alternative to this approach, we could first administer Dunbar's (2105) instrument to *one* group of CFIs to determine the group's baseline level of complacency prior to the

DOI: 10.4324/9781003308300-16

workshop, and then after the workshop is over, administer Dunbar's instrument a second time as a post-assessment. We would then compare the mean post-assessment score to the mean pre-assessment score. If the post-assessment mean is lower than the pre-assessment mean, then the workshop was effective. This approach employs a *within-groups design* because we are comparing means within the same group.

In the within-groups approach, CFIs are being measured twice at two different times—Time Period 1 is the pre-assessment, and Time Period 2 is the post-assessment—and CFIs' pre-assessment scores are being paired with their respective post-assessment scores. Because we are repeatedly measuring the same individuals on the same dependent variable under different treatment conditions or at different points in time, this approach is referred to as a repeated-measures design, and the focus is on examining differences between the multiple measurements.

Similar to a repeated-measures design is a matched-samples design, which also involves paired data, but consists of separate groups of individuals such that participants in one group are matched, one-to-one, with participants in a second group. The objective of a matched-samples design is to form equivalent groups relative to a particular variable(s) the researcher would like to control. For example, in an experimental study targeting flight performance, it might be beneficial to match participants on total number of flight hours or total number of hours as PIC. A matched-samples design also is referred to as a *paired-samples* or *matched-pairs* design, and both a repeated-measures and matched-samples design are sometimes referred to generically as a *related-samples* design. For our purposes, we restrict our discussion to a repeated-measures design that involves one sample with exactly two measurements of the dependent variable. Following are several examples of repeated-measures studies along with their advantages and disadvantages.

Examples of Repeated-Measures Design Studies

Measuring the Effect of Different Treatments

Consider a medical experiment designed to study the effects two different drugs have on blood pressure. Using a between-groups approach, we would randomly assign participants to one of the two drug groups, measure participants' blood pressure, and then compare the mean blood pressure between the groups. Using a repeated-measures approach, we would administer the two drugs to each participant and allow time for the effects of the first drug to wear off before the second drug is given. We would then compare the mean blood pressure between the treatments. In this latter case, the same participants are being measured twice—once after each treatment.

An example of an aviation-related repeated-measures design that involves different treatments is Lindseth et al. (2011), who examined the effects of diet on cognition and flight performance among pilots enrolled in a large collegiate aviation commercial pilot curriculum. Each pilot received 4 days of prepared meals consisting of high protein, high carbohydrate, high fat, and a control diet, and Lindseth et al. allowed 2 full weeks to elapse between treatments (p. 3). In their study:

- Each diet represented a treatment, so there were four separate treatments, one of which was the control diet.
- Each participant received each treatment.

- Each participant's cognition and flight performance were measured after each treatment.
- The researchers compared the effect each diet had on participants' cognition and flight performance.

Measuring "Change" after an Intervention

A *change score study* measures the effect of a treatment by first assessing participants' *baseline* prior to treatment and then administering the same assessment post-treatment to determine the extent to which participants' scores changed as a result of the treatment. This is exactly what we presented earlier with respect to CFIs' level of complacency. As another example, consider a study designed to determine how effective a risk aversion workshop is in helping GA pilots become more risk-aversive. To conduct this study, we would

- Pre-assess pilots' tendency for or attitudes toward engaging in risky flight behaviors.
- Administer the treatment in the form of a workshop designed expressly to make pilots less risk prone.
- Post-assess participants' tendency for or attitudes toward engaging in risky behaviors.
- Examine the change in participants' scores on the two assessments.

An example of an aviation-related change study is Hamilton (2016) who examined changes in cognition between pre- and post-treatment, which was a specialized alcoholism treatment program, among a group of airline pilots diagnosed with alcohol use disorder.

A Longitudinal Study

A *repeated-measures longitudinal study* would involve measuring the same group of individuals over a specific time period. For example, Taylor et al. (2007) annually assessed the flight performance of the same 118 GA pilots over a 3-year period. Their flight performance was measured in a simulator and scored in terms of executing air traffic controller communications, traffic avoidance, scanning cockpit instruments, executing an approach to landing, and a flight summary score. Taylor et al. then compared pilots' flight performance scores across the 3-year period to determine if any tangible pattern emerged. For example, did flight performance improve each year, remain the same, or decline?

A Repeated-Measures Correlational Study

A repeated-measures design also can be implemented as a correlational study. For example, let's assume we would like to know how aviation students in their senior year of college feel about their current stress level compared to what they perceived their stress level was as a freshman. To conduct this study, we would:

- Randomly select a group of students who are seniors in a 4-year college aviation degree program.

- Using a scale from 1 to 10 where 1 = No Stress and 10 = Extremely High Stress, ask participants to report what they recall their stress level was during the first year they were in the program, and what they perceive is their current stress level.
- Measure the relationship between the two sets of scores using Pearson *r*.

In this example, a single group of participants is reporting information relative to two different time periods, and we are examining the relationship between the information acquired from the two time periods. Thus, there is no manipulation as there was in the first two examples. Because the same group of participants is being measured twice, and because we are examining the relationship between the measures in the absence of any manipulation, this is a *repeated-measures correlation study.*

A Combined Within-Groups and Between-Groups Comparison Study

It also is possible to mix a within-groups study with a between-groups study. For example, Yesavage et al. (1994) compared the flight performance of older pilots vs. younger pilots (between-groups) at two different time periods that were 10 months apart (repeated-measures). They also examined the effects of alcohol vs. placebo in the two age groups and at two points in time: acute intoxication (blood alcohol content = .10) and 8 hours after drinking alcohol. This type of study is known as a *repeated-measures comparative group design* because two independent groups are being measured more than once at different time periods and comparisons on a dependent variable are being made both within and between groups.

Advantages of Repeated-Measures Designs

Perhaps the single most important advantage of repeated-measures studies is they afford the researcher control of individual differences among participants. Individual differences are probably the largest source of variation in most human-subjects research studies. When left uncontrolled, as in a completely randomized design, they comprise part of the error term. In repeated-measures designs, participants serve as their own control. Thus, it is possible to identify the variance due to individual differences and separate it from the error term. This leads to a more precise analysis. It also leads to higher statistical power because participants serve as their own control, and therefore, there is no between-groups variability. Repeated-measures designs also are more economical than completely randomized designs because they provide considerable savings in the number of participants required for a given study. Because each participant is receiving all treatment conditions, a smaller sample size is needed. Repeated-measures designs also enable researchers to study phenomena or examine effects across time.

Potential Problems and Disadvantages of Repeated-Measures Designs

Repeated-measures designs also have several disadvantages. When the same group of participants is being tested more than once on the same dependent variable, there is the potential for *carry-over effects.* Carry-over effects refer to situations in which treatments administered earlier in a sequence continue to affect participants' behavior while they are being administered subsequent treatments. For example, in Lindseth et al.'s (2011) diet study presented earlier, participants were assessed four different times (once after each targeted diet) on the same dependent variables (cognition and flight performance). Focusing on flight performance, the concept of carry-over effects means pilots' flight performance after the first diet conceivably could have influenced their flight performance after the second diet, and this second assessment could have further influenced pilots' flight performance when they were assessed after the third and fourth diets.

This influence of carry-over effects could be beneficial or detrimental. For example, if repeated exposure to the same dependent measure after each treatment provides a *practice effect*, then participants' performance on each subsequent measure of the DV most likely will improve. In Lindseth et al.'s (2011) diet study, this means that if each subsequent flight performance improved pilots' skills, then this performance might not be due to the targeted diet but instead to practice effect. It also is possible that participants could get tired or bored from being repeatedly assessed on the same dependent measure, which could cause performance to become inferior or decline after subsequent measures. This is known as a *fatigue effect*. Thus, in Lindseth et al.'s diet study, it also is possible that pilots became fatigued by the time their flight performance was assessed a second, third, and fourth time. Lindseth et al. attempted to mitigate carry-over effects by including a 2-week "lapse" period between diets.

11.3 Statistical Inferences Involving the Repeated-Measures *t* Test

The *t* Statistic for Repeated-Measures

To analyze data from a repeated-measures study, we use the repeated-measures *t* test, which also is called the *paired-samples t test, related-samples t test, matched-pairs t test,* and *dependent-samples t test*. This *t* test is conceptually similar to the single-sample *t* test from Chapter 5 and the independent-samples *t* test from Chapter 8, which both have the same generic structure:

$$t = \frac{\text{Sample Statistic} - \text{Population Parameter}}{\text{Standard Error}}$$

Because a repeated-measures design involves a single sample, the repeated-measures *t* test is derived from the single-sample *t* test. For example, recall from Chapter 5 that the single-sample *t* test compares the mean of a single sample to a hypothesized mean from the parent population, and the formula is

$$t = \frac{M - \mu}{SE_M} = \frac{M - \mu}{SD / \sqrt{n}}$$

where M is the mean of the sample, μ is the mean of the hypothesized population, SE_M is the estimated standard error of the mean, and n is the sample size. When applied to the repeated-measures *t* test, the numerator will be the *mean difference between the paired scores*, denoted M_D, and the estimated standard error will be with respect to this mean difference, denoted SE_D. Thus, the general formula for the repeated-measures *t* test is

$$t = \frac{M_D - \mu_D}{SE_D}$$

Let's now examine how this test statistic is used for statistical inferences.

Hypothesis Testing for a Repeated-Measures Study

Consider the set of hypothetical data given in Table 11.1, which represents GA pilots' pre- and post-workshop scores from a risk aversion workshop study that was designed to help GA pilots become more risk-aversive. Higher scores reflect a tendency toward risk-taking behavior, and lower scores reflect a tendency toward risk-aversive behavior. The objective of the study is to determine the effectiveness of the workshop. For example, if the mean post-workshop score is less than the mean pre-workshop score, then we could claim that the workshop was effective in helping GA pilots become less tolerant of engaging in risk-taking behavior. One statistical strategy for examining the effectiveness of this workshop is to conduct a hypothesis test, which is an indirect approach to statistical inference. Following are the four steps to hypothesis testing applied to the repeated-measures risk aversion workshop data from Table 11.1.

Step 1. Identify the targeted parameter, and formulate null/alternative hypotheses. The targeted parameter is the mean, μ, and we are interested in determining if in the population there is a significant difference between the means of the paired scores (i.e., μ_D). The reader is reminded that although we will be examining the difference between the pre-workshop scores (Y_1) and the post-workshop scores (Y_2), we will infer the mean difference to the population, which is why hypothesis testing is considered an indirect approach to statistical inference. As a result, the null and alternative hypotheses are as follows:

Table 11.1 Fictitious Data to Demonstrate the Computation of the Repeated-Measures t Test

Participant	Y_1 = Pre-Workshop Scores	Y_2 = Post-Workshop Scores	D = Difference Scores $(Y_2 - Y_1)$
1	19	19	0
2	19	11	−8
3	15	20	5
4	15	22	7
5	18	16	−2
6	21	19	−2
7	21	21	0
8	10	22	12
9	20	15	−5
Sum	158	165	7
M	17.55	18.33	0.78
SD	3.61	3.67	6.22

H_0: $\mu_D = 0$. There is no significant difference in the mean pre- and post-workshop scores.

H_1: $\mu_D < 0$. There is a significant difference in the mean pre- and post-workshop scores. More specifically, we believe the workshop will be effective, which will result in a significantly smaller post-workshop mean when compared to the pre-workshop mean. (*Note:* Because difference scores are calculated as $Y_2 - Y_1$, the mean difference must be negative for the workshop to be effective. This is why we use a one-tailed test to the left.)

Step 2. Determine the test criteria. The test criteria involve three components:

Level of Significance (α). We will set alpha to $\alpha = .05$.

Test Statistic. The test statistic is the repeated-measures t.

$$t = \frac{M_D - \mu_D}{SE_D}$$

Because the null hypothesis assumes there will be no significant difference between the means of the paired scores (i.e., $\mu_D = 0$) in the population, this formula reduces to

$$t = \frac{M_D}{SE_D} = \frac{M_D}{SD / \sqrt{n}}$$

where M_D is the mean of the sample of D scores, SE_D is the estimated standard error of the distribution of D scores, and n is the sample size.

The Critical Boundary. The critical t value is acquired from Table 2/Appendix A. Given $n = 9$, $df = 8$, and $\alpha = .05$, the corresponding critical t value for a one-tailed test to the left is $t(8) = -1.86$. This is shown in Figure 11.1(a).

Step 3. Collect data, check assumptions, run the analysis, and report the results.

Collect Data. The data are provided in Table 11.1.

Check the Assumptions. The assumptions for the repeated-measures t test are the same as for the single-factor t test.
- Independence. Although the pre- and post-workshop scores were obtained from the same individuals, the scores *within* each condition (pre and post) were from different individuals and hence were independent of one another. Therefore, this assumption is satisfied.
- Normality. The p value for the Shapiro–Wilk Goodness of Fit Test for the residuals of the difference scores is $p = .802$, which means that this assumption is met for the given data set.

Run the Analysis and Report the Results. In practice, we will run the analysis using our statistics program and record the results. However, for instructional purposes, we manually calculate the t value below. Observe from Table 11.1 there are three columns: $Y_1 =$ Pre-Workshop Scores, $Y_2 =$ Post-Workshop Scores, and $D =$ Difference Scores ($Y_2 - Y_1$). Focusing on the distribution of the difference scores, $M_D = 0.78$, $SD_D = 6.22$, and $n = 9$. Applying these data to the repeated-measures t statistic:

$$t = \frac{M_D}{SE_D} = \frac{M_D}{SD / \sqrt{n}} = \frac{0.78}{6.22 / \sqrt{9}} = \frac{0.78}{6.22 / 3} = \frac{0.78}{2.07} = 0.37$$

(a)

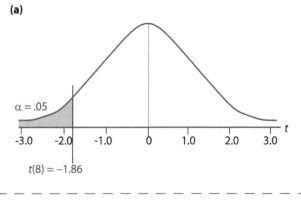

$\alpha = .05$

$t(8) = -1.86$

(b)

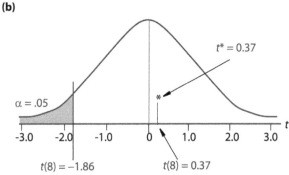

$t^* = 0.37$

$\alpha = .05$

$t(8) = -1.86$ $t(8) = 0.37$

Figure 11.1 (a) Illustration of critical *t* value for the data in Table 11.1, and (b) location of calculated *t* value with respect to the critical *t* value.

Step 4. Decide whether to reject or fail to reject the null hypothesis, and make a concluding statement relative to the RQ.

Decision. The calculated t^* is $t(8) = 0.37$. As shown in Figure 11.1(b), t^* is greater than $t_{critical} = -1.86$ and hence lies outside the critical region. Therefore, the decision is *fail to reject H₀*.

Concluding Statement. Relative to the preset significance level of $\alpha = .05$, the mean difference between the pre- and post-workshop scores was $M_D = 0.78$, which is positive and close to 0. This implies that the workshop, on average, had no effect reducing pilots' tendency toward risk-taking behavior. Therefore, the sample data do not provide sufficient evidence that the workshop was effective.

Confidence Interval for a Repeated-Measures Study

As discussed in previous chapters, the general form of a CI is as follows:

$$Population\ Parameter = Point\ Estimate \pm (SE)(Critical\ t\ Score)$$

When applied to the repeated-measures t statistic:

- The point estimate is the mean difference between the paired scores, M_D.
- SE is the standard error with respect to this mean difference, SE_D.
- The critical t score is acquired from Table 2/Appendix A for $df = n - 1$ relative to a two-tailed test.

As a result:

$$95\% \text{ CI } (\mu_D) = [a, b]$$

where

- a is the lower bound and is equal to $(M_D) - (SE_D)(t_{critical})$
- b is the upper limit and is equal to $(M_D) + (SE_D)(t_{critical})$

To illustrate the construction of the 95% CI for the repeated-measures t statistic, we will use the data set given in Table 11.1 where $M_D = 0.78$ and $SE_D = 2.07$. Because CIs are based on a two-tailed test, the critical t score for $df = 8$ is $t_{critical} = 2.306$. As a result, the corresponding 95% CI is as follows:

- Lower Limit: $M_D - (SE_D)(t_{critical}) = 0.78 - (2.07)(2.306) = 0.78 - 4.77 = -4.00$
- Upper Limit: $M_D + (SE_D)(t_{critical}) = 0.78 + (2.07)(2.306) = 0.78 + 4.77 = 5.55$

Therefore, the 95% CI for $\mu_D = [-4.00, 5.55]$. Based on the sample data, we believe with 95% certainty that in the population of GA pilots, when exposed to the treatment in the form of the given risk aversion workshop, the mean difference between pilots' pre- and post-workshop scores will be between −4.00 and 5.55. In other words, if we were to collect 100 random samples of size $n = 9$ from the same population of GA pilots, 95 of these samples would have a mean difference between the paired scores that is between −4 and 5.55. Because 0 is within this interval, it is possible that the mean difference between the paired scores could be zero. This means that the result of a corresponding hypothesis test with null hypothesis H_0: $\mu_D = 0$ would lead to a decision of fail to reject H_0, which is exactly what we found earlier.

11.4 Using the Repeated-Measures *t* Test in Research: A Guided Example

We now provide a guided example of a research study that involved the use of the repeated-measures t test. The research context is an assessment of GA pilots' perceptions of what constitutes a risky situation. Hunter (2002, p. 3) defined risk perception as "the recognition of the risk inherent in a situation." To implement this study, a random selection of GA pilots was obtained from several local flight schools, and pilots' risk perceptions were pre-assessed for baseline purposes using Hunter's (2006) Risk Perception–Other instrument. This instrument consists of 17 scenarios that describe various aviation situations. The scenarios are written in the third person, and participants rate the level of risk they perceive for the given situation from the perspective of the pilot described in the scenario, not for themselves. An example of a scenario follows:

> *A line of thunderstorms blocks the route of flight, but a pilot sees that there is a space of about 10 miles between two of the cells. He can see all the way to clear sky on the other side of the thunderstorm line, and there does not seem to be any precipitation along the route, although it does go under the extended anvil of one of the cells. As he tries to go between the storms, he suddenly encounters severe turbulence and the aircraft begins to be pelted with hail.*

The instrument initially was designed with a response scale of $1 = $ low risk and $100 = $ high risk. For the current study, though, the researchers used a Likert-type 5-point response scale with $1 = $ low risk and $5 = $ high risk. Thus, scores could range from 17 (a pilot perceives all 17 scenarios as being low risk) to 85 (a pilot perceives all 17 scenarios as being high risk), with higher scores reflecting a higher perception of a risky situation.

After completing the instrument as a pre-assessment, the pilots attended a 15-hour workshop that was designed to make them more cognizant of risky situations in aviation. The workshop was held every Friday from 1 p.m. to 4 p.m. for five consecutive weeks. At the end of the workshop, pilots' risk perceptions were post-assessed using the same instrument. A copy of the data acquired from this study is given in the file, Ch11_Guided Example Data. We will structure this guided example into three distinct parts: (a) pre-data analysis, (b) data analysis, and (c) post-data analysis.

Pre-Data Analysis

Research Question and Operational Definitions

The overriding research question is: "What is the effect of the risk perception workshop with respect to increasing GA pilots' perceptions of risky situations?" The key terms that require definitions include the *risk perception workshop*, which was defined earlier, and *risky situations*, which refer to the 17 scenarios given in Hunter's (2006) Risk Perception–Other instrument. The dependent variable is pilots' perceptions of a risky situation, which is measured by Hunter's instrument.

Research Hypothesis

The mean difference between the mean pre- and post-workshop scores will be positive, which reflects that the workshop was effective in increasing pilots' perceptions (i.e., making them more aware) of what constitutes a risky situation.

Research Methodology

The research methodology that would best answer this question is *repeated-measures*. We select a sample of GA pilots, pre-assess their risk perceptions using Hunter's (2006) Risk Perception–Other instrument, have them participate in the 15-hour risk perception workshop, and then post-assess their risk perceptions using the same instrument.

Sample Size Planning (Power Analysis)

To determine the minimum sample size needed, we consult Statistics Kingdom's (2022a) online calculator and enter the following:

- Tails: Right (H_1: $\mu > \mu_0$)
- Distribution: Student's t
- Significance level (α): .05
- Effect: Medium
- Effect size: 0.5

- Digits: 4
- Sample: One sample
- Power: .8
- Effect type: Cohen's d

The result is $n^* = 27$ for power $= .8118$. The given data set has $n = 66$ participants, which suggests we will have a greater than 81.18% of correctly rejecting H_0 based on a medium effect size. We use a one-tailed test to the right because we believe the workshop will *increase* GA pilots' level of risk perception. This implies that the mean difference between their pre- and post-assessment scores (i.e., Difference = postscore – prescore) will be positive, which indicates the workshop changed their perception of a situation from low risk to high risk.

Alternatively, we could use *G*Power* (Faul et al., 2007, 2009) with the following inputs:

- Test family = t tests.
- Statistical test = Means: Difference between two dependent means (matched pairs).
- Type of power analysis = A priori: Compute required sample size—given α, power, and effect size.
- Input parameters are Tails = One, Effect size $dz = 0.5$ (medium effect), α error prob = .05, and Power = .80.

This also returns a minimum sample size of $n = 27$.

Data Analysis

We now analyze the given data set relative to a hypothesis test involving the repeated-measures t

Step 1: Identify the targeted parameter, and formulate null/alternative hypotheses. The targeted parameter is the mean, and the null/alternative hypotheses are:

H_0: $\mu_D = 0$. There will be no significant change in GA pilots' risk perception. The mean difference, D, between the mean pre- and post-workshop scores will be zero.

H_1: $\mu_D > 0$. There will be a significant change in GA pilots' risk perception. The mean difference, D, between the mean pre- and post-workshop scores will be positive, which reflects an increase in their perception of risky situations.

Step 2: Determine the test criteria.

Test Statistic. The test statistic is t.

Level of Significance. The preset alpha level is $\alpha = .05$.

Boundary of the Critical Region. There is no entry in Table 2/Appendix A for $n = 66$, $df = 65$, $\alpha = .05$, and a one-tailed test to the right. Therefore, we consult Omni Calculator's (n.d.) online critical t-value calculator, which yields a critical t value of $t(65) = 1.67$. This is illustrated in Figure 11.2(a).

Step 3: Collect data, check assumptions, run the analysis, and report the results.

Collect the Data. The collected data are in the file Ch_11 Guided Example Data.

Check Assumptions. There are two assumptions.
- Independence. We assume the scores within each treatment condition are independent. It is important to recognize that although each pre- and post-assessment score is from the same individual, and hence there is a dependency, the scores *within* each treatment condition are independent.
- Normality. The normality assumption applies to the distribution of the difference scores, where $D = Y_2 - Y_1$. The p value for the Shapiro–Wilk Goodness of Fit Test for the residuals of the difference scores is $p = .663$, which means that this assumption is met for the given data set.

Run the Analysis and Report the Results. The results from our statistics program are as follows:
- $M_{Post} = 62.05$, $M_{Pre} = 39.11$.
- The mean difference is $M_D = 62.05 - 39.11 = 22.94$. Thus, the workshop increased pilots' risk perception by an average of approximately 23 points.

(a)

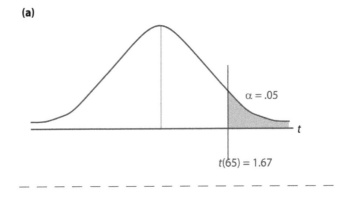

(b)

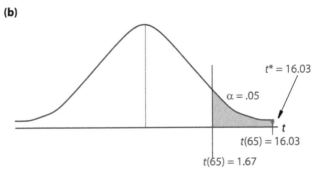

Figure 11.2 (a) Illustration of critical t value for Chapter 11's Guided Example, and (b) the location of the calculated t value with respect to the critical t value.

- The calculated t^* statistic is $t(65) = 16.03$, which is greater than t critical $= 1.671$, and therefore falls in the critical region.
- The corresponding p value is $p < .0001$.

Step 4: Decide whether to reject or fail to reject the null hypothesis, and make a concluding statement relative to the RQ.

Decision. The calculated $t^* = 16.03$ lies in the critical region as shown in Figure 11.2(b). Therefore, the decision is *reject the null hypothesis*.

Concluding Statement. The risk perception workshop was effective in making GA pilots significantly more aware of aviation situations that are risky. Post-workshop scores increased an average of nearly 23 points when compared to pre-workshop scores, indicating a higher perception of risk in aviation situations. Furthermore, this near 23-point difference was statistically significant, $t(65) = 16.03$, $p < .0001$.

Post-Data Analysis

We now examine the effect size, 95% CI, post hoc power, and plausible explanations.

Effect Size

Recall there are two ways to measure the effect size involving a t test (see Chapter 5): Cohen's estimated d, which is with respect to standard deviation, and r^2, which is with respect to explained variance.

Cohen's estimated d. When applied to a repeated-measures t test, Cohen's estimated d is relative to the distribution of the difference scores, $D = Y_2 - Y_1$.

$$\text{Estimated } d = \frac{M_D}{SD_D} = \frac{22.94}{11.63} = 1.97$$

Thus, when comparing pre- and post-workshop scores, the risk perception workshop increased participants' level of risk perception by nearly two full standard deviations, which is considered a large treatment effect.

Explained variance (r^2). To determine how much of the variability in the difference scores is being explained by the treatment, which was the risk perception workshop, we use the r^2 formula (see Chapter 5):

$$r^2 = \frac{t^2}{t^2 + df} = \frac{16.03^2}{16.03^2 + 65} = \frac{256.96}{256.96 + 65} = \frac{256.96}{321.96} = 0.798$$

Thus, approximately 80% of the variability in the difference scores is being explained by the risk perception workshop. Based on Cohen's (1988) metric-free units, this is considered a large effect. Another way of interpreting this is to say, "If you tell me your pre-workshop score, then I will have 80% of the information needed to perfectly predict your post-workshop score."

The 95% Confidence Interval

The 95% CI for the mean difference reported from our statistics program is [20.1, 25.8]. Thus, 95% of the time we can expect the true mean difference between the mean pre- and post-workshop scores in the population to range anywhere from 20.1 points to 25.8 points higher than the mean pre-workshop scores. Another way of looking at this is to say that if we were to randomly select 100 random samples of size $n = 66$ from the same population of GA pilots, then 95 of these samples would have a mean post-workshop score that is between 20.1 and 25.8 points higher than the mean pre-workshop scores. Furthermore, because the null hypothesis states that the mean difference score will be 0, and because 0 is not within the 95% CI, we would reject the corresponding null hypothesis. Lastly, based on the width of the 95% CI, the corresponding accuracy in parameter estimation (AIPE) is moderately strong relative to the spread in scores, and hence, the sample data are a good source for accurately estimating the true difference in mean scores in the population between pre- and post-workshops.

Power (Post Hoc)

To determine the power of the study, we consult Statistics Kingdom's (2022b) online power calculator and enter the following inputs:

- Tails: Right (H_1: $\mu > \mu_0$)
- Distribution: T
- Significance level (α): .05
- Effect type: Standardized effect size
- n_1: 66

- Digits: 4
- Sample: One sample
- Effect: Large
- Effect size: 1.97

The result is a power of 1.0. Because we do not believe in a power probability of 1.0, we report that the power of our hypothesis test was greater than .99. Thus, there is a greater than 99% probability that we made the correct decision to reject H_0, and that the large effect found in the sample also exists in the population.

Alternatively, we could use *G*Power* (Faul et al., 2007, 2009) as we did earlier but change the type of power analysis to "Post hoc: Compute achieved power—given α, sample size, and effect size." We also change the input parameters to reflect the actual effect size (*ES* = 1.97) and sample size ($n = 66$). When these changes are made, the power of the study is 1.00. Once again, because nothing in statistics is 100% certain, we report that power > .99.

Plausible Explanations for the Results

One plausible explanation is the design used. We chose to implement the study as a repeated-measures design, which eliminated the single source of variability, namely, individual differences, including participants' age, level of education, and flight experience. A second plausible explanation is sample size. We had approximately two and one-half times as many participants needed based on sample size planning. When you combine a relatively large sample size with the absence of individual differences via a repeated-measures design, you increase the likelihood of finding a significant effect. A third plausible explanation is the workshop itself. Perhaps the manner in which the workshop was implemented—3 hours each Friday afternoon for 5 weeks—contributed to the large mean difference score. There is some theory to support this notion. For example, the *spacing effect theory* posits that distributed practice is more effective than massed practice. Thus, if the workshop had been conducted as a weekend event such as 8 hours on Saturday and 7 hours on Sunday, it is conceivable that the mean change score might not have been as large. Finally, keep in mind the potential effect of outliers relative to the difference scores. Although the boxplot associated with the difference scores shows no outliers, an outlier analysis using Jackknife distances yields five outliers: Cases 43, 50, 52, 55, and 63. When these cases are excluded from the analysis, the results are, $t(59) = 19.07$, $p = 1.0000$. Thus, the presence of the outliers had little effect on the final results. Nevertheless, the reader is reminded that it is critical to conduct an outlier analysis as part of data analysis.

Chapter Summary

1. In a repeated-measures design, a single group receives all the treatments and performance on the dependent variable is then measured after each treatment has been administered. This design is in contrast to a between-subjects design where two or more independent groups receive different treatments and then comparisons on the dependent variable are made between groups.
2. In a repeated-measures design, participants serve as their own control, which leads to a more precise analysis because it removes individual differences. When left uncontrolled, individual differences are one of the largest sources of variation in most research studies.
3. Advantages of a repeated-measures design include the need for fewer participants and the ability to examine changes across time. Potential problems include carry-over effects (both practice effects and fatigue effects).
4. The repeated-measures t test statistic is used to analyze data from a repeated-measures design that involves exactly two sets of scores where the focus is on examining the mean difference between the paired data. The general structure of the repeated-measures t test is the same as the single-sample t except the mean and standard error are relative to the difference scores (i.e., $Y_2 - Y_1$).
5. Conducting a hypothesis test and computing/interpreting corresponding CIs, *ESs*, and power for a repeated-measures study are performed the same as for a single-sample t test except now the focus is on the distribution of the difference scores.

Vocabulary Check

Baseline	Repeated-measures comparative group design
Carry-over effects	Repeated-measures correlation study
Change score study	Repeated-measures design
Fatigue effects	Repeated-measures longitudinal study
Matched-samples design	Repeated-measures t test
Mean of the difference scores (M_D)	Standard error of the difference scores (SE_D)
Paired scores	Within-subjects design
Practice effects	

Review Exercises

A. Check Your Understanding

In 1–10, choose the best answer among the choices provided.

1. When a repeated-measures *t* test is used to compare pre- and post-assessment scores, the following relationship with respect to sample size (*n*) is expected:
    a. $n_{Pre} > n_{Post}$
    b. $n_{Pre} = n_{Post}$
    c. $n_{Pre} < n_{Post}$
    d. n_{Pre} does not have to equal n_{Post}

2. A flight instructor is developing a new test to assess flight students' knowledge of the NAS. As part of the test construction process, he develops two parallel forms of the instrument. He will then administer both forms to a group of flight stusdents and compare the mean scores from the two sets of data. An appropriate statistical procedure is
    a. independent-samples *t* test.
    b. single-factor ANOVA.
    c. repeated-measures *t* test.
    d. correlation analysis.

3. Which of the following is an example of paired data where a repeated-measures *t* test would be appropriate to compare the means of the two sets of scores?
    a. Scores from married couples with respect to a study designed to examine participants' attitudes toward children.
    b. Scores from experimental and control groups with respect to a study designed to determine the effectiveness of a new treatment.
    c. Departure delay times in minutes for a specific airline at a specific airport with respect to a study designed to examine an airline's overall mean departure delays.
    d. The number of GA pilots who answered "yes" or "no" to the statement, "I prefer glass cockpits over round dials" in a study designed to assess GA pilots' cockpit preferences.

4. The concept of "individuals serve as their own control" with respect to a repeated-measures design means:
    a. The same group of individuals is assigned to both treatment and control groups simultaneously.
    b. Individuals receive all treatments, which eliminates individual differences such as age, IQ, and education among participants.
    c. Individuals within a group are in control of what treatment they receive.
    d. Individuals who receive a treatment are in control over how the treatment will be administered, measured, and compared.

5. Let's assume you conducted a study that measured the reaction times in seconds at two different points in time involving a group of GA pilots who were all the same age: the year when they turned 40 and the year when they turned 50. Let's further assume that the mean difference in the two sets of data was +3 seconds and the 95% CI was [0.5, 5.5]. These results suggest:
    a. There was a significant improvement in pilots' reaction times over the 10-year period.

b. There was no significant change in pilots' reaction times over the 10-year period.

c. There was a significant deterioration in pilots' reaction times over the 10-year period.

d. Pilots' reaction times increased on average $5.5 - 0.5 = 5$ seconds.

6. Let's assume that an aviation researcher is interested in examining the effectiveness of two different programs (P1 and P2) designed to improve flight students' learning strategies. To conduct this study, she randomly selects a group of flight students at her university who are enrolled in a Part 141 program and then randomly assigns them to either the P1 group or the P2 group. Both groups are administered Panadero et al.'s (2021) Deep Learning Strategies Questionnaire (DLS-Q) as a pre-assessment prior to the implementation of the new programs, and again as a post-assessment after the completion of the programs. To determine which program was more effective in improving students' learning strategies, which *t* test would be most appropriate?

a. Independent-samples *t* test that would compare the differences in the post-assessment scores.

b. Repeated-measures *t* test that would compare the difference scores for all participants as a single sample.

c. Independent-samples *t* test that would compare the difference scores between the two groups.

d. A *t* test is not appropriate because there are four sets of scores: pre- and post-assessment scores for participants in the P1 and P2 groups.

7. Which research setting describes a study that uses a repeated-measures design?

a. You seek to compare the mean annual salary between a randomly selected group of male Part 121 pilots with that of a randomly selected group of female Part 121 pilots.

b. You seek to replicate Part a above, but this time both groups are randomly selected from the same airline.

c. You seek to compare the number of runway incursions for a group of pilot–copilot pairs by examining the relationship between the number of incursions committed by pilots to those committed by copilots.

d. You seek to compare the mean number of runway incursions involving a group of GA pilots measured at the end of the current year with the mean number of runway incursions of the same group of pilots measured 5 years later.

8. Let's assume you conducted a study to determine the effect "noise" has on flight students' performance, which was measured in a simulator and defined as the mean deviation from the glidepath in degrees. The same group of students was first exposed to a "silent" condition in which the CFI was quiet throughout the flight instruction, followed by a "noisy" condition in which the CFI constantly asked students questions related to the flight and required students to respond to the questions. Students completed 10 trials under each condition. Suppose the results were $M_{Silent} = 1.7$, $M_{Noisy} = 3.7$, $t(8) = -3$, $p = .0171$. Which of the following describes how carry-over effects could have impacted the results?

a. Students completed 10 trials before they underwent the noisy condition, which led to a practice effect that significantly improved their performance.

b. Students completed 10 trials before they underwent the noisy condition, which led to a fatigue effect that significantly degraded their performance.

    c.   The ordering of the conditions was responsible for the results. If students were exposed to the noisy condition first followed by the silent condition, then the results would be different.

    d.   Based on the results, it does not appear that carry-over effects impacted the results because the mean difference was not statistically significant.

9. A repeated-measures design also is referred to as a paired-samples design because a first set of scores is paired with a second set of scores. Which of the following is NOT an example of paired data and hence could not be analyzed using a repeated-measures *t* test?

    a.   Scores acquired from the same group of individuals measured under two separate conditions.

    b.   Scores acquired from two separate groups of unrelated individuals who have been matched on a specific attribute such as age or IQ.

    c.   Scores acquired from separate groups formed by assigning naturally occurring pairs of individuals—one person to each group—such as married couples, siblings, and identical twins.

    d.   Scores acquired from two separate groups of unrelated individuals who have been randomly assigned to each group.

10. Let's assume a researcher collected data using a repeated-measures design that consisted of two measurements of performance on the dependent variable but inadvertently analyzed the data using an independent-samples *t* test. Which would be an accurate statement relative to this analysis?

    a.   The results would reflect those of a bivariate correlation analysis because both variables would consist of numerical data.

    b.   The results would be the same as those from using a repeated-measures *t* test.

    c.   The analysis could not be completed because there would not be a group membership variable.

    d.   The analysis could not be completed because there would not be an independent variable.

B. Apply Your Knowledge

Use the following research description and the corresponding data set to conduct the activities given in Parts A–C (see also Section 11.4).

This research study is with respect to an aviation maintenance, repair, and overhaul (MRO) facility, and the objective is to assess the effect a stress management initiative has on reducing employees' stress level. The rationale for the study is grounded in the belief that employee stress could lead to increased feelings of rushing to complete a maintenance task, frustration, fatigue, and complacency, all of which could lead to maintenance errors. To implement the study, a researcher pre-assessed employees' stress level using the Stress subscale of Fogarty's (2005) Maintenance Environment Survey. This 9-item subscale assesses participants' perceived feelings and consequences about their stress and what contributes to it. A sample item is, "The demands of my work interfere with my home and family life." Items are scored using a traditional 5-point Likert scale ranging from 1 = Strongly Disagree to 5 = Strongly Agree. Thus, overall scores could range from 9 to 45, with higher aggregate scores

reflecting higher levels of stress. Following the pre-assessment, employees engaged in various stress reduction activities for 1 month. These included a 30-minute
extended lunch break each day, access to free onsite mental health counseling,
a 1-hour after-work social event every Friday afternoon, and a 30-minute stress-
reduction strategies workshop that was scheduled at various times each day to
accommodate employees' different schedules. At the end of the study period, employees were post-assessed using the same instrument. The corresponding data are
located in the file Ch_11 Exercises Part B Data. Your assignment is to import this data
set into your statistical software program and analyze the data using a repeated-
measures *t* test.

A. Pre-Data Analysis

1. Specify the research questions and corresponding operational definitions.
2. Specify the research hypotheses.
3. Determine the appropriate research methodology/design and explain why it is
   appropriate.
4. Conduct an a priori power analysis to determine the minimum sample size needed.
   Compare this result to the size of the given data set and explain what impact
   the size of the given sample will have on the results relative to the minimum size
   needed.

B. Data Analysis

1. Conduct a hypothesis test via a repeated-measures *t* test by applying all four steps of
   hypothesis testing as presented in Sections 11.3 and 11.4.

C. Post-Data Analysis

1. Determine and interpret the estimated effect size using Cohen's *d*.
2. Determine and interpret the effect size relative to explained variance (r^2).
3. Determine and interpret the power of the study from a post hoc perspective.
4. Determine and interpret the 95% confidence interval, including its precision and
   AIPE.
5. Present at least two plausible explanations for the results.

References

Cohen, J. (1988). *Statistical power analysis for the behavioral sciences* (2nd ed.). Lawrence Erlbaum
   Associates.
Dunbar, V. L. (2015). *Enhancing vigilance in flight instruction: Identifying factors that contribute to
   flight instructor complacency* (Publication No. 3664585) [Doctoral dissertation, Florida Institute of
   Technology]. ProQuest Dissertations and Theses Global.
Faul, F., Erdfelder, E., Buchner, A., & Lang, A.-G. (2009). Statistical power analyses using G*Power 3.1:
   Tests for correlation and regression analyses. *Behavior Research Methods, 41*, 1149–1160. [Download
   software at http://www.gpower.hhu.de]. https://www.psychologie.hhu.de/fileadmin/redaktion/
   Fakultaeten/Mathematisch-Naturwissenschaftliche_Fakultaet/Psychologie/AAP/gpower/GPow-
   er31-BRM-Paper.pdf

Faul, F., Erdfelder, E., Lang, A.-G., & Buchner, A. (2007). G*Power 3: A flexible statistical power analysis program for the social, behavioral, and biomedical sciences. *Behavior Research Methods, 39,* 175–191. [Download software at http://www.gpower.hhu.de]. https://www.psychologie.hhu.de/fileadmin/redaktion/Fakultaeten/Mathematisch-Naturwissenschaftliche_Fakultaet/Psychologie/AAP/gpower/GPower3-BRM-Paper.pdf

Fogarty, G. J. (2005). Psychological strain mediates the impact of safety climate on maintenance errors. *International Journal of Applied Aviation Studies, 5*(1), 53–63. https://eprints.usq.edu.au/434/1/IJAAS_paper_Mar_05_revised.pdf

Hamilton, H. C. (2016). *Airline pilots in recovery from alcoholism: A quantitative study of cognitive change* [Doctoral dissertation, Walden University]. Walden Dissertations and Doctoral Studies. https://scholarworks.waldenu.edu/dissertations/1516/

Hunter, D. R. (2002). *Risk perception and risk tolerance in aircraft pilots.*(Report No. DOT/FAA/AM-02/17). Washington, DC: Federal Aviation Administration. https://rosap.ntl.bts.gov/view/dot/57933

Lindseth, G. N., Lindseth, P. D., Jensen, W. C., Petros, T. V., Helland, B. D., & Fossum, D. L. (2011). Dietary effects on cognition and pilots' flight performance. *International Journal of Aviation Psychology, 21*(13), 269–282. https://doi.org/10.1080/10508414.2011.582454

Omni Calculator (n.d.). *The critical value calculator.* https://www.omnicalculator.com/statistics/critical-value

Panadero, E., Alonso-Tapia, J., García-Pérez, D., Fraile, J., Sánchez Galán, J. M., & Pardo, R. (2021). Deep learning self-regulation strategies: Validation of a situational model and its questionnaire. *Revista de Psicodidáctica, 26*(1). https://ojs.ehu.eus/index.php/psicodidactica/article/download/23347/20729

Statistics Kingdom (2022a). *Z-test and T-test sample size calculator.* https://www.statskingdom.com/sample_size_t_z.html

Statistics Kingdom (2022b). *Normal, T—Statistical power calculators.* https://www.statskingdom.com/32test_power_t_z.html

Taylor, J. L., Kennedy, Q., Noda. A., & Yesavage, J. A. (2007). Pilot age and expertise predict flight simulator performance. *Neurology, 68*(9), 648–654. https://www.ncbi.nlm.nih.gov/pmc/articles/PMC2907140/

Yesavage, J. A., Dolhert, N., & Taylor, J. L. (1994). Flight simulator performance of younger and older aircraft pilots: Effects of age and alcohol. *Journal of the American Geriatrics Society, 42*(6), 577–582. https://doi.org/10.1111/j.1532-5415.1994.tb06852.x

Part F

Part F

Nonparametric Statistics: Working with Frequency Data

.

12 The Chi-Square Statistic

Student Learning Outcomes

After studying this chapter, you will be able to do the following with respect to analyzing frequency data using chi-square:

1. Distinguish between a parametric vs. a nonparametric test.
2. Perform and interpret the results of a chi-square test for goodness of fit.
3. Perform and interpret the results of a chi-square test for independence.
4. Engage in pre-data analysis, data analysis, and post-data analysis activities with respect to a chi-square test for goodness of fit and a chi-square test for independence as described in Sections 12.3 and 12.5, respectively.

12.1 Chapter Overview

The statistical strategies presented in the previous chapters are known as parametric tests because they involve confidence intervals and hypothesis tests with respect to population parameters such as the mean, correlation coefficient, and regression coefficient. We now focus our attention on nonparametric tests, which do not examine parameters but instead involve examining frequency data. In a nonparametric test, the dependent variable is either nominal or ordinal, and outliers usually are not a concern because they are unlikely to arise in ordinal or nominal data. Nonparametric tests also do not require scores on the outcome variable to be normally distributed, and they typically do not require an assumption of homogeneity of variances. In this chapter, we introduce two commonly used nonparametric tests: the chi-square test for goodness of fit and the chi-square test for independence. The former is sometimes called one-way chi-square, and the latter, two-way chi-square. As part of our discussion, we present information about the chi-square distribution, the chi-square test statistic, corresponding assumptions, effect sizes, and hypothesis testing. We conclude each presentation with guided examples of research studies that use the chi-square test for goodness of fit and the chi-square test for goodness independence, respectively.

12.2 One-Way Chi-Square: The Test for Goodness of Fit

The Concept of the Chi-Square Test for Goodness of Fit

In the most general sense, chi-square, denoted by the square of the Greek letter chi, χ^2, is a nonparametric statistical strategy that tests whether sets of frequencies follow certain

DOI: 10.4324/9781003308300-18

patterns. For example, Torres et al. (2011) examined the relationship between various human factor errors and the occurrence of runway incursions. As part of their study, they reviewed reports of runway incursions submitted to the Aviation Safety Reporting System (ASRS) and NTSB between January 2005 and March 2009. From these reports, they recorded the number of runway incursions that were attributed to various human factor errors such as situational awareness, miscommunications, and airport markings. For example, in Table 12.1, which contains the top five categories cited in Torres et al. but with hypothetical frequencies for instructional purposes, we can see that 90 runway incursions were attributed to situational awareness errors, 70 runway incursions were attributed to miscommunications, etc. In the vocabulary of chi-square, each factor is considered a category, and the number of incursions associated with each category represents the observed frequencies, denoted O.

When examined from a hypothesis test perspective, Torres et al. (2011) applied a chi-square test for goodness of fit, which also is known as one-way chi-square, because it involves a single variable (human factor errors) with multiple categories. This is analogous to a single-factor ANOVA, which involves one factor with multiple levels. To perform this analysis, Torres et al. hypothesized that the proportion of incursions would be the same across all categories, and compared the observed frequencies to what they expected relative to their hypothesis.

If we apply Torres et al.'s expectation to the five categories in Table 12.1, then each category is expected to have one-fifth, or 20%, of the total frequencies. Given that $N = 250$, this means that each category is *expected* to have .20 × 250 = 50 incursions. This is illustrated in Table 12.2. These hypothesized frequencies are referred to as expected frequencies, denoted E, because they are what is expected if the hypothesis is true. As a result, the chi-square test for goodness of fit is used to determine if a frequency distribution obtained from sample data is consistent with a claimed or established distribution. In short, it is used to compare observed frequencies to expected frequencies.

The Chi-Square Distribution and Critical Values

The chi-square test is based on the chi-square distribution, which is illustrated in Figure 12.1. This distribution is not symmetrical, but instead its shape depends on the degrees of freedom, similar to the t and F distributions, and hence it reflects a family of distributions. The chi-square distribution has one degree of freedom, with $df = C - 1$ where $C =$ the number of categories. Thus, the degrees of freedom for the runway incursions example summarized in Table 12.2 are $df = 5 - 1 = 4$ because there are five categories. Observe from

Table 12.1   Number of Runway Incursions per Human Factor Errors

Category	Observed Frequencies (O)
Situational Awareness	90
Miscommunication	70
Distraction	40
Airport Markings	30
Complex Taxiways	20

Note. N = 250.

Figure 12.1 that as *df* increases, the chi-square distribution becomes more symmetrical. Further observe that the numerical values of chi-square always are nonnegative (i.e., ≥ 0).

To determine critical chi-square values relative to *df* and alpha (α), we consult Table 6/Appendix A. For example, as illustrated in Figure 12.2(a), given α = .05 and *df* = 4, the critical chi-square value is $\chi^2(4) = 9.488$ because the targeted proportion of area under the curve is 5% to the right. However, if the targeted proportion of area under the curve is .90 and *df* = 19, then as shown in Figure 12.2(b), the critical chi-square value is $\chi^2(19) = 11.651$, because 90% of the curve lies to the right of this boundary value. In the event that the targeted *df* is not listed in Table 6/Appendix A, we use Omni Calculator's (n.d.) online critical value calculator for chi-square.

The Test Statistic for the Chi-Square Test for Goodness of Fit

The formula for the chi-square test statistic is

$$\chi^2 = Sum\left[\frac{(O_i - E_i)^2}{E_i}\right]$$

where the numerator is determined by calculating the difference between the observed (O_i) and expected (E_i) frequencies for each category, squaring this difference, and then dividing this squared difference by the expected frequency. The final chi-square value is

Table 12.2 Observed and Expected Number of Runway Incursions per Human Factor Errors Based on a Hypothesized Even Distribution

Category	Observed Frequencies (O_i)	Expected Proportion	Expected Frequencies (E_i)
Situational Awareness	90	0.2	0.2 × 250 = 50
Miscommunication	70	0.2	0.2 × 250 = 50
Distraction	40	0.2	0.2 × 250 = 50
Airport Markings	30	0.2	0.2 × 250 = 50
Complex Taxiways	20	0.2	0.2 × 250 = 50

Note. N = 250.

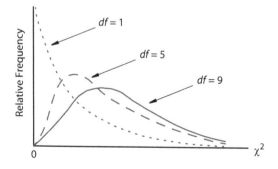

Figure 12.1 Illustration of the chi-square (χ^2) distribution.

the overall sum of these quotients. This is illustrated below using the data from Table 12.3, which is an expansion of Table 12.2 for the runway incursion example.

$$
= \frac{(90-50)^2}{50} + \frac{(70-50)^2}{50} + \frac{(40-50)^2}{50} + \frac{(30-50)^2}{50} + \frac{(20-50)^2}{50}
$$

$$
= \frac{(40)^2}{50} + \frac{(20)^2}{50} + \frac{(-10)^2}{50} + \frac{(-20)^2}{50} + \frac{(-30)^2}{50}
$$

$$
= \frac{1600}{50} + \frac{400}{50} + \frac{100}{50} + \frac{400}{50} + \frac{900}{50}
$$

$$
= 32 + 8 + 2 + 8 + 18
$$

$$
= 68
$$

As a result, the calculated chi-square value is $\chi^2(4) = 68.0$. Given the critical chi-square $\chi^2(4) = 9.488$ from Figure 12.2(a), if we were conducting a hypothesis test where $H_0 =$ the observed frequencies do not differ significantly from the hypothesized frequencies, then we would reject H_0. This is illustrated in Figure 12.3.

Assumptions for the Chi-Square Test for Goodness of Fit

In addition to the requirement that data be measured on a nominal or ordinal scale, the chi-square test has two primary assumptions: independence and minimum cell size.

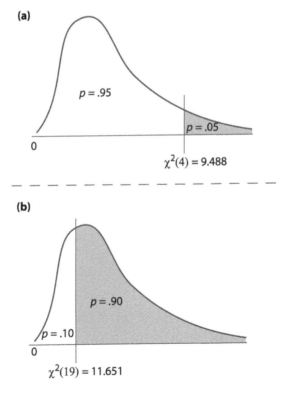

Figure 12.2 Illustration of how to use Table 6/Appendix A.

Table 12.3 Observed and Expected Number of Runway Incursions per Human Factor Errors Based on a Hypothesized Even Distribution and Configured to Follow the Chi-Square Test Statistic Formula

Category	$(O_i - E_i)$	$(O_i - E_i)^2$	$\dfrac{(O_i - E_i)^2}{E_i}$
Situational Awareness	$90 - 50 = 40$	$40^2 = 1600$	$1600/50 = 32.0$
Miscommunication	$70 - 50 = 20$	$20^2 = 400$	$400/50 = 8.0$
Distraction	$40 - 50 = -10$	$-10^2 = 100$	$100/50 = 2.0$
Airport Markings	$30 - 50 = -20$	$-20^2 = 400$	$400/50 = 8.0$
Complex Taxiways	$20 - 50 = -30$	$-30^2 = 900$	$900/50 = 18.0$
			Sum: 68.0

Note. N = 250. See also Table 12.2.

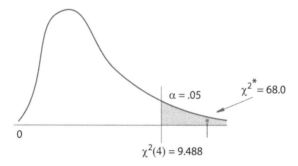

Figure 12.3 Illustration of a critical χ^2 value vs. the corresponding calculated χ^2 value.

Independence

The independence assumption requires that each *observed* frequency be independent of all the other observations. This means that an observed frequency cannot be placed in more than one category. For example, in our version of Torres et al. (2011)'s runway incursions study, a runway incursion error cannot be attributed to both situational awareness and airport markings. It must one or the other, but not both.

Minimum Cell Size

The minimum cell size assumption refers to the minimum number of *expected* frequencies in each category, or cell. When testing for this assumption, two general rules of thumb are acceptable and either may be applied: (a) the expected frequency for any cell must be at least 5, or (b) at least 80% of the cells must have an expected frequency greater than 5. In the event that this assumption is not satisfied, then alternative strategies such as combining categories or using Fisher's exact probability test should be considered.

Applying these assumptions to our version of Torres et al.'s (2011) runway incursions study, we presume that the five error categories associated with human factors listed in Table 12.2 are mutually exclusive. We also observe from Table 12.2 that all of the categories have an expected frequency of 50, which is greater than 5.

Effect Size for the Chi-Square Test for Goodness of Fit

The effect size for the chi-square test for goodness of fit is Cohen's *w*

$$w = \sqrt{\frac{\chi^2}{N}}$$

Cohen's *w* varies between 0 and 1. As *w* approaches 1, the effect size becomes larger, and as *w* approaches 0 the effect size becomes smaller. Cohen (1988, pp. 224–225) also suggested metric-free units for interpreting the magnitude of an effect: small ($w = 0.10$), medium ($w = 0.30$), and large ($w = 0.50$). Applying Cohen's *w* to the hypothetical data given in Table 12.3 where $\chi^2(4) = 68.0$ and $N = 250$:

$$w = \sqrt{\frac{\chi^2}{N}} = \sqrt{\frac{68}{250}} = \sqrt{0.272} = 0.52$$

Thus, based on Cohen's guidelines, this represents a large effect.

12.3 Using the Chi-Square Test for Goodness of Fit in Research: A Guided Example

We now provide a guided example of a research study that involves the use of the chi-square test for goodness of fit. The context of this study is airline passengers' seat preferences relative to four choices: aisle, exit row, middle, and window. We will assume that in the population of airline passengers, the general belief is that the targeted seat categories have equal preferences. A survey of $N = 200$ airline passengers with similar flight experiences was conducted, and the raw data are provided in the file Ch_12 Guided Example Data for Chi-Square Test for Goodness of Fit. Let's now analyze these data to see if their respective proportions are consistent with the general belief regarding seat preferences. We will structure this guided example into three distinct parts: (a) pre-data analysis, (b) data analysis, and (c) post-data analysis.

Pre-Data Analysis

Research Question and Operational Definitions

The overriding research question is: "What is the difference in the proportion of seat preferences between the observed and expected frequencies among airline passengers?" The key terms that require definitions—aisle, exit row, middle, and window seats—should be self-evident, but we provide operational definitions for completeness.
- Aisle seats are at the end of a row adjacent to the aisle. For example, in a 3 + 3 configuration—three seats to the right and three seats to the left of the aisle, where the seats are labeled from right-to-left A-B-C and D-E-F—aisle seats are labeled C and D.
- Exit row seats are in the same row as the emergency exit doors of the aircraft. Furthermore, regardless of the seat's location—window, middle, or aisle—it is still considered an exit row seat.
- Middle seats have at least one seat on either side of them. For example, in a 3 + 3 configuration, the middle seats are labeled B and E.

- Window seats are next to a window. For example, in a 3 + 3 configuration, the window seats are labeled A and F.

Research Hypothesis

The proportion of seat preferences between what was observed and what is expected will be different.

Research Methodology

The research methodology is a *one-way chi-square design* and involves a single nominal variable (airline seat preferences) with four categories (aisle, exit row, middle, and window). The objective is to determine if what was observed "fits" with what is expected, and therefore the study requires a chi-square design for goodness of fit.

Sample Size Planning (Power Analysis)

To determine the minimum sample size needed, we consult Statistics Kingdom's (2022a) online calculator and enter the following inputs relative to the selection of "Chi-Squared Sample size":

- Test: Goodness of fit
- Significance level (α): .05
- Effect: Medium
- Categories: 4
- Digits: 4
- Power: .8
- Effect size (w): 0.3

The result is $N^* = 122$ for power = .803. This means that the sum of all the observed frequencies for all categories must be at least 122. For the current example, we have four categories, and the total number of observations is 200. Thus, we have a sufficient number of observations to have at least an 80.3% chance of correctly rejecting H_0 with respect to a medium effect size ($w = 0.3$).

Alternatively, we could use *G*Power* (Faul et al., 2007, 2009) with the following inputs:

- Test family = χ^2 tests.
- Statistical test = Goodness-of-fit tests: Contingency tables.
- Type of power analysis = A priori: Compute required sample size—given α, power, and effect size.
- Input parameters: Effect size $w = 0.3$, α error prob = .05, and Power = .80, and $df = 3$ (recall that $df = C - 1$ where $C =$ the number of categories).

This also results in a minimum sample size of $N = 122$.

Data Analysis

We now analyze the given data set relative to a hypothesis test involving the chi-square test for goodness of fit.

Step 1: Formulate null/alternative hypotheses. The null/alternative hypotheses are:

H_0: The proportion of frequencies across the four seat categories are equal: 25%–25%–25%–25%.

H_1: The proportion of frequencies across the four seat categories are different. We believe that the proportion of frequencies from the sample data—where Aisle = 70/200 = 35%, Exit Row = 40/200 = 20%, Middle = 30/200 = 15%, and Window = 60/200 = 30%—will be significantly different than the hypothesized split.

Step 2: Establish the test criteria.

Test Statistic. The test statistic is χ^2 and $df = 4 - 1 = 3$.

Level of Significance. The preset alpha level is $\alpha = .05$.

Boundary of the Critical Region. The t boundary value for the critical region is acquired from Table 6/Appendix A for $\chi^2(3, \alpha = .05) = 7.815$. This is illustrated in Figure 12.4(a).

Step 3: Collect data, check assumptions, run the analysis, and report the results.

Collect Data. The data are provided in the file as indicated earlier, and the corresponding frequency distribution table is provided in Table 12.4.

Check Assumptions. There are two assumptions.

- Independence. The seat categories are mutually exclusive, and participants provided their single, highest preference. Therefore, this assumption is met.

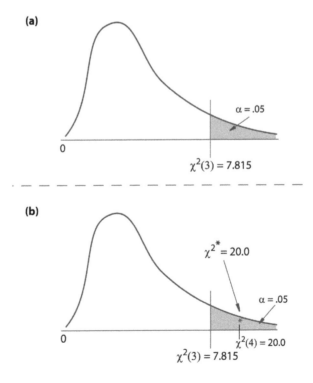

Figure 12.4 (a) Illustration of the critical chi-square value for Chapter 12's Guided Example of goodness of fit test, and (b) the location of the calculated chi-square value with respect to the critical value.

Table 12.4 Frequency Distribution Table for the Guided Example Involving Chi-Square Test for Goodness of Fit

Seat Category	Observed Frequencies (O)
Aisle	70
Exit Row	40
Middle	30
Window	60

Note. N = 200.

- Minimum cell size. As shown in Table 12.5, the expected frequency of each cell is 50, which is greater than 5. Therefore, this assumption is met.

Run the Analysis and Report the Results. Although we will use our statistics program to calculate the χ^2 test statistic, a summary of the long-hand calculations associated with the data in Table 12.5 is provided for instructional purposes. A summary of these calculations is provided in Table 12.6.

$$\chi^2 = \text{Sum}\left[\frac{(O_i - E_i)^2}{E_i}\right]$$

$$= \frac{(70-50)^2}{50} + \frac{(40-50)^2}{50} + \frac{(30-50)^2}{50} + \frac{(60-50)^2}{50}$$

$$= \frac{(20)^2}{50} + \frac{(-10)^2}{50} + \frac{(-20)^2}{50} + \frac{(10)^2}{50}$$

$$= \frac{400}{50} + \frac{100}{50} + \frac{400}{50} + \frac{100}{50}$$

$$= 8 + 2 + 8 + 2$$

$$= 20$$

Thus, the calculated chi-square test statistic is $\chi^{2*} = 20.0$. As shown in Figure 12.4(b), the calculated χ^2 statistic falls inside the critical region, and as reported from our statistics program, $p = .0002$. Observe from these calculations that aisle and middle seats (the first and third categories) each contribute 8 units, or 40%, to the overall χ^2.

Step 4: **Decide whether to reject or fail to reject the null hypothesis, and make a concluding statement relative to the RQ.**

Decision. Because the calculated $\chi^{2*} = 20.0$ lies in the critical region, the decision is to reject the null hypothesis.

Concluding Statement. The proportions of observed frequencies of the respective seat preferences as expressed by sample participants are significantly different from what was hypothesized. The distribution of observed frequencies did not fit the distribution of expected frequencies, and hence the analysis did not confirm an equal split across the four airline seat categories with respect to the seat preferences expressed by participants. Thus, airline passengers do not have equal preferences with respect to the four types of airline seats that were targeted.

Table 12.5 Observed and Expected Frequencies for the Guided Example Involving Chi-Square Test for Goodness of Fit

Seat Category	Observed Frequencies (O_i)	Expected Proportion	Expected Frequencies (E_i)
Aisle	70	0.25	$0.25 \times 200 = 50$
Exit Row	40	0.25	$0.25 \times 200 = 50$
Middle	30	0.25	$0.25 \times 200 = 50$
Window	60	0.25	$0.25 \times 200 = 50$

Note. N = 200.

Table 12.6 Observed and Expected Frequencies for Airline Seat Preferences for the Guided Example Involving Chi-Square Test for Goodness of Fit

Seat Category	$(O_i - E_i)$	$(O_i - E_i)^2$	$\dfrac{\left(O_i - E_i\right)^2}{E_i}$
Aisle	$70 - 50 = 20$	$20^2 = 400$	$400/50 = 8.0$
Exit Row	$40 - 50 = -10$	$-10^2 = 100$	$100/50 = 2.0$
Middle	$30 - 50 = -20$	$-20^2 = 400$	$400/50 = 8.0$
Window	$60 - 50 = 10$	$10^2 = 100$	$100/50 = 2.0$
			Sum: 20.0

Note. N = 200. See also Table 12.5.

Post-Data Analysis

We now examine effect size, power, and plausible explanations.

Effect Size

Applying Cohen's w:

$$w = \sqrt{\frac{\chi^2}{N}} = \sqrt{\frac{20}{200}} = \sqrt{0.10} = 0.316$$

Thus, based on Cohen's guidelines, we have a medium effect.

Power (Post Hoc)

To determine the power of the study we conduct a post hoc power analysis using Statistics Kingdom's (2022b) online chi-square statistical power calculator with the following inputs:

- Tails: Goodness of fit
- Significance level (α): .05
- Effect: Medium
- Categories: 4
- Digits: 4
- Sample size: 200
- Effect size (w): 0.316

The result is a power of .9749. Thus, there is a 97.5% probability that we made the correct decision in rejecting H_0.

Alternatively, we could use *G*Power* (Faul et al., 2007, 2009) but change the type of power analysis to "Post hoc: Compute achieved power—given α, sample size, and effect size." We also change the input parameters to reflect the actual effect size ($w = 0.316$) and sample size ($N = 200$). When these changes are made, the power of the study is .9748885.

Plausible Explanations for the Results

One plausible explanation for the results is sample size. Based on our sample planning, we needed at least 122 observations to have an 80% chance of correctly rejecting the null hypothesis with respect to a medium effect ($w = 0.3$). The current study met this minimum sample size, and the magnitude of the effect was indeed medium ($w = 0.316$). A second plausible explanation is with respect to the even split we hypothesized for the expected proportions. If we had given greater preferences to aisle and window seats so that the expected split was $\frac{1}{3} : \frac{1}{5} : \frac{1}{10} : \frac{1}{3}$ instead of the hypothesized split of $\frac{1}{4} : \frac{1}{4} : \frac{1}{4} : \frac{1}{4}$ then we might not have found a significant effect.

12.4 Two-Way Chi-Square: The Test for Independence

The Concept of the Chi-Square Test for Independence

In the chi-square test for goodness of fit, we compare the *relative proportions* of frequencies for a distinct number of categories obtained from sample data to the corresponding *theoretical proportions* of frequencies in the population. For example, in the Guided Example of Section 12.3, we compared sample data about airline passengers' seat preferences to a hypothetical model that was based on the presumption that the distributions across the four targeted seat categories would be equal. The distribution of participants' seat preferences represented the relative proportions, the hypothesized distribution reflected what we expected and represented the theoretical proportions, and the underlying research question was, "Does the sample frequency distribution 'fit' what we expect from the population frequency distribution if the null hypothesis were true?" The study also involved a single categorical variable (seat preference), and we collected frequency data on this variable and classified the data into specific seat categories.

We now extend this concept to research studies that involve *two* categorical variables where frequency data are collected on two separate variables, and the goal will be to determine if the two variables are related to each other. As an example, consider Table 12.7, which contains a summary of the responses to the question, "Do you support, oppose, or have no opinion on the FAA's minimum 1500-hour requirement for newly hired Part 121 pilots?" to which flight students and CFIs responded. In the context of chi-square, this summary table is referred to as a contingency table.

Note the following from Table 12.7:

- The table's structure is the same as the *means summary table* we presented in our discussion of factorial ANOVA in Chapter 10: It consists of rows and columns, where the rows represent the levels of the "row variable," and the columns represent the levels of the "column variable." For example, in Table 12.7, the row variable is Flight Status, which has two levels, Flight Student and CFI, and the column variable is Opinion, which has three levels, Support, Oppose, and No Opinion.

- Unlike a means summary table, the cells of a contingency table do *not* consist of group means, but instead consist of frequencies, or *counts*. For example, 29 flight students and 14 CFIs support the FAA requirement, 36 flight students and 24 CFIs oppose it, and 15 flight students and 2 CFIs have no opinion.

- Each cell entry represents paired data. For example, the first cell entry (n = 29) represents the frequency associated with the data pair Flight Students–Support. This notion of paired data is similar to bivariate correlation from Chapter 6.

To analyze the data in Table 12.7, we use a statistical strategy called the chi-square test for independence, which is similar to other measures of association such as the Pearson *r*. The chi-square test for independence, which also is called two-way chi-square because it involves two variables, tests the null hypothesis that the *proportion of frequencies in the categories that represent the levels of one variable are the same for those at all levels of the other variable.* For the FAA 1500-hour requirement example, the underlying question would be, "Do the proportions of flight students who respectively support, oppose, or have no opinion differ significantly from proportions of CFIs who respectively support, oppose, or have no opinion?" Notice that this underlying question is similar to the question we posed with respect to the concept of interactions in Chapter 10: "Do the levels of one variable operate the same or differently under the levels of the second variable?" Once again, the difference is we are working with distributions of frequency data, not group means.

The Test Statistic for the Chi-Square Test for Independence

The test statistic for the chi-square test for independence is the same as that for the chi-square test for goodness of fit:

$$\chi^2 = \text{Sum } \frac{(O_i - E_i)^2}{E_i}$$

When applying the formula, it is beneficial to recognize that the entries in the contingency table represent observed frequencies (O_i) for each paired category. To get the

Table 12.7 Contingency Table for FAA 1500-Hour Requirement Study

		Support	Oppose	No Opinion	Row Totals
		— Response to Question (Opinion)			
Flight Status	*Flight Student*	29	36	15	80
	CFI	14	24	2	40
	Column Totals	43	60	17	N = 120

corresponding expected frequencies (E_i), we have to compute the ratio of an observed frequency and the product of the respective row and column totals as

$$E_{ij} = \frac{T_i \times T_j}{N}$$

where i represents the row and j represents the column. For example, using the data from Table 12.7, the expected frequency for the CFI–Oppose cell (i.e., Row 2, Column 2), and denoted E_{22}, is the product of Row 2's total (40) and Column 2's total (60) divided by $N = 120$:

$$E_{22} = \frac{T_i \times T_j}{N} = \frac{40 \times 60}{120} = \frac{2400}{120} = 20$$

Thus, when applying the chi-square test statistic formula for this cell, the observed frequency is $O_{22} = 24$ and the corresponding expected frequency is $E_{22} = 20$. Although we will use our statistics program to calculate the χ^2 test statistic, a summary of the hand calculations associated with the data in Table 12.7 is provided for instructional purposes, and a copy of the raw data that corresponds to this example is provided in the file Ch_12 General Example Data for Chi-Square Test for Independence. The reader is encouraged to confirm the expected frequencies given in these calculations.

$$\chi^2 = \text{Sum}\left[\frac{(O_i - E_i)^2}{E_i}\right]$$

$$= \frac{(29 - 28.7)^2}{28.7} + \frac{(36 - 40)^2}{40} + \frac{(15 - 11.3)^2}{11.3} + \frac{(14 - 14.3)^2}{14.3} + \frac{(24 - 20)^2}{20} + \frac{(2 - 5.7)^2}{5.7}$$

$$= \frac{(0.3)^2}{28.7} + \frac{(-4)^2}{40} + \frac{(3.7)^2}{11.3} + \frac{(-0.3)^2}{14.3} + \frac{(4)^2}{20} + \frac{(-3.7)^2}{5.7}$$

$$= \frac{0.09}{28.7} + \frac{16}{40} + \frac{13.69}{11.3} + \frac{0.09}{14.3} + \frac{16}{20} + \frac{13.69}{5.7}$$

$$\approx 0.003 + 0.4 + 1.21 + 0.006 + 0.8 + 2.4$$

$$= 4.819$$

From an instructional perspective, these hand calculations enable us to observe each cell's contribution to the overall chi-square. For example, the first cell (Flight Student–Support) contributed nearly nothing (0.003 units) to the overall chi-square of 4.819, but the third cell (Flight Student–No Opinion) contributed 1.21 units, or 25%.

The degrees of freedom for the chi-square test for independence are equal to the product of one less than the number of rows (R) in the contingency table and one less than the number of columns (C). That is,

$$df = (R - 1)(C - 1)$$

For the current example involving the opinions of flight students and CFIs about the FAA's 1500-hour requirement, $df = (2 - 1)(3 - 1) = (1)(2) = 2$. From Table 6/Appendix A, the corresponding critical value is $\chi^2(2) = 5.991$ for $\alpha = .05$. Because the calculated chi-square value is less than the boundary value of the critical region (4.82 < 5.99), we would fail to reject the corresponding null hypothesis.

The Null Hypothesis for the Chi-Square Test for Independence

With respect to hypothesis testing, the chi-square test for independence tests the null hypothesis that the variables are independent of each other. For this to be true, the *proportion* of participants in the different levels (categories) of one variable should be the same regardless of a participant's position on the other variable. For example, in our running example, if "Flight Status" is unrelated to "Opinions," then the *proportion* of flight students and CFIs should be the same across the three opinion categories (Support, Oppose, No Opinion). The corresponding null and alternative hypotheses would then be expressed as follows:

H_0: The distribution of opinions to the FAA's 1500-hour requirement has the same proportions for flight students and for CFIs. In other words, the two levels of Flight Status are independent of the three levels of Opinions.

H_1: The proportions are significantly different between flight students and CFIs.

To illustrate this, we can see from the contingency table given in Table 12.7 that the observed frequencies are not uniform across all the levels of opinion for flight students and CFIs. However, if we focus our attention on the *proportions* of those responding in each category as summarized in Table 12.8, observe the following:

- Of the 120 participants, two-thirds ($n = 80$) were flight students and one-third ($n = 40$) were CFIs.
- 29 of 80 flight students (36.25%) and 14 of 40 CFIs (35%) "support" the requirement.
- 36 of 80 flight students (45%) and 24 of 40 CFIs (60%) "oppose" the requirement.
- 15 of 80 flight students (18.75%) and 2 of 40 CFIs (5%) have "no opinion."

Table 12.8 Contingency Table for FAA 1500-Hour Requirement Study with Proportions (p)

Flight Status	Support		Oppose		No Opinion		Row Total	
	n	p^a	n	p^a	n	p^a	n	p^b
Flight Student	29	0.3625	36	0.4500	15	0.1875	80	2/3
		0.6744		0.6000		0.8824		
		0.2417		0.3000		0.1250		
CFI	14	0.3500	24	0.6000	2	0.0500	40	1/3
		0.3256		0.4000		0.1176		
		0.1167		0.2000		0.1670		
Column Total	43	0.3583	60	0.5000	17	0.1417	120	100%

Note. N = 120.
$^a p$ = The base for the three proportions given for each of the six primary cells is relative to the row n, column n, and overall n. For example, in the Flight Student-Support cell, the proportions reflect 29/80, 29/43, and 29/120, respectively. $^b p$ = the base is the overall total.

When examined from this perspective, the respective percentages of flight students' and CFIs' opinions differ by 1% for "Support," 15% for "Oppose," and 13.75% for "No Opinion." Although these differences in proportion are *numerically* different, they are not *statistically* different, which was confirmed by the calculated test statistic of $\chi^2(2)$ = 4.82 for α = .05 vs. the critical value of $\chi^2(2)$ = 5.99. As a result, we would conclude that based on the sample data, a person's flight status (flight student or CFI) is *independent* of his or her opinion of the FAA's 1500-hour requirement, and vice versa, $\chi^2(2, N = 120) = 4.82, p > .05$.

As an alternative to expressing H_0 and H_1 from an independence perspective, we also could express the null and alternative hypotheses from a correlation perspective because the concept of independence presumes that the variables are unrelated. When considered from this perspective, the corresponding null and alternative hypotheses would be:

H_0: There is no significant relationship between flight status and opinions.
H_1: There is a significant relationship between flight status and opinions.

Thus, similar to the Pearson *r*, the null hypothesis of the chi-square test for independence posits there is no relationship between the variables (i.e., they are independent of each other), whereas the alternative hypothesis posits the variables are related to each other.

Before we close this section, let's return to Table 12.8 for a moment. In addition to reporting the proportions with respect to the row totals, we also included proportions with respect to the column and overall totals. For example:

- Of the 43 participants who reported they "support" the FAA 1500-hour requirement, 29 (67.44%) were flight students, and 14 (32.56%) were CFIs.
- The Flight Student–Support group represented approximately 25% of the overall sample (29/120).
- Independent of their opinions, two-thirds of the sample (80/120) were flight students and one-third (40/120) were CFIs.
- Independent of their flight status, the largest opinion group was "oppose," which constituted 50% (60/120) of the sample.

Although it is not necessary to report all of these proportions, we choose to do so because it provides the reader a more complete picture of the analysis.

Assumptions for the Chi-Square Test for Independence

The assumptions for the chi-square test for independence are the same as those for the goodness of fit test presented earlier.

- *Independence.* Each observation (i.e., each observed cell of a contingency table) must be independent. Thus, the levels of one variable cannot be dependent on any of the levels of the second variable.
- *Minimum cell sizes.* The expected frequency of any cell should not be less than 5, or at least 80% of the expected frequency cells should be greater than 5.

Effect Size for the Chi-Square Test for Independence

Cohen's *w*, which we used to measure the effect size for the chi-square test for goodness of fit, also can be used to measure the effect size for the chi-square test for independence. The most widely reported effect size for the chi-square test for independence, though, is *Cramer's V*, and the general rule of thumb is to use Cramer's *V* if the design is larger than 2 × 2.

$$V = \sqrt{\frac{\chi^2}{N \times m}}$$

where

- χ^2 = the calculated chi-square statistic
- N = the total number of observations
- m = (the smaller of *R* or *C*) *minus* 1, where R = the number of rows and C = the number of columns in the contingency table.

Applying Cramer's *V* to the FAA 1500-hour requirement example, which involves a 2 × 3 design (2 rows, 3 columns), with $\chi^2 = 4.82$ and $N = 120$, the smaller of *R* vs. *C* is *R*, which is equal to 2. Therefore, $m = (R - 1) = (2 - 1) = 1$, and *V* is:

$$V = \sqrt{\frac{\chi^2}{N \times m}} = \sqrt{\frac{4.82}{120 \times 1}} = \sqrt{\frac{4.82}{120}} = \sqrt{0.04} = 0.20$$

To interpret the effect size, we use Cohen's (1988) labels of "small," "medium," and "large" as summarized in Table 12.9. As a result, based on these labels, the magnitude of the effect for the current example is somewhere between small and medium.

12.5 Using the Chi-Square Test for Independence in Research: A Guided Example

We now provide a guided example of a research study that involves the use of the chi-square test for independence. The context of this guided example is an extension of Torres et al.'s (2011) runway incursions study. In addition to examining the relationship between human factor errors and runway incursions, Torres et al. also examined the relationship between weather conditions and runway incursions. Because we did not have

Table 12.9 Cohen's Effect Sizes for Chi-Square Test for Independence

m^a	Small Effect	Medium Effect	Large Effect
1	0.10	0.30	0.50
2	0.07	0.21	0.35
3	0.06	0.17	0.29

Note. $^a m$ = the smaller of *R* or *C minus* 1, where R = the number of rows and C = the number of columns of the contingency table. For example, in a 2 × 3 design, $R = 2$ and $C = 3$. Therefore, $m = (R - 1) = (2 - 1) = 1$.

Torres et al.'s raw data, we instead examined the FAA's (2022) runway incursions database and randomly selected $N = 560$ cases between January 2008 and August 2013 that involved weather conditions and the different types of runway incursion errors. A copy of the raw data is given in the file Ch_12 Guided Example Data for Chi-Square Test for Independence, and the corresponding contingency table is presented in Table 12.10 as a convenience to the reader. We will structure this guided example into three distinct parts: (a) pre-data analysis, (b) data analysis, and (c) post-data analysis.

Pre-Data Analysis

Research Question and Operational Definitions

The overriding research question is: "What is the relationship between weather conditions and runway incursions?" The key terms and variables that require definitions include *weather conditions* and *runway incursions*. In the context of the current study, weather conditions are defined as visual meteorological conditions (VMC) and instrument meteorological conditions (IMC). VMC refers to weather conditions under which pilots are able to fly an aircraft with sufficient visibility that allows them to maintain visual separation from terrain and other aircraft, and IMC refers to weather conditions under which pilots must fly primarily using the aircraft's instruments (International Civil Aviation Organization, ICAO, 2005).

As for runway incursions, the Federal Aviation Administration (FAA, n.d.) defines a runway incursion as "Any occurrence at an aerodrome involving the incorrect presence of an aircraft, vehicle or person on the protected area of a surface designated for the landing and takeoff of aircraft" (para. 1). The FAA also classifies runway incursions as operational incidents (OIs), which are attributed to air traffic control and also include operational errors (OEs) and operational deviations (ODs); pilot deviations (PDs), which are attributed to pilots; and vehicle/pedestrian deviations (V/PDs), which are attributed to individuals such as drivers of food service vehicles or baggage handlers.

Research Hypothesis

Weather conditions and runway incursions will not be independent of each other. In other words, runway incursions will be related to weather conditions.

Table 12.10 Contingency Table for the Guided Example for Chi-Square Test for Independence Involving Weather Conditions and Runway Incursions with Observed Frequencies

	Runway Incursions[a]			
Weather Conditions	OE	PD	V/PD	*Total*
Instrument Meteorological Conditions (IMC)	8	13	5	26
Visual Meteorological Conditions (VMC)	251	230	53	534
Total	259	243	58	560

Note. [a]OE = Operational Errors, which consists of Operational Incidents, Operational Deviations, and Other; PD = Pilot Deviations; and V/PD = Vehicle/Pedestrian Deviations.

Research Methodology

The research methodology is a 2×3 chi-square design, which involves two nominal variables where the first variable, Weather Conditions, has two levels, and the second variable, Runway Incursions, has three levels. The objective is to determine if the levels of one factor are related to or independent of the levels of the second factor. We also may consider the research methodology as correlational because we are examining a relationship between variables.

Sample Size Planning (Power Analysis)

To determine the minimum sample size needed, we prepared Table 7a ($\alpha = .05$) and Table 7b ($\alpha = .01$) in Appendix A for small, medium, and large effect sizes relative to Cohen's labels from Table 12.9. Although not comprehensive, the tables provide guidance for establishing sample sizes and determining post hoc power. For instance, in the current example, we have a 2×3 design, which means that $m = 1$ in Table 12.9. Therefore, if $\alpha = .05$, power $= .80$, and *ES* is medium, then $N = 107$ from the second sub-table of Table 7a/Appendix A.

Alternatively, we could use *G*Power* (Faul et al., 2007, 2009) similar to what we did for the Guided Example in Section 12.3, which involved the chi-square test for goodness of fit. Given a medium effect size of $w = 0.3$, $\alpha = .05$, and $df = (2 - 1)(3 - 1) = 2$: $N = 108$. If using *G*Power*, though, recognize that the effect size is with respect to w.

Data Analysis

We now analyze the given data set relative to a hypothesis test involving the chi-square test for independence.

Step 1: Formulate null/alternative hypotheses. The null/alternative hypotheses are:

> H_0: There is no significant relationship between weather conditions and runway incursions: The two levels of weather conditions are independent of the three levels of runway incursions.
>
> H_1: There is a significant relationship between weather conditions and runway incursions.

Step 2: Determine the test criteria.

> *Level of Significance.* The preset alpha level is $\alpha = .05$.
>
> *Test Statistic.* The test statistic is χ^2 and $df = (2 - 1)(3 - 1) = 2$.
>
> *Boundary of the Critical Region.* The boundary of the critical region acquired from Table 7a/Appendix A is $\chi^2(2, \alpha = .05) = 5.991$. This is illustrated in Figure 12.5(a).

Step 3: Collect data, check assumptions, run the analysis, and report the results.

> *Collect Data.* The data are given in the file as indicated earlier, and the corresponding contingency table is presented in Table 12.10.
>
> *Check Assumptions.* There are two assumptions.

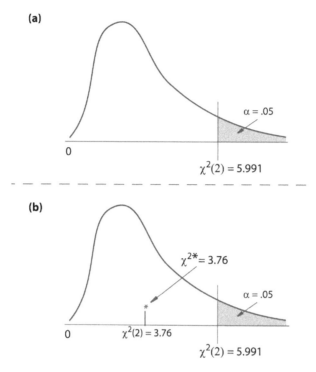

Figure 12.5 (a) Illustration of the critical chi-square value for Chapter 12's Guided Example of test for independence, and (b) the location of the calculated chi-square value with respect to the critical value.

- Independence. The two weather conditions, IMC and VMC, are mutually exclusive as are the three types of runway incursions. Therefore, this assumption is satisfied.
- Minimum cell sizes. As we will demonstrate shortly, the expected frequency for the IMC–V/PD cell is 2.69, which is less than 5. However, because five of the six categories (83.3% > 80%) have expected frequencies greater than 5, this assumption is satisfied.

Run the Analysis and Report the Results. The results of this analysis from our statistics program are:

- The calculated χ^2 test statistic for $df = 2$ and $N = 560$ is $\chi^{2*} = 3.76$, which lies outside the critical region as shown in Figure 12.5(b).
- The corresponding p value is $p = .1526$.

Hand Calculations. Independent of reporting the results from our statistics program, we also demonstrate the manual calculation χ^2. To do this, though, we need to first determine the expected frequencies. Observe from Table 12.10 that 259 of the 560 cases (46.25%) were OEs, 243 of the 560 cases (43.39%) were PDs, and 58 of the 560 cases (10.36%) were V/PDs. Given that the null hypothesis claims that the distribution is the same for the two categories of weather conditions, we use these proportions to determine what we would expect the frequencies to be if the null hypothesis were true:

- 46.25% of the 26 IMC cases are expected to be OEs: $.4625 \times 26 = 12.03$
- 43.39% of the 26 IMC cases are expected to be PDs: $.4339 \times 26 = 11.28$

- 10.36% of the 26 IMC cases are expected to be V/PDs: .1036 × 26 = 2.69
- 46.25% of the 534 VMC cases are expected to be OEs: .4625 × 534 = 246.98
- 43.39% of the 534 VMC cases are expected to be PDs: .4339 × 534 = 231.70
- 10.36% of the 534 VMC cases are expected to be V/PDs: .1036 × 534 = 55.32

Table 12.11 contains a summary of these calculations and includes both the observed frequencies (*O*) and the expected frequencies (*E*). We now apply the formula.

$$\chi^2 = \text{Sum}\left[\frac{(O_i - E_i)^2}{E_i}\right]$$

$$= \frac{(8-12.03)^2}{12.03} + \frac{(13-11.28)^2}{11.28} + \frac{(5-2.69)^2}{2.69} + \frac{(251-246.98)^2}{246.98} + \frac{(230-231.7)^2}{231.7} + \frac{(53-55.32)^2}{55.32}$$

$$= \frac{(-4.03)^2}{12.03} + \frac{(1.72)^2}{11.28} + \frac{(2.31)^2}{2.69} + \frac{(4.02)^2}{246.98} + \frac{(-1.7)^2}{231.7} + \frac{(-2.32)^2}{55.32}$$

$$= \frac{16.2409}{12.03} + \frac{2.9584}{11.28} + \frac{5.3361}{2.69} + \frac{16.1604}{246.98} + \frac{2.89}{231.7} + \frac{5.3824}{55.32}$$

$$= 1.35 + 0.26 + 1.98 + 0.065 + 0.012 + 0.097$$

$$= 3.764$$

Thus, our hand calculation yields $\chi^2 = 3.774$, which is off by 0.004 units compared to what was provided by our statistics program. Once again, recall that the hand calculations enable us to see each cell's contribution to the overall chi-square. For example, the third cell (IMC–V/PD) had the largest contribution at 1.98 units, or 53%, to the overall chi-square of 3.774.

Step 4: Decide whether to reject or fail to reject the null hypothesis, and make a concluding statement relative to the RQ.

Decision. Because the calculated $\chi^{2*} = 3.76$ lies outside the critical region as shown in Figure 12.5(b), the decision is *fail to reject the null hypothesis.*

Concluding Statement. There is no significant relationship between weather conditions and runway incursions. The distribution of frequencies of OEs, PDs, and V/PDs is statistically uniform with respect to the meteorological weather conditions of IMC and VMC.

Table 12.11 Complete Contingency Table for the Weather Conditions–Runway Excursions Study for the Guided Example for the Chi-Square Test for Independence

| | Runway Incursions[b] | | | | | |
| | OE | | PD | | V/PD | |
Weather Conditions[a]	O	E	O	E	O	E
IMC	8	12.03	13	11.28	5	2.69
VMC	251	246.98	230	231.70	53	55.32

Note. [a]IMC = instrument meteorological conditions, VMC = visual meteorological conditions.
[b]OE = operational errors, which include operational deviations and operational incidents; PD = pilot deviations; and V/PD = vehicle/pedestrian deviations. O = observed frequencies and E = expected frequencies.

Post-Data Analysis

We now determine, report, and interpret the corresponding effect size and power. We also discuss plausible explanations for the results.

Effect Size

As presented earlier, the effect size for the chi-square test for independence is Cramer's V:

$$V = \sqrt{\frac{\chi^2}{N \times m}} = \sqrt{\frac{3.76}{560 \times 1}} = \sqrt{\frac{3.76}{560}} = \sqrt{0.0067} = 0.08$$

Based on Cohen's guidance, this is a small effect size. As a point of information, because $m = 1$, the effect sizes given by Cramer's V and w will be equal.

Power (Post Hoc)

To determine the power of the study, we focus on the second sub-table of Table 7a/Appendix A for $df = 2$ (2×3) design. Because the effect size was $V = 0.08$, which is a "small" effect, we scan the first row of this sub-table and look for a sample size that is close to $N = 560$. This is between 496 and 621, which equates to power values of .50 and .60, respectively. As a result, we estimate the power of the study to be somewhere between .50 and .60, but because the effect size was smaller than .10, the power most likely is smaller than .50. Thus, there is less than a 50% chance that the small effect found in the sample truly exists in the population.

   Alternatively, because in the current example Cramer's V is equal to w, we could use *G*Power* (Faul et al., 2007, 2009) as we did earlier for the goodness of fit test. Based on inputs of $w = 0.08$, $\alpha = .05$, total sample size = 560, and $df = 2$, the corresponding power is .3768.

Plausible Explanations for the Results

Two immediate plausible explanations should come to mind. The first is related to the power of the study. Because the effect size was so small, this implies that we did not have a sufficiently large sample size (recall the basketball in a haystack vs. a needle in the haystack concept). Consulting the second sub-table of Table 7a/Appendix A for guidance, to find a "small" effect size ($w = 0.10$) for $df = 2$ (2×3 design) with power = .80, we would need a sample size of at least $N = 964$. If we use *G*Power* (Faul et al., 2007, 2009) and enter an effect size of $w = 0.08$, then the minimum sample size needed is $N = 1,506$. Thus, one plausible explanation for the results is we did not have a sufficiently large sample size.

   A second plausible explanation is with respect to the sample data. Observe from the contingency table given in Table 12.10 that of the 560 runway incursion cases, only 26 (4.64%) occurred under IMC, which means that 95.36% occurred under VMC. This begs the question: "Is weather really a variable in this study or is it a constant?" Lastly, observe from Table 12.11, which contains both observed and expected frequencies, that the expected frequency for the IMC–V/PD cell is 2.69, which is less than 3. Harris (1998, p. 466) cautioned against using a chi-square test if the expected frequency of any cell falls below 3. Her recommendation is to use an alternative strategy such as combining categories or applying

Fisher's exact probability test. Thus, it is possible that the results might be spurious be-cause of this one cell.

Chapter Summary

1. A chi-square test is a nonparametric inferential statistics procedure used for analyzing frequency data. The focus is on determining if the proportion of observed frequen-cies measured on a nominal-level variable is significantly different than the propor-tion of frequencies that are expected to occur.

2. The chi-square test for goodness of fit is known as one-way chi-square because it involves a single nominal variable with multiple levels, or categories. A one-way chi-square is used to determine if the proportion of observed data is consistent with or "fits" with what is expected.

3. The chi-square test for independence is known as two-way chi-square because it involves two nominal variables each with multiple levels, or categories. A two-way chi-square is used to determine if the two variables are related. If the variables are un-related, then this means that the levels of one variable are independent of the levels of the second variable. This objective is accomplished by comparing the proportion of observed frequencies with the proportion of frequencies that would be expected if the variables were independent of each other.

4. The chi-square distribution is nonsymmetrical and nonnegative and consists of a family of distributions governed by the degrees of freedom, similar to the t and F distributions. The degrees of freedom are $df = C - 1$, where C = the number of cat-egories, for the chi-square test for goodness of fit, and $df = (R - 1)(C - 1)$, where R = the number of rows and C = the number of columns, for the chi-square test for independence.

5. The chi-square test statistic is $\chi^2 = \text{Sum}\left[\dfrac{(O_i - E_i)^2}{E_i}\right]$, where O_i are the observed frequencies and E_i are the expected frequencies. The data for a chi-square test are summarized in a frequency distribution table (test for goodness of fit) or a contin-gency table (test for independence). The assumptions for chi-square are independ-ence and minimum cell sizes. Cohen's w is used to determine the effect size for one-way chi-square, and Cramer's V is used to determine the effect size for two-way chi-square.

Vocabulary Check

Category	Expected frequency (E)
Cell	Independence assumption
Chi-square distribution	Minimum cell size assumption
Chi-square test for goodness of fit	Nonparametric test
Chi-square test for independence	Observed frequency (O)
Chi-square test statistic	One-way chi-square
Cohen's w	Paired data
Contingency table	Parametric test
Cramer's V	Two-way chi-square

Review Exercises

A. Check Your Understanding

In 1–10, choose the best answer among the choices provided.

1. One of the differences between one-way chi-square and two-way chi-square is:
   a. One-way chi-square involves scores on one variable whereas two-way chi-square involves scores on two variables.
   b. One-way chi-square has one *df* like a *t* test, whereas two-way chi-square has two *df*s like an *F* test.
   c. The calculated chi-square statistic always will be higher for two-way chi-square than for one-way chi-square.
   d. The chi-square distribution for one-way chi-square is different than the distribution for two-way chi-square.
2. If a chi-square test for goodness of fit results in a large chi-square value that is statistically significant, then this means that the
   a. observed frequencies are consistent with the expected frequencies.
   b. observed frequencies are inconsistent with the expected frequencies.
   c. sample data are a good fit to the null hypothesis.
   d. level of significance had to be extremely small.
3. Which of the following situations is true for the chi-square test for goodness of fit?
   a. A small number of categories most likely will lead to a large χ^2.
   b. A large number of categories most likely will lead to a small χ^2.
   c. A large number of categories most likely will lead to a large χ^2.
   d. The number of categories has no effect on the size of χ^2.
4. If the observed and expected frequencies are equal, then the χ^2 test statistic will be
   a. positive
   b. zero
   c. negative
   d. infinitely large
5. A group of 100 flight students and 100 nonflight students was asked to reply to the question, "In general do you enjoy reading a college textbook?" The results are given below. What would be the appropriate statistical procedure to apply to these data?

	Yes	No	Sometimes
Flight Students	30	45	25
Nonflight Students	50	30	20

   a. χ^2 test for Goodness of Fit
   b. *t* test for independent sample means
   c. ANOVA
   d. χ^2 test for Independence
6. An aviation researcher examined whether airport executives (AEs) and aviation maintenance technicians (AMTs) also had a PPL. A random sample yielded the following data:

Group	f
A. AEs w/PPL	25
B. AEs w/o PPL	10
C. AMTs w/PPL	45
D. AMTs w/o PPL	20

Data analysis resulted in $\chi^2 = 10.42$. If the researcher expected a ratio of 3:2:3:2 for Groups A–D, respectively, are the observed results *inconsistent* with the expected ratio for $\alpha = .05$?

a. Yes, because $\chi^{2*} > \chi^2$ critical.
b. Yes, because $\chi^{2*} < \chi^2$ critical.
c. No, because $\chi^{2*} < \chi^2$ critical.
d. Yes, because $\chi^{2*} > \chi^2$ critical.

7. Referencing Item 6 above, are the observed results inconsistent with the expected ratio at the 1% level of significance?

a. Yes, because $\chi^{2*} > \chi^2$ critical.
b. Yes, because $\chi^{2*} < \chi^2$ critical.
c. No, because $\chi^{2*} < \chi^2$ critical.
d. Yes, because $\chi^{2*} > \chi^2$ critical.

8. Referencing Item 6 above, suppose the researcher organized and analyzed the data from the perspective shown below. Would the analysis yield the same chi-square value as reported in Item 6?

	PPL	
Occupation	Yes	No
Airport Executive	25	10
Aviation Maintenance Technician	40	25

a. Yes, because the data and corresponding proportion of frequencies are the same.
b. Yes, because both analyses involve a chi-square test.
c. No, because one uses a chi-square test for goodness of fit and the other uses a chi-square test for independence.
d. No, because the degrees of freedom will be different.

9. Let's assume that an aviation researcher hypothesized that the proportion of runway incursions would be uniform across four different runway incursion categories: (a) those attributed to air traffic control (OE/OD/OI), (b) those attributed to pilots (PD), (c) those attributed to vehicles and pedestrians (V/PD), and (d) those that were classified as "other" or "unknown" (OTH/UNK). Let's further assume that a 10-year analysis of runway incursions from the FAA database resulted in sample sizes of $N = 2990, 98644, 3507$, and 773, respectively, for the four categories, and the results of data analysis were $\chi^2(3, \alpha = .05) = 10638.95, p < .0001$. What can you conclude?

a. There is a significant difference in the proportion of frequencies of runway incursions across the four categories.

b. The sample data are consistent with the researcher's hypothesis.

c. The results are flawed because the proportion of frequencies are heavily biased toward pilots.

d. There must be an error in the calculation of χ^2 because χ^2 cannot be that large.

10. Three major airports, as listed by the U.S. Bureau of Transportation Statistics (BTS, 2022), were randomly selected and the number of flight delays was recorded with respect to BTS' categories as shown below. The data reflected all flight delays between January 2022 and April 2022, inclusive, and included all airlines. The results of data analysis were $\chi^2(8) = 19463.71$, $p = 0$, Cramer's $V = 0.06$. Which of the following is the most appropriate conclusion?

		Flight Delays			
Airport	Air Carrier	Weather	NAS	Security	Late-Arrival Aircraft
JFK	299583	24047	133716	1887	318224
ORD	445262	79515	243477	2803	433251
SEA	208778	33182	99921	1710	179482

a. Because $p = 0$, there is no significant relationship between airports and flight delays.

b. Because $\chi^2 = 19463.71$, the distribution of flight delays has the same proportions across the three airports.

c. Because $df = 8$, there is a significant difference in the proportion of flight delays across the three airports.

d. Because Cramer's $V = 0.06$, the effect size is too small to find a significant effect.

B. Apply Your Knowledge: Chi-Square Test for Goodness of Fit

Use the following research description and the corresponding data set to conduct the activities given in Parts A–C (see also Section 12.3).

This research study is an extension of the guided example given in Section 12.3. Let's assume that Delta Air Lines reports the following information about seat preferences among U.S. travelers who took a Delta flight to/from at least one domestic airport during the past 3 months:

- 30% prefer an aisle seat
- 20% prefer a window seat
- 20% prefer an exit row seat
- 10% prefer a middle seat
- 10% prefer to sit in Delta Comfort Plus
  10% prefer to sit in the main cabin

Let's further assume that you randomly surveyed 400 Delta passengers with similar flight experiences and asked them to specify their primary seat preference. A copy of these data is contained in the file titled "Ch_12 Exercises Part B Data." For the reader's convenience these data are summarized in the table below.

Seat Preference	f
Aisle	125
Window	77
Exit Row	90
Middle	31
Delta Comfort Plus	42
Main Cabin	35

A. Pre-Data Analysis

1. Specify the research questions and corresponding operational definitions.
2. Specify the research hypotheses.
3. Determine the appropriate research methodology/design and explain why it is appropriate.
4. Conduct an a priori power analysis to determine the minimum sample size needed. Compare this result to the size of the given data set and explain what impact the size of the given sample will have on the results relative to the minimum size needed.

B. Data Analysis

1. Conduct a hypothesis test using the chi-square test for goodness of fit by applying all four steps of hypothesis testing as presented in Section 12.3.

C. Post-Data Analysis

1. Determine and interpret the estimated effect size using Cohen's w.
2. Determine and interpret the power of the study from a post hoc perspective.
3. Present at least two plausible explanations for the results.

C. Apply Your Knowledge: Chi-Square Test for Independence

Use the following research description and the corresponding data set to conduct the activities given in Parts A–C (see also Section 12.5).

The data file "Ch_12 Exercises Part C Data" contains a subset of the runway incursions data acquired from the FAA's runway incursions database for five airports that had the highest frequency of reported runway incursions between January 2008 and August 2013. Your assignment is to analyze these data to determine if there is a relationship between airport location and runway incursions. The airports are Hartsfield–Jackson Atlanta International (ATL), Charlotte Douglas International (CLT), Phoenix Deer Valley (DVT), Los Angeles International (LAX), and O'Hare International (ORD). Furthermore, the runway incursions include Operational Errors (OEs), which consist of operational incidents, operational deviations, and "Other"; Pilot Deviations (PDs); and Vehicle/Pedestrian Deviations (V/PDs).

A. Pre-Data Analysis

1. Specify the research questions and corresponding operational definitions.
2. Specify the research hypotheses.
3. Determine the appropriate research methodology/design and explain why it is appropriate.
4. Conduct an a priori power analysis to determine the minimum sample size needed. Compare this result to the size of the given data set and explain what impact the size of the given sample will have on the results relative to the minimum size needed.

B. Data Analysis

1. Conduct a hypothesis test using the chi-square test for independence by applying all four steps of hypothesis testing as presented in Section 12.5.

C. Post-Data Analysis

1. Determine and interpret the estimated effect size using Cramer's *V*.
2. Determine and interpret the power of the study from a post hoc perspective.
3. Present at least two plausible explanations for the results.

References

Bureau of Transportation Statistics (2022). *Airline on-time statistics and delay causes.* https://www. transtats.bts.gov/ot_delay/ot_delaycause1.asp

Cohen, J. (1988). *Statistical power analysis for the behavioral sciences* (2nd ed.). Lawrence Erlbaum Associates.

Faul, F., Erdfelder, E., Buchner, A., & Lang, A.-G. (2009). Statistical power analyses using G*Power 3.1: Tests for correlation and regression analyses. *Behavior Research Methods, 41*, 1149–1160. [Download software at http://www.gpower.hhu.de]. https://www.psychologie.hhu.de/fileadmin/redaktion/Fakultaeten/Mathematisch-Naturwissenschaftliche_Fakultaet/Psychologie/AAP/gpower/GPower31-BRM-Paper.pdf

Faul, F., Erdfelder, E., Lang, A.-G., & Buchner, A. (2007). G*Power 3: A flexible statistical power analysis program for the social, behavioral, and biomedical sciences. *Behavior Research Methods, 39*, 175–191. [Download software at http://www.gpower.hhu.de]. https://www.psychologie.hhu.de/fileadmin/redaktion/Fakultaeten/Mathematisch-Naturwissenschaftliche_Fakultaet/Psychologie/AAP/gpower/GPower3-BRM-Paper.pdf

Federal Aviation Administration (2022). *FAA runway safety office—Runway incursions (RWS).* https://www.asias.faa.gov/apex/f?p=100:28:::NO:28

Harris, M. B. (1998). *Basic statistics for behavioral science research* (2nd ed.). Allyn and Bacon.

Omni Calculator (n.d.). *The critical value calculator.* https://www.omnicalculator.com/statistics/critical-value

Statistics Kingdom (2022a). *Chi-squared sample size calculator.* https://www.statskingdom.com/sample_size_chi2.html

Statistics Kingdom (2022b). *Chi-squared statistical power calculator.* https://www.statskingdom.com/34test_power_chi2.html

Torres, K. R, Metscher, D. S., Smith, M. (2011). A correlational study of the relationship between human factor errors and the occurrence of runway incursions. *International Journal of Professional Aviation Training & Testing Research, 5*(1), 3–25. https://ojs.library.okstate.edu/osu/index.php/IJAR/article/view/8111/7457

Appendix

Statistics Tables

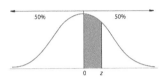

Table 1 Area Under the Standard Normal Curve Between the Mean and Z (see Figure A1)

Figure A1 Figure for Table 1.

z	|| Second decimal digit of z									
	0	1	2	3	4	5	6	7	8	9
0.0	.0000	.0040	.0080	.0120	.0160	.0199	.0239	.0279	.0319	.0359
0.1	.0398	.0438	.0478	.0517	.0557	.0596	.0636	.0675	.0714	.0753
0.2	.0793	.0832	.0871	.0910	.0948	.0987	.1026	.1064	.1103	.1141
0.3	.1179	.1217	.1255	.1293	.1331	.1368	.1406	.1443	.1480	.1517
0.4	.1554	.1591	.1628	.1664	.1700	.1736	.1772	.1808	.1844	.1879
0.5	.1915	.1950	.1985	.2019	.2054	.2088	.2123	.2157	.2190	.2224
0.6	.2257	.2291	.2324	.2357	.2389	.2422	.2454	.2486	.2517	.2549
0.7	.2580	.2611	.2642	.2673	.2704	.2734	.2764	.2794	.2823	.2852
0.8	.2881	.2910	.2939	.2967	.2995	.3023	.3051	.3078	.3106	.3133
0.9	.3159	.3186	.3212	.3238	.3264	.3289	.3315	.3340	.3365	.3389
1.0	.3413	.3438	.3461	.3485	.3508	.3531	.3554	.3577	.3599	.3621
1.1	.3643	.3665	.3686	.3708	.3729	.3749	.3770	.3790	.3810	.3830
1.2	.3849	.3869	.3888	.3907	.3925	.3944	.3962	.3980	.3997	.4015
1.3	.4032	.4049	.4066	.4082	.4099	.4115	.4131	.4147	.4162	.4177
1.4	.4192	.4207	.4222	.4236	.4251	.4265	.4279	.4292	.4306	.4319

(Continued)

Second decimal digit of z

z	0	1	2	3	4	5	6	7	8	9
1.5	.4332	.4345	.4357	.4370	.4382	.4394	.4406	.4418	.4429	.4441
1.6	.4452	.4463	.4474	.4484	.4495	.4505	.4515	.4525	.4535	.4545
1.7	.4554	.4564	.4573	.4582	.4591	.4599	.4608	.4616	.4625	.4633
1.8	.4641	.4649	.4656	.4664	.4671	.4678	.4686	.4693	.4699	.4706
1.9	.4713	.4719	.4726	.4732	.4738	.4744	.4750	.4756	.4761	.4767
2.0	.4772	.4778	.4783	.4788	.4793	.4798	.4803	.4808	.4812	.4817
2.1	.4821	.4826	.4830	.4834	.4838	.4842	.4846	.4850	.4854	.4857
2.2	.4861	.4864	.4868	.4871	.4875	.4878	.4881	.4884	.4887	.4890
2.3	.4893	.4896	.4898	.4901	.4904	.4906	.4909	.4911	.4913	.4916
2.4	.4918	.4920	.4922	.4925	.4927	.4929	.4931	.4932	.4934	.4936
2.5	.4938	.4940	.4941	.4943	.4945	.4946	.4948	.4949	.4951	.4952
2.6	.4953	.4955	.4956	.4957	.4959	.4960	.4961	.4962	.4963	.4964
2.7	.4965	.4966	.4967	.4968	.4969	.4970	.4971	.4972	.4973	.4974
2.8	.4974	.4975	.4976	.4977	.4977	.4978	.4979	.4979	.4980	.4981
2.9	.4981	.4982	.4982	.4983	.4984	.4984	.4985	.4985	.4986	.4986
3.0	.4987	.4987	.4987	.4988	.4988	.4989	.4989	.4989	.4990	.4990
3.1	.4990	.4991	.4991	.4991	.4992	.4992	.4992	.4992	.4993	.4993
3.2	.4993	.4993	.4994	.4994	.4994	.4994	.4994	.4995	.4995	.4995
3.3	.4995	.4995	.4995	.4996	.4996	.4996	.4996	.4996	.4996	.4996
3.4	.4997	.4997	.4997	.4997	.4997	.4997	.4997	.4997	.4997	.4998
3.5	.4998	.4998	.4998	.4998	.4998	.4998	.4998	.4998	.4998	.4998
3.6	.4998	.4998	.4999	.4999	.4999	.4999	.4999	.4999	.4999	.4999

Note. For z ≥ 3.7, use .5000.

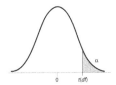

Table 2 Critical Values of Student's *t* Distribution, which Equate to a Particular Proportion of the Area Under the Curve for a One-Tailed Test (see Figure A2)

Figure A2 Figure for Table 2.

			p (One-Tailed Probabilities)				
df	.10	.05	.025	.01	.005	.001	.0005
1	3.078	6.314	12.706	31.821	63.657	318.309	636.619
2	1.886	2.920	4.303	6.965	9.925	22.328	31.600
3	1.638	2.353	3.182	4.541	5.841	10.214	12.924
4	1.533	2.132	2.776	3.747	4.604	7.173	8.610
5	1.476	2.015	2.571	3.365	4.032	5.894	6.869
6	1.440	1.943	2.447	3.143	3.707	5.208	5.959
7	1.415	1.895	2.365	2.998	3.499	4.785	5.408
8	1.397	1.860	2.306	2.896	3.355	4.501	5.041
9	1.383	1.833	2.262	2.821	3.250	4.297	4.781
10	1.372	1.812	2.228	2.764	3.169	4.144	4.587
11	1.363	1.796	2.201	2.718	3.106	4.025	4.437
12	1.356	1.782	2.179	2.681	3.055	3.930	4.318
13	1.350	1.771	2.160	2.650	3.012	3.852	4.221
14	1.345	1.761	2.145	2.624	2.977	3.787	4.140
15	1.341	1.753	2.131	2.602	2.947	3.733	4.073
16	1.337	1.746	2.120	2.583	2.921	3.686	4.015
17	1.333	1.740	2.110	2.567	2.898	3.646	3.965
18	1.330	1.734	2.101	2.552	2.878	3.610	3.922
19	1.328	1.729	2.093	2.539	2.861	3.579	3.883
20	1.325	1.725	2.086	2.528	2.845	3.552	3.850
21	1.323	1.721	2.080	2.518	2.831	3.527	3.819
22	1.321	1.717	2.074	2.508	2.819	3.505	3.792
23	1.319	1.714	2.069	2.500	2.807	3.485	3.768
24	1.318	1.711	2.064	2.492	2.797	3.467	3.745
25	1.316	1.708	2.060	2.485	2.787	3.450	3.725
26	1.315	1.706	2.056	2.479	2.779	3.435	3.707
27	1.314	1.703	2.052	2.473	2.771	3.421	3.689
28	1.313	1.701	2.048	2.467	2.763	3.408	3.674
29	1.311	1.699	2.045	2.462	2.756	3.396	3.660
30	1.310	1.697	2.042	2.457	2.750	3.385	3.646
40	1.303	1.684	2.021	2.423	2.704	3.307	3.551
60	1.296	1.671	2.000	2.390	2.660	3.232	3.460
120	1.289	1.658	1.980	2.358	2.617	3.160	3.373
200	1.286	1.653	1.972	2.345	2.601	3.131	3.340
250	1.285	1.651	1.969	2.341	2.596	3.123	3.330
1000	1.282	1.646	1.962	2.330	2.581	3.098	3.300
∞	1.282	1.645	1.960	2.326	2.576	3.091	3.291

Table 3 Critical Values for Pearson *r*

N	df = n − 2	α (One-Tailed) .10	.05	.01	α (Two-Tailed) .10	.05	.01
4	2	.800	.900	.980	.900	.950	.990
5	3	.687	.805	.934	.805	.878	.959
6	4	.608	.729	.882	.729	.811	.917
7	5	.551	.669	.833	.669	.755	.875
8	6	.507	.621	.789	.621	.707	.834
9	7	.472	.582	.750	.582	.666	.798
10	8	.443	.549	.715	.549	.632	.765
11	9	.419	.521	.685	.521	.602	.735
12	10	.398	.497	.658	.497	.576	.708
13	11	.380	.476	.634	.476	.553	.684
14	12	.365	.457	.612	.457	.532	.661
15	13	.351	.441	.592	.441	.514	.641
16	14	.338	.426	.574	.426	.497	.623
17	15	.327	.412	.558	.412	.482	.606
18	16	.317	.400	.542	.400	.468	.590
19	17	.308	.389	.529	.389	.456	.575
20	18	.299	.378	.515	.378	.444	.561
21	19	.291	.369	.503	.369	.433	.549
22	20	.284	.360	.492	.360	.423	.537
23	21	.277	.352	.482	.352	.413	.526
24	22	.271	.344	.472	.344	.404	.515
25	23	.265	.337	.462	.337	.396	.505
26	24	.260	.330	.453	.330	.388	.496
27	25	.255	.323	.445	.323	.381	.487
28	26	.250	.317	.437	.317	.374	.479
29	27	.245	.311	.430	.311	.367	.471
30	28	.241	.306	.423	.306	.361	.463
35	33	.222	.283	.392	.283	.335	.430
40	38	.207	.264	.367	.264	.312	.402
45	43	.195	.248	.346	.248	.294	.378
50	48	.184	.235	.328	.235	.279	.361
60	58	.168	.214	.300	.214	.254	.330
70	68	.155	.198	.278	.198	.236	.305
80	78	.145	.185	.260	.185	.220	.286
90	88	.136	.174	.244	.174	.207	.269
100	98	.129	.165	.232	.165	.196	.256
120	118	.118	.151	.212	.151	.179	.234

Note. Adapted from Kachigan (1991).

Table 4 Power Table for Pearson *r*

				One-Tailed Test (a = .05)					
					r				
Power	.10	.20	.30	.40	.50	.60	.70	.80	.90
.25	96	26	13	8	6	5	4	4	3
.50	271	68	31	18	12	8	6	5	4
.60	360	90	40	23	14	10	7	6	4
.70	470	117	51	29	18	12	9	6	5
.75	537	133	59	32	20	14	19	7	5
.80	616	153	67	37	23	15	11	8	5
.85	716	177	77	42	26	17	12	8	6
.90	853	211	92	50	31	20	14	10	6
.95	1077	266	115	63	38	25	17	11	8
.99	1569	386	167	91	55	36	24	16	10
				Two-Tailed Test (a = .05)					
.25	167	42	20	12	8	6	5	4	3
.50	385	96	42	24	15	10	7	6	4
.60	490	122	53	29	18	12	9	6	5
.70	616	153	67	37	23	15	10	7	5
.75	692	172	75	41	25	17	11	8	6
.80	783	194	85	46	28	18	12	9	6
.85	895	221	97	52	32	21	14	10	6
.90	1047	259	113	62	37	24	16	11	7
.95	1294	319	139	75	46	30	19	13	8
.99	1828	450	195	105	64	40	27	18	11

Table 5a Critical F Values (α = .05)

df (Den.)	1	2	3	4	5	6	7	8	9	10	15	20	30	40	50	150	9999
									df (Num.)								
1	161.40	199.50	215.71	224.58	230.16	233.99	236.77	238.88	240.54	241.88	245.95	248.01	250.10	251.14	251.77	253.46	254.30
2	18.51	19.00	19.16	19.25	19.30	19.33	19.35	19.37	19.38	19.40	19.43	19.45	19.46	19.47	19.48	19.49	19.50
3	10.13	9.55	9.28	9.12	9.01	8.94	8.89	8.85	9.81	8.79	8.70	8.66	8.62	8.59	8.58	8.54	8.53
4	7.71	6.94	6.59	6.39	6.26	6.16	6.09	6.04	6.00	5.96	5.86	5.80	5.75	5.72	5.70	5.65	5.63
5	6.61	5.79	5.41	5.19	5.05	4.95	4.88	4.82	4.77	4.74	4.62	4.56	4.50	4.46	4.44	4.39	4.36
6	5.99	5.14	4.76	4.53	4.39	4.28	4.21	4.15	4.10	4.06	3.94	3.87	3.81	3.77	3.75	3.70	3.67
7	5.59	4.74	4.35	4.12	3.97	3.87	3.79	3.73	3.68	3.64	3.51	3.44	3.38	3.34	3.32	3.26	3.23
8	5.32	4.46	4.07	3.84	3.69	3.58	3.50	3.44	3.39	3.35	3.22	3.15	3.08	3.04	3.02	2.96	2.93
9	5.12	4.26	3.86	3.63	3.48	3.37	3.29	3.23	3.18	3.14	3.01	2.94	2.86	2.83	2.80	2.74	2.71
10	4.96	4.10	3.71	3.48	3.33	3.22	3.14	3.07	3.02	2.98	2.85	2.77	2.70	2.66	2.64	2.57	2.54
11	4.84	3.98	3.59	3.36	3.20	3.09	3.01	2.95	2.90	2.85	2.72	2.65	2.57	2.53	2.51	2.44	2.40
12	4.75	3.89	3.49	3.26	3.11	3.00	2.91	2.85	2.80	2.75	2.62	2.54	2.47	2.43	2.40	2.33	2.30
13	4.67	3.81	3.41	3.18	3.03	2.92	2.83	2.77	2.71	2.67	2.53	2.46	2.38	2.34	2.32	2.24	2.21
14	4.60	3.74	3.34	3.11	2.96	2.85	2.76	2.70	2.65	2.60	2.46	2.39	2.31	2.27	2.24	2.17	2.13
15	4.54	3.68	3.29	3.06	2.90	2.79	2.71	2.64	2.59	2.54	2.40	2.33	2.25	2.20	2.18	2.10	2.07
20	4.35	3.49	3.10	2.87	2.71	2.60	2.51	2.45	2.39	2.35	2.20	2.12	2.04	1.99	1.96	1.88	1.84
30	4.17	3.32	2.92	2.69	2.53	2.42	2.33	2.27	2.21	2.16	2.01	1.93	1.84	1.79	1.76	1.67	1.62
40	4.08	3.23	2.84	2.61	2.45	234	2.25	2.18	2.12	2.08	1.92	1.84	1.74	1.69	1.66	1.56	1.51
50	4.03	3.18	2.79	2.56	2.40	2.29	2.20	2.13	2.07	2.03	1.87	1.78	1.69	1.63	1.60	1.50	1.44
150	3.90	3.05	2.66	2.43	2.27	2.16	2.07	2.00	1.94	1.89	1.73	1.64	1.54	1.58	1.44	1.31	1.22
9999	3.84	3.00	2.60	2.37	2.21	2.10	2.01	1.94	1.88	1.83	1.67	1.57	1.46	1.39	1.35	1.20	1.03

Note. Critical values were created manually using Soper (2022b) and then checked against Kachigan (1991, Table V, p. 293).

Table 5b Critical *F* Values ($\alpha = .01$)

df (Den.)	\multicolumn{17}{c}{df (Num.)}
	1	2	3	4	5	6	7	8	9	10	15	20	30	40	50	150	9999
1	4052	4999.50	5403	5625	5764	5859	5928	5981	6022	6056	6157	6209	6261	6287	6302	6345	6366
2	98.50	99.00	99.17	99.25	99.30	99.33	99.36	99.37	99.39	99.40	99.41	99.45	99.47	99.47	99.45	99.49	99.50
3	34.12	30.82	29.46	28.71	28.24	27.91	27.67	27.49	27.35	27.23	26.87	26.69	26.50	26.41	26.35	26.20	26.13
4	21.20	18.00	16.69	15.98	15.52	15.21	14.98	14.80	14.66	14.55	14.20	14.02	13.84	13.75	13.69	13.54	13.46
5	16.26	13.27	12.06	11.39	10.97	10.67	10.46	10.29	10.16	10.05	9.72	9.55	9.38	9.29	9.24	9.09	9.02
6	13.75	10.92	9.78	9.15	8.75	8.47	8.26	8.10	7.98	7.87	7.56	7.40	7.23	7.14	7.09	6.95	6.88
7	12.25	9.55	8.45	7.85	7.46	7.19	6.99	6.84	6.72	6.62	6.31	6.16	5.99	5.91	5.86	5.72	5.65
8	11.26	8.65	7.59	7.01	6.63	6.37	6.18	6.03	5.91	5.81	5.52	5.36	5.20	5.12	5.06	4.93	4.86
9	10.56	8.02	6.99	6.42	6.06	5.80	5.61	5.47	5.35	5.26	4.96.	4.81	4.65	4.57	4.52	4.38	4.31
10	10.04	7.56	6.55	5.99	5.64	5.39	5.20	5.06	4.94	4.85	4.56	4.41	4.25	4.17	4.12	3.98	3.91
11	9.65	7.21	6.22	5.67	5.32	5.07	4.89	4.74	4.63	4.54	4.25	4.10	3.94	3.86	3.81	3.67	3.60
12	9.33	6.93	5.95	5.41	5.06	4.82	4.64	4.50	4.39	4.30	4.01	3.86	3.70	3.62	3.57	3.43	3.36
13	9.07	6.70	5.74	5.21	4.86	4.62	4.44	4.30	4.19	4.10	3.82	3.66	3.51	3.43	3.38	3.24	3.17
14	8.86	6.51	5.56	5.04	4.69	4.46	4.28	4.14	4.03	3.94	3.66	3.51	3.35	3.27	3.22	3.08	3.00
15	8.68	6.36	5.42	4.89	4.56	4.32	4.14	4.00	3.89	3.80	3.52:	3.37	3.21	3.13	3.08	2.94	2.87
20	8.10	5.85	4.94	4.43	4.10	3.87	3.70	3.56	3.46	3.37	3.09	2.94	2.78	2.69	2.64	2.50	2.42
30	7.56	5.39	4.51	4.02	3.70	3.47	3.30	3.17	3.07	2.98	2.70	2.55	2.39	2.30	2.24	2.09	2.01
40	7.31	5.18	4.31	3.83	3.51	3.29	3.12	2.99	2.89	2.80	2.52	2.37	2.20	2.11	2.06	1.90	1.80
50	7.17	5.05	4.20	3.72	3.41	3.19	3.02	2.90	2.78	2.70	2.42	2.26	2.10	2.00	1.95	1.78	1.68
150	6.81	4.75	3.91	3.45	3.14	2.92	2.76	2.63	2.53	2.44	2.16	2.00	1.83	1.73	1.66	1.46	1.33
9999	6.63	4.61	3.78	3.32	3.02	2.80	2.64	2.51	2.41	2.32	2.04	1.88	1.70	1.59	1.52	1.25	1.00

Note. Critical values were created manually using Soper (2022b) and then checked against Kachigan (1991, Table V, p. 293).

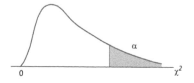

Figure A3 Figure for Table 6.

Table 6 Critical Values of the Chi-Square Distribution (see Figure A3)

	Proportion in Critical Region (shaded area to the right: α)									
df	.995	.99	.975	.95	.90	.10	.05	.025	.01	.005
1	–	–	0.001	0.004	0.016	2.706	3.841	5.024	6.635	7.879
2	0.010	0.020	0.051	0.103	0.211	4.605	5.991	7.378	9.210	10.597
3	0.072	0.115	0.216	0.352	0.584	6.251	7.815	9.348	11.345	12.838
4	0.207	0.297	0.484	0.711	1.064	7.779	9.488	11.143	13.277	14.860
5	0.412	0.554	0.831	1.145	1.610	9.236	11.070	12.833	15.086	16.750
6	0.676	0.872	1.237	1.635	2.204	10.645	12.592	14.449	16.812	18.548
7	0.989	1.239	1.690	2.167	2.833	12.017	14.067	16.013	18.475	20.278
8	1.344	1.646	2.180	2.733	3.490	13.362	15.507	17.535	20.090	21.955
9	1.735	2.088	2.700	3.325	4.168	14.684	16.919	19.023	21.666	23.589
10	2.156	2.558	3.247	3.940	4.865	15.987	18.307	20.483	23.209	25.188
11	2.603	3.053	3.816	4.575	5.578	17.275	19.675	21.920	24.725	26.757
12	3.074	3.571	4.404	5.226	6.304	18.549	21.026	23.337	26.217	28.300
13	3.565	4.107	5.009	5.892	7.042	19.812	22.362	24.736	27.688	29.819
14	4.075	4.660	5.629	6.571	7.790	21.064	23.685	26.119	29.141	31.319
15	4.601	5.229	6.262	7.261	8.547	22.307	24.996	27.488	30.578	32.801
16	5.142	5.812	6.908	7.962	9.312	23.542	26.296	28.845	32.000	34.267
17	5.697	6.408	7.564	8.672	10.085	24.769	27.587	30.191	33.409	35.718
18	6.265	7.015	8.231	9.390	10.865	25.989	28.869	31.526	34.805	37.156
19	6.844	7.633	8.907	10.117	11.651	27.204	30.144	32.852	36.191	38.582
20	7.434	8.260	9.591	10.851	12.443	28.412	31.410	34.170	37.566	39.997
21	8.034	8.897	10.283	11.591	13.240	29.615	32.671	35.479	38.932	41.401
22	8.643	9.542	10.982	12.338	14.041	30.813	33.924	36.781	40.289	42.796
23	9.260	10.196	11.689	13.091	14.848	32.007	35.172	38.076	41.638	44.181
24	9.886	10.856	12.401	13.848	15.659	33.196	36.415	39.364	42.980	45.559
25	10.520	11.524	13.120	14.611	16.473	34.382	37.652	40.646	44.314	46.928
26	11.160	12.198	13.844	15.379	17.292	35.563	38.885	41.923	45.642	48.290
27	11.808	12.879	14.573	16.151	18.114	36.741	40.113	43.195	46.963	49.645
28	12.461	13.565	15.308	16.928	18.939	37.916	41.337	44.461	48.278	50.993
29	13.121	14.256	16.047	17.708	19.768	39.087	42.557	45.722	49.588	52.336
30	13.787	14.953	16.791	18.493	20.599	40.256	43.773	46.979	50.892	53.672
40	20.707	22.164	24.433	26.509	29.051	51.805	55.758	59.342	63.691	66.766
50	27.991	29.707	32.357	34.764	37.689	63.167	67.505	71.420	76.154	79.490
60	35.534	37.485	40.482	43.188	46.459	74.397	79.082	83.298	88.379	91.952
70	43.275	45.442	48.758	51.739	55.329	85.527	90.531	95.023	100.425	104.215
80	51.172	53.540	57.153	60.391	64.278	96.578	101.879	106.629	112.329	116.321
90	59.196	61.754	65.647	69.126	73.291	107.565	113.145	118.136	124.116	128.299
100	67.328	70.065	74.222	77.929	82.358	118.498	124.342	129.561	135.807	140.169

Note. The proportions represent the areas to the right of the critical value. To look up an area to the left of the critical value, subtract the proportion from 1 and then use the corresponding column.

Table 7a Sample Size (*N*) for χ^2 Test for Independence to Detect Effect Size (*V*) for $\alpha = .05$ with Corresponding *df* and Contingency Table of Size (*R* × *C*)

w^a	.25	.50	.60	.70	.75	.80	.85	.90	.95	.99
					Power					

df = 1 (2 × 2)

w^a	.25	.50	.60	.70	.75	.80	.85	.90	.95	.99
0.10	165	384	490	617	694	785	898	1051	1300	1837
0.30	18	43	54	69	77	87	100	117	144	204
0.50	7	15	20	25	28	31	36	42	52	73

df = 2 (2 × 3)

0.10	226	496	621	770	859	964	1092	1265	1544	2140
0.30	25	55	69	86	95	107	121	141	172	238
0.50	9	20	25	31	34	39	44	51	62	86

df = 3 (2 × 4)

0.10	258	576	715	879	976	1090	1230	1417	1717	2352
0.30	29	64	79	98	108	121	137	157	191	261
0.50	10	23	29	35	39	44	49	57	69	94

df = 4 (3 × 3)

0.07	629	1310	1616	1976	2188	2437	2739	3143	3790	5151
0.21	70	146	180	220	243	271	304	349	421	572
0.35	25	52	65	79	88	97	110	126	152	206

Note: Sample sizes determined by $\dfrac{N_{.10}}{(100)w^2}$ where $N_{.10}$ reflects entries from Cohen (1988, Column .10 in Table 7.46 and Table 7.47, pp. 258–259).
[a] Effect sizes are from Table 12.9 for *df* = 1, 2, and 3 where *m* = 1, and for *df* = 4 where *m* = 2. The effect sizes are for *w*. The rows of each sub-table reflect "small," "medium," and "large effects," respectively.

Table 7b Sample Size (*N*) for χ^2 Test for Independence to Detect Effect Size (*V*) for $\alpha = .01$ with Corresponding *df* and Contingency Table of Size (R × C)

w^a	Power									
	.25	.50	.60	.70	.75	.80	.85	.90	.95	.99
					df = 1 (2 × 2)					
0.10	362	664	800	961	1056	1168	1305	1488	1781	2403
0.30	40	74	89	107	117	130	145	165	198	267
0.50	14	27	32	38	42	47	52	60	71	96
					df = 2 (2 × 3)					
0.10	467	819	975	1157	1264	1388	1540	1743	2065	2742
0.30	52	91	108	129	140	154	171	194	229	305
0.50	19	33	39	46	51	56	62	70	83	110
					df = 3 (2 × 4)					
0.10	544	931	1101	1297	1412	1546	1709	1925	2267	2983
0.30	60	103	122	144	157	172	190	214	252	331
0.50	22	37	44	52	56	62	68	77	91	119
					df = 4 (3 × 3)					
0.07	1239	2088	2457	2882	3131	3363	3769	4233	4965	6490
0.21	138	232	273	320	348	374	419	470	552	721
0.35	50	84	98	115	125	135	151	169	199	260

Note: Sample sizes determined by $\dfrac{N_{.10}}{(100)w^2}$ where $N_{.10}$ reflects entries from Cohen (1988, Column .10 in Table 7.46 and Table 7.47, pp. 258–259).
[a] Effect sizes are from Table 12.9 for *df* = 1, 2, and 3 where *m* = 1, and for *df* = 4 where *m* = 2. The effect sizes are for *w*. The rows of each sub-table reflect "small," "medium," and "large effects," respectively.

Answers to Part A Review Exercises

Item	Chapter											
	1	2	3	4	5	6	7	8	9	10	11	12
1.	d	d	c	d	a	d	c	b	a	b	b	a
2.	b	c	c	b	d	c	c	c	b	d	c	b
3.	c	a	d	a	b	b	b	b	c	a	a	c
4.	c	b	c	c	d	d	c	b	c	c	b	b
5.	d	d	b	d	b	a	a	a	d	c	c	d
6.	d	c	c	c	d	c	c	c	b	c	c	a
7.	a	b	a	c	c	c	a	c	d	c	d	c
8.	d	c	a	a	b	b	b	b	c	d	b	c
9.	a	d	d	d	a	d	b	c	b	c	d	a
10.	c	a	a	a	d	c	a	b	c	b	a	c

Index

Note: **Bold** page numbers refer to tables; *italic* page numbers refer to figures and page numbers followed by "n" denote endnotes.

accuracy in parameter estimation (AIPE) 85–6; used in examples 103, 105, 125, 146, 154, 175, 184, 199, 206, 288

accuracy of prediction (regression) 171

alpha level 92–3; approach to significance 92; and experimentwise error 213, 234; inflated 213–14, 218; testwise 213–14, 234; and Type I errors 100–1, 106

alternative hypothesis (H_1); chi-square test for goodness of fit 304; chi-square test for independence 310–11, 314; concept of 89, 91–3, 96; correlation 149, 152–3; factorial ANOVA 253–4, 257; independent-samples t test 199, 203; repeated-measures t test 281; regression 176, 179; single-factor ANOVA 226–7, 230, 235; single-sample t test 116, 119, 121–3; z test 98, 104–6

ANOVA *see* factorial ANOVA; single-factor ANOVA

area under a curve 56–62, 71, 80–1, 85, 92–3, 95, 96, 99, 115–16, 299

assumptions of: chi-square test for goodness of fit 300–1, 304, 318; chi-square test for independence 311, 314, 318; correlation 143, 149, 152; factorial ANOVA 253, 258, 265; independent-samples t test 193–5, 200, 203; regression 176, 180; repeated-measures t test 281, 286; single-factor ANOVA 227, 230–1; single-sample t test 119, 123; z test 98

bar graphs 28

bell-shaped distribution 33–4, 53–5, 68, 71

best fitting line *see* line of best fit

beta (β) 101

beta-alpha ratio (β:α) 101

between-groups design 192

between-groups variability 216

biased sample 7–8, 11

bivariate correlation *see* correlation analysis

bivariate regression *see* regression analysis

Bonferroni procedure 214

box plot 40

carry-over effects 278–9, 289

category for: chi square 298–9, 301, 310; frequency distributions 28–31, 37, 44; nominal variables 19

causal-comparative research 15

causation 150, 155

cause-and-effect 11, 15, 22, 150

cell of a matrix (factorial ANOVA) 241–2, 246, 249–50, 260

cell size (chi-square) 300–1, 305, 308–11, 315–18

census 16

Central limit theorem (CLT) 70

central tendency measures *see* measures of central tendency

change-score study 277, 289

chi-square distribution 298–9, 318

chi-square test for goodness of fit 297–307; assumptions 300; categories 298; critical region **300–1**, 304; degrees of freedom 298; effect size 302, 306; expected frequencies 298–9; guided example 302; minimum cell size 301; hypothesis test 304–5; observed frequencies 298–9; power analysis 303, 306–7; test statistic 299

chi-square test for independence 307–18; assumptions 311, 314–15; concept of 307; contingency table 307–8, **310**, **313**, **317**; critical region 315; degrees of freedom 309; effect size 312, 317; expected frequencies 309; guided example 312; hypothesis test 314–16; null hypothesis for 310–11; observed frequencies 308; power analysis 314, 317; test statistic 308

class intervals 31–2, 44

class width 31–2, 44
cluster random sampling 8
coefficient: correlation 14, 132, 145; regression 162
coefficient of determination (r^2) 125, 144–5, 153, 155, 172–3, 183, 232, 287
Cohen's d 102, 104, 106, 124–5, 127, 144, 204, 287
Cohen's f 179, 183, 233, 235, 262–3
Cohen's w 302, 306, 312, 318
column effect (for factorial ANOVA) 243, *245*, *250*, *259*
common cause 150, 155
confidence intervals 80; AIPE (*see* accuracy in parameter estimation); and correlation 145, 153; construction of 81; and factorial ANOVA 262–3; hypothesis tests (compared to) 104–5; and independent-samples t test 196, 198; interpreting 83; precision 85; and regression 173, 183; and repeated-measures t test 282, 287–8; and single-factor ANOVA 233; and single-sample t test 117, 125; width of 85
constant: regression (B_0) 162–4, 169; variable (compared to) 17, 317
contingency table 307–14, **316**, 317
continuous variables *18*
control group 13, 275
convenience sampling 8
correlation analysis 131–59; causation (compared to) 150; coefficient 14, 132, 145; coefficient of determination 144; confidence intervals 145, 153; correlational research (compared to) 131; critical *r* formula 147; critical *t* formula 148; curvilinear 136–7; effect size 144, 153; Fisher z transformation 146; formula 139; guided example 150; hypothesis testing 147, 148–9, 152; intermediate 137; line of best fit 134; negative 135; outliers in 142, *144*; Pearson *r* 137; perfect 137; phi-coefficient 139; point-biserial 138; positive 134; power 151, 154; scatter plots 133, *138*; significance of 142–3; Spearman correlation 138; standard error 149; strength of the relationship 137; t test statistic 148, 150; zero 137
correlational research 11, 14–15, 17, 22, 150–1, 155, 179, 277, 314
correlation coefficient (*r*) 14, 132, 145
Cramer's V 312, 317–18
Critical F 224
critical r 147
Critical z 92–93
critical region (for): chi-square 304–5, 310, 314–16; concept of 92, *93*–4; correlation 142, *143*, 149–50; factorial ANOVA 253; independent-samples t test 201, 204; repeated-measures t test 282, 286–7; regression 178, 182;

single-factor ANOVA *215*, 227, *228*, 231; single-sample t test 120, 123; z test 98, 105–6
critical *t* 147–8
cross-sectional survey 16, 22
cumulative frequency 28, **29**
cumulative relative frequency 28, **29**
curvilinear relationship 136–7

degrees of freedom (*df*) 115; chi-square test for goodness of fit 298, 318; chi-square test for independence 309, 318; factorial ANOVA 249, 253, 257; and F statistic 222–3; independent-samples t test **194**, 199, 203; repeated-measures t test 281; single-factor ANOVA 222–5, 227; single-sample t 115, 117–18, 126; and t statistic 115
denominator of F test statistic 217, 221, 224–5
dependent variable 13, 17, 125, 133, 176, 192, 197, 226, 232, 252, 276, 278, 284, 289, 297
derived score 3, 51
descriptive statistics 26; central tendency measures 34; dispersion measures 37; percent and percentile 36, 43
deviation scores 41, 53, 172, 219
dichotomous variables *18*, 21–2, 138–9
difference scores (*D*) for repeated-measures t test 279–80, **280**, 288
direction of a relationship 134, 137
direct relationship 135
disordinal interaction 244, 247, 252, 260–1, 264
distributions: bimodal 35; chi-square 298–9; frequency 27–31; F 223–4; graphical representations *33*; multimodal 35; normal 53, 55; of sample means of 64–5; shapes of **33**; skewed **33**, 34; standard normal 57; symmetrical **33**, 34; t 115–17; z 52–33
distribution shapes 33–4; bell-shaped 33; *j*-shaped *33*, 34; skewed *33*, 34, *37*; symmetrical *33*, 34, *37*, uniform 33; *u*-shaped 33

effect size: coefficient of determination (r^2) 125, 144, 153, 155, 287; for chi-square goodness of fit test 302–3, 306–7; for chi-square test for independence 312, 314, 317–18, 333–4; Cohen's d 102, *103*, 124; Cohen's f 179, 183–5, 233; Cohen's w 302, 306; concept of 102, *103*; for correlation (r^2) 144–5, 153; Cramer's V 312, 317–18; explained variance 125; for factorial ANOVA (partial η^2) 256–7, 261–3; for independent-samples t test 204–6; for regression (f^2) 179, 183–5; for repeated-measures t test 285, 287–8; for single-factor ANOVA 230, 232–5; for single-sample t test 124–5, 127

empirical probability *vs.* theoretical probability 60
empirical rule 58
equation of a line 160
errors: experimentwise 213, 234; sampling 11, 21–2, 63, 65, 67, 81, 171, 223; standard error of estimate 171–3, **177**, 182, 184–5; testwise 213–14, 234; Type I 100–1, 106; Type II 100–1, 106
error term **194**, 217, 247–8, 278
error variance 221
estimated *d see* Cohen's *d*
estimated population standard deviation 171
estimated standard error (for): independent-samples *t* test 193; repeated-measures *t* test 279–81; single-sample *t* test 114, 123
eta squared (η^2): factorial ANOVA 261; partial 261; single-factor ANOVA 232–3, 235
explained variance (r^2) 124–5, 127, 144, 153, 155, 183, 204–5, 232, 235, 261–2, 287
expected frequencies 298, **299**, 302, 305, **306**, 309, 315–18
experimental error 216–17, 247
experimental research 11; control group 13; manipulation and control 13; treatment group 13
experimentwise alpha 213, 234
experimentwise error 213, 234
ex post facto *see* causal-comparative research
extraneous factors 13, 15

factorial ANOVA 240–71; assumptions for 253, 258, 265; column effect 243, *243, 245, 250, 259*; concept of 240–1; confidence intervals 262–3; degrees of freedom 249, 253, 257; effect size 256–7, 261–3; formulas 248–9, 261; *F*-ratio 248; group-means table 242, *250*; guided example 255; hypothesis testing 252, 257; interaction effect *243, 245, 250, 259*; interactions 243–7; interpreting significance 251–2, 254–5, 259–60; logic and structure 247, *248*; main effects 248, 251–2, 254–6, 259, 263, 265; matrix representation 241, *242–3, 250, 259*; omnibus 240, 247, 250–4, 256–9, 263–5; power analysis 256–7, 263; research questions 252; row effect 242, *243, 245, 250, 259*; statistical hypotheses 252–3, 257; summary table **249**, 250, **251**; test statistic 247–9
factorial design 241
fatigue effect 279, 289
F distribution 223
F distribution table 330–1
Fisher *z* transformation 146
five-number summary 40
form of a relationship 136

F ratio: distribution of 223–4; error variance 221; factorial ANOVA 248; single-factor ANOVA 217, 221, 224–5; structure of 222–3
frequency distribution graphs: bar graph 28, *29*, 44; histograms 30; shapes of 33; skewed distribution *33, 34, 37*; symmetrical distribution *33, 37*
frequency distribution tables 27–8; for qualitative data 27; for quantitative data 28; grouped 31
frequency polygon 54

generalize 7, 10–11
goodness of fit test *see* chi-square test for goodness of fit
grand mean 66–9, 71–2, 114
grouped frequency distributions: class intervals 31; class width 31; guidelines for creating 31–2
group-means table 242, 244, 246, *250*, 254–5, 258, *259*, 260–1, 263–4
group membership concept 191–2
guided examples (involving): chi-square test for goodness of fit 302; chi-square test for independence 312; correlation 150; factorial ANOVA 255; independent-sample *t* test 201; regression 178; repeated-measures *t* test 283; single-factor ANOVA 229; single-sample *t* test 121

histogram 30
homogeneity of variances assumption 195
homoscedasticity of the residuals 176
hypotheses: alternative 89, *90*; directional *vs.* nondirectional *90*; null 88, *90*; research 88, *90*; research *vs.* alternative 89; statistical 88, *90*; writing 90, *91*
hypothesis testing 86; alpha level 92; alpha-level approach 92; for chi-square test for goodness of fit 304–5; for chi-square test for independence 314–16; confidence intervals (compared to) 104–5; for correlation 147, 148–9, 152; critical region 92; decision criteria 94, 96; decision errors 100–2; effect size 102; factorial ANOVA 252, 257; four steps of 98; guided examples 121, 150, 178, 201, 229, 255, 283, 302, 312; independent-samples *t* test 199, 202; logic of 87; non-critical region 92; one-tailed test 93; *p*-value approach 95; power of (concept) 100–1; repeated-measures *t* test 280–2, 285–7; regression 175, 179; single-factor ANOVA 225, 230; single-sample *t* test 119; statistical significance of 92; *t* statistic 114; two-tailed test 93, 116; Type I errors 100–1, 106; Type II errors 100–1, 106; involving the mean 98; *z* score test statistic 98

independent events 65
independent samples concept 191–2
independent-samples *t* statistic 193, 201
independent-samples *t* test 191–211;
    assumptions 193–5, 200; between-group
    design 192; confidence interval 196, 198;
    effect size 204–6; formulas for **194**; group
    membership concept 191–2; guided
    example 201; homogeneity of variances 195;
    hypothesis test 199, 202; Levene test 195,
    200, 204, 207, 227, 231, 253, 258; pooled
    variance 193–4, 197; power analysis 202,
    206; repeated-measures *t* test (compared
    to) 279–80; separate variances 193–4, 198;
    single-sample *t* test (compared to) 192–3;
    standard error 195; test statistic 193, 201;
    Welch's *t* statistic 195
independent observations 193–4
independent variable 12, 17, 133, 150, 196, 232
individual differences 216–17, 247, 278, 288–9
inferential statistics 77, 99, 233, 318
inflated alpha 213–14, 218
interactions 243–7; concept of 243–7, *245*;
    disordinal 244, *245*; effect *243, 245, 250, 259*;
    graph of *246, 251*, 260; interpreting 247, 250,
    254–5; numerical representation 243, *245, 250,
    259*; ordinal 244; zero 244, *245*
intermediate correlation 137
interpreting: confidence intervals 83, 146, 174,
    198; correlation coefficient 14, 137; effect
    size 102, 124–5, 145, 155, 179, 183, 203–4,
    232, 262, 287, 302, 312, 317; interactions
    247; regression coefficient 168; regression
    constant 169; significance (factorial ANOVA)
    251–2, 254–5, 259–60
interquartile range (*IQR*) 39
interval estimate 79–80, 105
interval measurement scale 20–2
interval of predictability 168
intervention studies 13, 22, 277
inverse relationship 136, 151–2

j-shaped distribution *33*, 34
Jackknife distances 143, 152, 155, 289
judgment sampling *88*

least-squares criterion 162
level of significance *see* significance level
Levene test 195
Likert response scale 6; scores from 20
line of best fit 134
linear regression *see* regression analysis
linear relationship *138*
logic of ANOVA 215
longitudinal survey 16

main effects (of factorial ANOVA) 248, 251–2,
    254–6, 259, 263, 265

manipulation and control 13
margin of error 16
matched-samples design 278
matching groups 209
matrix representation (factorial ANOVA) 241,
    242–3, 250, 259
mean 10, 36; distribution of sample *64*, 65–72,
    77–8, *79*, 81–2, 98, 114; grand 66–7; median
    and mode (compared to) 37; of population
    10, 37; of sample 10, 37; of sample means
    66–7; standard error of the 67
mean difference score (*M_D*) 279–81, **280**
measurement scales 19; interval 20; nominal
    19; ordinal 19; quasi-interval 20; ratio 20
measures of central tendency 34; mode 35;
    median 36; mean 36
measures of dispersion 37; *see also* measures of
    variation
measures of position 36, 43; median 36, 39;
    percentiles 36, 43; quartiles 39, 43
measures of variation 39; box plot 40; five-
    number summary 40; interquartile range
    (*IQR*) 39; range 39; standard deviation 41
median 36; mean and mode (compared to) 37;
    see also measures of position
minimum cell size assumption (chi-square)
    300–1, 305, 311, 315, 318
mode 35

negative correlation 135
negatively skewed distributions 34
nominal measurement scale 19
noncritical region 92, 96, 99
nonparametric test 297, 318
normal curve 55–6, 58, 60, *61*, 71, 80, 81, 92, 325
normal distribution 34, 53, 55–60, 69–72, 78,
    81–2, 92, 105, 114–15, *116*, 126, 145–6, 178,
    223, 225, 235
normality assumption 226
normality of residuals 181; *see also* Shapiro-Wilk
    test for goodness of fit
null hypothesis (*H_0*): chi-square test for
    goodness of fit 303–4, 307; chi-square test
    for independence 310–11, 314; concept of
    88–9, 91–3, 96; correlation 147, 149, 152–4;
    factorial ANOVA 252–3, 257; independent-
    samples *t* test 195, 199, 202, 204, 206–7;
    repeated-measures *t* test 281–3, 285, 288;
    regression 175, 178–9, 181–2, 184; single-
    sample *t* test 116, 119, 120–3, 126; single-
    factor ANOVA 213–14, 225, 226, 228, 230–1,
    235; *z* test 98, 100, 104–6

observational studies 11, *13*, 22, 150
observed frequencies (*O_i*) 298, 300, 303, 305,
    306, 308, 316, 318
observed scores 165, **167**, 168, 173, 185
omega squared 262

omnibus test 214, 218, 225–6, 228, 230–4, 240, 247, 250–4, 256–9, 263–5
one-tailed test 93, 116, 119, 149–51, 176, 281, 285, 286
one-way chi-square *see* chi-square test for goodness of fit
operational definitions 4
ordinal interaction 244–5, 250, 261, 264
ordinal measurement scale 19–20, 22, 138, 297, 300
ordinary least squares (OLS) regression 162; *see also* regression analysis
outliers 34, 142, *144*
overall mean *see* grand mean

paired data 133–4, 149, 155, 276, 289, 308
paired scores 139, 163, 280–1, 283
pairwise comparisons 212
parameter estimation 77; confidence interval 79; point estimate 78; interval estimate 78
parametric tests 297
parent population 7, 9–11, 16, 21; and sampling distribution of the mean 63, 66, 69–70, 72
partial eta squared 232, 261, 282
Pearson correlation (*r*) 137; *see also* correlation
percentage of variance (*r²*) *see* coefficient of determination
percentages 28, 57, *59*
percentile 36, 43–4
perfect correlation 137
perfect prediction 168
phi-coefficient 139
point-biserial correlation 138
point estimate 78–82, 104, 145, 148, 173, 176–7, 196, 282–3
pooled variances *t* test 193–4, 197
population 6; generalize to 7, 10–11; mean (μ) 10; parent 7; representative 7; sample (compared to) 7; standard deviation (σ) 42
positive correlation 134
positively skewed distributions 34, 223, 235
post hoc tests: factorial ANOVA 255, 259; single-factor ANOVA 226–7
power 101, 151, 154
power analysis: a priori (sample size planning) 103, 122, 151, 179, 202, 230, 256–7, 285; concept of 7; post hoc 125, 154, 184, 206, 234, 263, 288
practical significance 96–7
practice effect 279, 289
precision (of a CI) 85–6, 105, 146, 175
predicted scores 165, *166*
prediction equation 161–2, 182; accuracy of 171
probability 56; distribution of sample means 66, 68; empirical *vs.* theoretical 60; independent event 65; normal distribution 59; and proportion 56

proportion, percentage, and probability 56–60
purpose statement 4, 6
p-value 92, 95, *97, 99*, 106, 195, 214
pyramiding effect of Type I errors 213

Q_1 39–40, 43, 45
Q_3 39–40, 43–5
qualitative: data 27; studies 3; variable 18
quantitative: data 28; research methodologies 13; studies 3; variable 18
quartiles 43, 45
quota sampling *8*

r (Pearson correlation coefficient) *see* correlation coefficient
r² *see* coefficient of determination
random assignment 13
random sample 7
random selection 7
range 39
ratio measurement scale 20
ratio variable *18*
raw scores 43, 51–3, 65, 71, 88, 114, 121
regression analysis 160–188; accuracy of 171; coefficient (*B*) 162; confidence intervals 173, 184; constant (B_0) 162–4, 169; effect size 183–5; equation 161, 182; equation of a line 160; guided example 178; homoscedasticity 176; hypothesis testing 175, 179; interpreting 168–170; interval of predictability 168; least-squares criterion 162; line 161; observed scores *vs.* predicted scores 165; ordinary least squares (OLS) 162; perfect prediction 168; power analysis 179, 184; prediction equation 161–2, 182; residual plot 176, *182*; residuals ($y - \hat{y}$) 165; root mean square error (*RMSE*) 172; slope-intercept form 161; standard deviation of residuals 171–4; standard error of estimate 172–3, 184; statistical significance 171; *t* statistic 176–7; *y* hat 162
regression equation 161–2, 164–72, **173**, 175, **177**, 181–5
regression lines 161–71, 173, 175–6, 178–9, *182*, 185–5
relative frequency 28, **29**, **31**, 58, 60
repeated-measures comparative group design 278
repeated-measures correlation study 277–8
repeated-measures longitudinal study 277
repeated-measures research design 276
repeated measures *t* test 275–94; advantages of 278; assumptions 281, 286; carry-over effects 278–9, 289; concept of 275; confidence intervals 282, 287–8; disadvantages of 278; effect size 287; estimated standard error 281; fatigue effect 279, 289; guided example 283; hypotheses for 281; hypothesis test 280–82, 285–7; mean difference scores 279–81, **280**;

power analysis 285, 288; practice effect 279, 289; sample size 278; *t* statistic 279–80

representative sample 7, 10–11

research: hypothesis 88–91, 105, 122, 132, 141, 151, 163, 170, 202, 256, 284, 303, 313; methodology 11; objective 4; process 3, *5*; question (RQ) 4

research methodologies 11; causal-comparative research 15; correlational 14; experimental 11; survey 15

research questions (factorial ANOVA) 252

residual plots *181*

residuals ($y - \hat{y}$) 165

reverse causality 150, 155

rho (ρ) 145–7, 155

root mean square (*RMS*) 43

root mean square error (*RMSE*) 172

row effect 242, *243*, *245*, *250*, *259*

row means 242, 264–5

sample 6; assignment *14*; of sample means 64; selection 6–7, *14*; statistic 10

sample size planning 85–6, 103, 105–6, 122, 151, *152*, 179, 202, 230, 233, 256, 285, 303, 314

sample variance (SD^2) 42, 219

sampling distribution; concept of 63; of sample means 64, *79*

sampling: error 11, 21–2, 63, 65, 67, 81, 171, 223; from a normal distribution 70; from a non-normal distribution 70; strategies 7, 8, 10; with replacement 66, 72

scatter plots 133, *138*

separate variances *t* test 193–4, 198

Shapiro-Wilk test for goodness of fit 153, 181, 194, 200, 203, 226–7, 231, 253, 258, 281, 286

significance *see* statistical significance

significance level 92–3, 120, 149, 178, 196, 201, 282

simple random sampling *8*

single-factor ANOVA 212–39; assumptions 227, 231; between-groups variability 216; Bonferroni procedure 214; concept of 212–13, 219; confidence intervals 233; degrees of freedom 222–3; effect size 230, 232–5; *F* distribution 223; *F* distribution table 330–1; *F* ratio 217, 221, 224–5; formulas 221–2; guided example 229; hypothesis testing 225, 230; logic of 215; mean squares (*MS*) 219; normality assumption 226; post hoc tests 226–7; power analysis 230, 234; statistical hypotheses for 226, 230; summary table structure *220*, 221–3; test statistic for 217, 221, 224–5; testwise *vs.* experimentwise alpha 213; Tukey's HSD test 227–8, **229**, 231–2; within-groups variability 216

single factor with *n* levels 192, 196

single-sample *t* test 114–130: assumptions 119, 123; confidence intervals 117, 125; effect size 124; estimated standard error 114; guided example 121; hypothesis testing 119; independent-samples *t* test (compared to) 192–3; power analysis 122, 125; repeated-measures *t* test (compared to) 279–280; *t* distribution 115; test statistic 113

skewed distributions *33*, *34*, *36*–7, 44

slope 161–4, 168–71, 174–5, 178–9, 182–3

slope-intercept form 161–2

smooth curves 56–7

snowball sampling *8*

Spearman correlation 138

square root 42–3, 69–70, 78, 141–2, 172, 204, 219, 225

SS see sum of squares

$SS_{between}$ 220–3, 233, **244**, 261–2

SS_{within} 220–3, 233, 247–8, **249**, 261

standard deviation: concept of 41–3, 44–45, 52–3; and effect size 102, *103*; sampling distribution of sample means 64–5, 67–71; standard normal distribution of 57–8

standard error: concept of 67, 69, 71–2; and confidence intervals 78, 81–3, 85–6; and effect size 102; and independent-samples *t* test 193, **194**, 196–98, 201, 207; and Pearson *r* 145, 148; and regression 171–73, **174**, 176–77, 184–85; and repeated-measures *t* test 279–81, 283, 289; and single-sample *t* test 117–119, 126

standard error of estimate ($SE_{Residuals}$) 172–4

standard error of difference scores (SE_D) 279–281, 283

standard error of the mean 67, 114

standard error of the regression coefficient (SE_B) 173

standard normal curve 60, *61*, 71, 325

standard normal distribution 57

standard scores 52, 71; *see also z* scores

standardized distributions 53

statistical hypotheses 88, 226, 230

statistically equivalent groups 13

statistical power *see* power

statistical significance 89, 92, 142–3, 171; α-level approach 92; *p*-value approach 95; practical significance (compared to) 96–7

statistical tables: chi-square distribution 332; *F* distribution 330–1; Pearson correlation *r* 328; power for Pearson *r* 329; sample size and power for chi-square 333–4; *t* distribution 327; *z* distribution 325

stratified random sampling *8*

strength of a relationship 137, 160

Student's *t* distribution 114; *see also t* distribution

Student's *t* test *see t* test

sum of squared deviations (*SS*) **68**, 221
survey research 15; census 16; cross-sectional 16; longitudinal 16; margin of error 16; survey instrument (compared to) 16
symmetrical distributions *33*, 34–5, *37*
systematic sampling *8*

t distribution 115; table 327
t test: ANOVA (compared to) 224–5; assumptions 193; independent-samples 191; repeated-measures 279–80; single-sample 113; test statistic 193
tables: *see* statistical tables
tail of a distribution 34, 37, 93, 117, 214
test for independence *see* chi-square test for independence
test statistic (for): chi-square test for goodness of fit 299; chi-square test for independence 308; correlation 148, 150; factorial ANOVA 247–9; independent-samples *t* test 193, 201; regression 176–7; repeated-measures *t* test 279–80; single-factor ANOVA 217, 221, 224–5; single-sample *t* test; *z* test 52
testing hypothesis *see* hypothesis testing
testwise alpha level 213–14, 234
total degrees of freedom (*df*$_{Total}$) 220, 222
total sum of squares (*SS*$_{Total}$) 220–1, 232, **249**, 262
total variability 216–17
treated population 87, 99, 102, 105, 120, 122–3, 125
treated sample 87, 92–4, 96–100, 104, 120
treatment effect 88–9, 100–2, *103*, 108, 120, 205, 216–17, 223, 226, 245, 287
treatment group 13, 275
Tukey's Honestly Significant Difference (HSD) 227–8, **229**, 231–2, 255, 260
two-factor design 241, 243, 247, 255
two-tailed test 93, 116
two-way ANOVA *see* factorial ANOVA
two-way chi-square *see* chi-square test for independence
Type I error 100–1
Type II error 100–1

uniform distribution 33
unit normal table *see z* table
untreated population 87, 91–2, 96–7, 99, 102, *103*, 120
u-shaped distribution 33

variability 38, 68, 115, 125, 171, 205, 215–19, 248, 278, 287–8
variables 17; constant (compared to) 17; continuous 19; dependent 17; dichotomous *18*; discrete *18*; extraneous 13, 15, 150; independent 17; outcome 17; predictor 17; qualitative 18; quantitative 18
variance 42; between-groups 217, 220–1, 235, 247–8, 252, 265; error 221; population 42; sample 42; within-groups 217, 221, 225, 235, 247, 248
variation 38

Welch's *t* statistic 195
within-groups degrees of freedom (*df*$_{Within}$) 220, 222–3, 227, 230, 240, 248–9
within-groups sum of squares *see SS*$_{Within}$
within-groups variance 216
within-subjects design *see* repeated-measures *t* test

$\hat{y}$ (*y*-hat) 162
y intercept 161–2, 164, 182

zero correlation 137
zero interaction 244, *245*
z scores: distribution of sample means *64*, 65–72; distribution table 325; formula 52; and hypothesis testing 98; and inferential statistics 77–109; location in a distribution *61*; and normal distribution 53, *59*; Pearson correlation 145–7; purposes 53; raw scores (compared to) 53; standard deviation 53, 57; standardizing distributions with 53; statistic 52; *t* statistic (compared to) 113–14; unit normal table 325
z table 325
z test statistic (for hypothesis testing) 98

Milton Keynes UK
Ingram Content Group UK Ltd.
UKHW052028141024
449569UK00017B/734